Polymer Thin Films

Edited by
Abbass A. Hashim

Polymer Thin Films,
Edited by Abbass A. Hashim

Republished by InTech

Janeza Trdine 9, 51000 Rijeka, Croatia

Polymer Thin Films,
Edited by Abbass A. Hashim
p. cm.
ISBN 978-953-307-059-9

For EU product safety concerns:
IN TECH d.o.o., Prolaz Marije Krucifikse Kozulić 3, 51000 Rijeka, Croatia,
info@intechopen.com or visit our website at intechopen.com.

Meet the editor

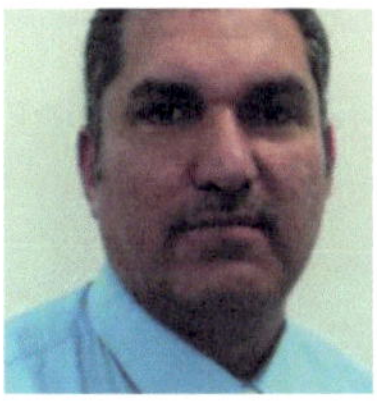

Abbass A. Hashim currently carrying out research in several areas of interest, mainly covering work on chemical and bio-sensors, semiconductor thin film device fabrication and characterisation. At the moment I have very strong interest in radiation environmental pollution and bacteriology treatment. The teams of researchers are working very hard to bring novel results in this field. I am also a member of the team in charge for the supervision of Ph.D. students in the fields of development of silicon based planar waveguide sensor devices, study of inelastic electron tunnelling in planar tunnelling nanostructures for sensing applications and development of organotellurium(IV) compounds for semiconductor applications. I am a specialist in data analysis techniques and nanosurface structure. I have served as the editor for many books, been a member of the editorial board in science journals, have published many papers and hold many patents.

Preface

In the past decade, polymer has generated much interest internationally for its potential to solve a wide variety of industry problem in nanotechnology and electronics devices. Several dozen companies are now designing and selling polymer units globally for a wide variety of expanding markets. Several research centres are also improving and developing polymer thin film, polymer fundamental studies and polymer techniques for preparation and application.

In recent times greater focus has been placed upon polymer thin films which plays an increasingly important role in technological applications ranging from coatings, adhesives and lithography to organic light emitting diodes and various organic material based devices, including sensors and detectors.

The physical properties of materials, small molecule, atomic or long chain polymers, confined to sufficiently small dimensions by external "walls" are generally difficult to predict because they manifest the influence of a confinement as well as the influence of interfacial interactions between the material (chemical) constituents and the external "walls." To this end, polymer films below a certain thickness range often exhibit physical properties that differ substantially from intrinsic bulk behavior. This is due, in part, to the increasing influence of entropic effects (confinement and chain "packing") and interfacial interactions as the film thickness decreases.

This textbook is written for researchers who have chosen to study an area that appeals to them and one in which they show a genuine interest. It is targeted together to form the knowledge pathway to obtain an advanced polymeric research in polymer thin film.

The approach adopted is user friendly with case study explained to reinforce essential point. The authors of this book are experienced researchers and have used this experience to include a wealth of particular investigations and results that will appeal to all research workers. These are presented so that they are suitable for use as application demonstrations or for assessment research and planning. Whereas appropriate results are included to assist the objectives research investigation.

This book provides a timely overview of a current state of knowledge of the use of polymer thin film for important technological applications. Polymer thin film book covers the scientific principles and technologies that are necessary to implement the use of polymer electronic device. A wide-ranging and definitive coverage of this emerging field is provided for both academic and practicing scientists. The book is intended to enable readers with a specific background, e.g. polymer nanotechnology, to become acquainted with other specialist aspects of this multidisciplinary field.

Following this preface, part A covers the fundamental of the key aspect related to the development and improvement of polymer thin film technology and part B covers more advanced aspects of the technology are dealt with nano-polymer layer which provide an up-to-date survey of current research directions in the area of polymer thin film and its application skills.

Part A Overview: Fundamental key research

Part A of the book is an overview of the fundamental key aspects of physical and chemical development research and case studies in the recent years. The section contains 8 different chapters discussing the physical and chemical properties of polymer thin film for a promising future application. The first chapter by Abbass Hashim describes the recent development of polymer electronic nano-devices due to its dielectric properties. The dielectric properties of polymer are governed by the position, direction and the length of the dipolar group with respect to the main chain. Simulation model can show the energy bands and the stable energy levels available for the specific polymer dipoles. The specific dipole group depends on the temperature range, the polymer structure and the solid state of the main polymer. Applying the simulation technique descried in this chapter can observed the energy band values specifically and separately. The dielectric relaxations can be clearly detected and the temperature ranges for the relaxation processes can be obtained.

The second chapter of part A by R. A. Minamisawa, et al. provides an overview on several materials and methods that have been proposed to enhance properties in porous membranes, exploring from the polymer to the microelectronic technologies. They reported their own recent development in the fabrication of PFA fluorpolymer porous thin film membranes by ion-induced physical etching using a stencil mask technique. PFA fluorpolymers are candidates for advanced filtration membranes for being chemically inert, potentially resistant to biofouling because of its lowest adhesion coefficient. The third chapter by Bin Wang and Ritesh N. Vyas gives an overview of the earliest study of the use of EIS in studying membrane resistivity and capacitance where phospholipid bilayers were deposited on LbL polyelectrolyte surface. EIS has been used to characterize the stability and insulating properties of LbL films that are stabilized by different approaches like cross-linking and thermal conversion. The charge transfer resistance and film resistance are obtained directly from the filling of experimental impedance spectrum to Randle's equivalent circuits. The results obtained in this chapter contain information for mass transfer through thin membranes. A model also developed for membrane with particle components has broader applications. The conditions considered in this chapter can be easily applied to many situations in both industry and basic science.

The fourth chapter by Nikitenko V.R. and Tameev A.R. considers most of the recent studies of charge transport focusing on behaviour of carrier mobility. The chapter is focusing on less studied problem of dispersion of charge carriers in space. The objective was emphasizing that a carrier's non-equilibrium manifestations are much wider than effects of dispersive transport. This chapter provides options for analytic modelling and correct determination of material's parameters from data of time-of-flight and transient electroluminescence measurements. Charge transport in disordered organic layers has been intensively investigated in recent years both experimentally and theoretically. The next chapter by Anton Georgiev,et al. describes the preparation of polyimide thin film by physical vapour deposition and comment on their application as a pure material or a thin layer matrix for producing nanocomposite layers. Their superb properties, such as a low dielectric constant, high thermal and photo-stability, high chemical resistance and high optical transmittance predetermine their wide-spread applications as a cast and layers used as insulators, protective or capsulation layers, mechanical or diffusion barriers, in opto- and microelectronics.

Chapter six is written by Jan Weszka covering the aspect of polyazomethine thin film prepared by chemical vapour deposition and thermal vacuum evaporation. The chapter shows that the structure, morphology and optical spectra of polyazomethine thin films are depend on the deposition's technological conditions. The optical spectra reveal features resembling optical spectra of polyparaphenylene vinylene. The expectation is that the low transport rate of monomers to the substrate of the setup used is favourable for small growth rates. The optical spectra reveal features are being resultant from the distribution length of the conjugated fragment in various chains within the whole thin film volume.

The seventh chapter by S. Saravanan, et al. investigates the influence of radiation on polyaniline structure and properties. Irradiated thin film of polyaniline with swift heavy ions has been studied. Polyaniline has been prepared by RF plasma of aniline. The effect of SHI on the structural, optical and electrical properties is investigated in this chapter. The study focused on the comparison of FTIR spectral results with the standard data and based on the analysis of the tentative structure for the pristine polyaniline and the irradiated polyaniline. The change in optical band gaps for irradiated samples was evaluated and found that the optical band gap reduces with increase of fluency. Electrical studies were carried out on the irradiated thin films and the results were compared with the pristine thin films.

The chapter eight by I. Apostol, et al. presents some results about single step surface relief modulation of polymeric films in order to create integrated structures in materials. The study polymeric materials are photoresist and azopolymers. The tridimensional surface relief modulation is presented in two points; (i) technical possibilities to obtain surface relief gratings in a single step process on polymeric materials (ii) physic-chemical processes responsibility for single step surface modulation under the action of light and time stability. A general accepted technique for micro and nanoscale processing technologies in microelectronics is lithography. The exposure of a photoresist to a light pattern is followed by a developing stage. This is a two step method, which is time consuming non-localized and in the second stage is using non-ecological solvents.

Part B Overview: application key research

Part B of the book is an overview of the some of the most interesting aspects of polymer thin film development research and applications in the recent years. The section contains 8 different chapters discussing the promising future application. The chapter nine by Xi Zhang and Guanglu Wu covers the possibility of implementing layer-by layer (LBL) assembly as one of the most powerful technique in fabricating multilayer thin film. The unconventional LbL assembly includes supramolcular assembly in solution and LbL deposition at liquid-solid interface. This method brings not only new supermolecular structures but also functions. Therefore, it can be regarded as one of the multi-level assembly. The conventional or unconventional LbL method can be employed, each method has its own scope of applications as well as limitations. The combination of these two methods may facilitate the assembly of thin film materials with complex and elaborate structures for the integration of functionalities.

The combination of layered nanostructures and functional assemblies are summarized and discussed in this chapter.

The chapter ten by Wenjun Zheng discuss the surface processing which caused changes in chemicophysical and physcochemical properties of polymer surface. The changes are including many interesting surface phenomena and impose a number of interesting aspects for scientific research and lead to engineering applications. The surface properties of polyimide thin films are of prime importance for elucidating mechanisms behind surface phenomena. This chapter covers the mechanical rubbing breaks of the two dimensional topographical uniformity of polyimide surface and causes changes in the surface energy of the thin film. The rubbed polyimide thin film, the hydrophilicity of the surface towards the rubbing direction is different from that in the direction against the rubbing direction. The surface anisotropy in the rubbed polyimide surface is thought to be created due to an orientational arrangement of polyimide chains at the surface. The next chapter by Tatyana I. Shabatina and Gleb B. Sergeev covers the used of low temperature enlarger the possibilities of nanochemistry and opens new prospects in creation of film materials with new conducting, protecting and sensor properties. Low temperature and matrix isolation methods are used for stabilization of high energetic and very active metal species as atoms, clusters and nano-particles. The chemical activity and properties of reaction product is due to the using metal atoms, clusters and nanoparticles or in another word the effect of the reaction particles size. This effect is the intrinsic feature of nanochemistry. The effective approach of cryochemistry is a developed aimed on stabilization of metal atoms, dimmers, trimmers and higher clusters and metal particles in inert gas matrices and polymer films and by certain organic substances layers at different substrates. The low temperature and controlled condensation of reagent vapors allowed us to obtain and stabilized metal particles of 1nm and less in size.

The chapter twelve by Albert Chin gives an overview of the organic thin film transistor (OTFT). The advantages of pentances-based OTFT are low cost, small weight and visible-light transparency, for potential use in applications such as organic displays, flexible displays and low cost integrated circuit (IC). The key factor of the higher operation current OLED and high resolution display is the high transistor current of OTFT. The transistor current can be achieved by increasing the mobility of pentacene and gate dielectric using high-k dielectric. The study goal in this chapter is to achieve high speed, large drive current and low power consumption for OTFT ICs.

The chapter thirteen by Masaru Nagai describes the concept of the emitting light of the dye molecules. The dye molecules absorb light and then they emit light at different red shifted wavelength. There is a wide range of application for the dye molecules in laser system and displays. This chapter is focusing on the high efficient color-conversion polymer film and their application to full-color, active-matrix (AM), organic light emitting diode (OLED) displays. These polymers are promising color-conversion materials that can be used to fabricate efficient CCM-OLED devices. The chapter fourteenth by Jun Ki Kim and Kyunghwan Oh introduces the principles, fabrication procedure, and characterization of beam propagation and beam patterns from linear and circular azo-polymer surface relief grating (SRC). The chapter discussed the experimental and theoretical background. Micro/nano scale phase front inscription techniques were investigated for flexible beam shaping on polymer thin layer. The numerical simulation of diffraction patters out of azo-polymer layer on the fibre is also analyzed. The proposed inscription technology based on polymer thin layer coating properties of the light and its applicability in integrated optical components as well as subsystem for optical communications.

The chapter fifteenth by G. Telipan, et al. describes the physical properties of linear polymer and organic molecules. The physical properties are strongly modified when the polymer contains an associating end groups. The monomer units kept together by non-covalent interactions called supramolecular. A organo-siloxane supramolecular polymer was obtained, starting from 4,4'-bipyridine as an acceptor and silicon-containing carboxylic acids as hydrogen-donor molecules. Thin and thick film sensors were made by spin coating techniques on alumina substrate. The sensor device was exposed to CO2 and NOx and the respond voltage measured as a function of time. The chapter discussed the theoretical aspects about equivalent scheme circuit of active sensors. Chapter sixteenth by T Kohoutek, et al. provides a study of the multilayer dielectric film with sufficient differences in refractive index. The refractive index played an important role in designing highly effective planar optical elements, namely mirrors and filters. The required thickness of the dielectric multilayer film and the optical constant meet the Bragg resonance condition formed one dimensional photonic bandgaps. The dielectric multilayers are widely used as highly effective reflectors and filters in current optical devices. The bandgap structure can be predicted according to the theory of light propagation through stratified dielectric media and their optical properties then adjusted appropriately. The chapter focused on the fabrication of Ge-Se/PS dielectric and Au/Ge-Se/PS meta/ dielectric reflector from amorphous chalcogenide and polymer films with the optical reflectivity higher than $r>99\%$ near $\lambda\sim1550$ nm using low temperature and inexpensive deposition techniques.

The chapter seventeenth by Cristina Riggio, et al. talks about the drug accessibility to the central nervous system (CNS) which is limited by the blood-brain barrier. Many techniques to deliver therapeutics to the CNS are in used including osmotic pumps and silicone reservoirs. This chapter outlined different applications of polymers, usually intended as agent acted to improve the bio-distribution of the desired drugs. They highlighted the huge possibilities offered by the thin film technology applied to the concepts of drug delivery and drug targeting. The chapter also obtained an efficient combination of physical and chemical features of an innovative neuronal interface based on a carbon nanotubes array. The result achieved by this study indicates that polymer technology could be efficient embedded in CNS array acting as drug delivery system at cellular level.

The last chapter by Hye Ri Yang and Chong Hoon Kwak presents the determination of the optical nonlinearties of azo dye doped polymer film by means of holographic gratings as a scalar effect and the photo-induced birefringenece as a vector effect. They have measured the diffraction efficiency of the holographic gratings and the photo-induced birefringence. The diffraction is caused by a linear polarized pump beam as a function of time for various laser beam intensities and azo dye concentrations. The authors found that the real time behaviours of both of the diffraction efficiencies and the photo-induced birefringence reveal the stretched exponential kinetics. A three state model for photoisomerization is proposed to analyse the stretched exponential kinetic behaviours.

We believe that the material presented in the book of polymer thin film should not only help readers to find out more about this new and challenging subject, but also act as a useful reference in the future.

In addition, this book is eminently suitable for those who are studying MSc and PhD in polymer thin film field applications.

March 2010

Abbass A. Hashim

Contents

Preface VII

1. Polymer dipoles relaxation and potential energy (New Simulation Model) 001
Abbass A. Hashim

2. Advanced PFA thin porous membranes 015
R. A. Minamisawa, R. L. Zimmerman, C. Muntele and D. ILA

3. Ion Transfer in Layer-by-Layer Films 029
Bin Wang and Ritesh N. Vyas

4. Non-equilibrium charge transport in disordered organic films 047
Vladimir Nikitenko and Alexey Tameev

5. Preparation of Polyimide Thin Films by Vapour Deposition and Solid State Reactions 071
Anton Georgiev, Erinche Spassova, Jacob Assa and Gencho Danev

6. Thin films of aromatic polyazomethines 093
Weszka J.

7. Investigations on pristine and swift heavy ion irradiated plasma polymerized aniline thin films 111
S. Saravanan, M. R. Anantharaman, S. Venkatachalam and D. K. Avasthi

8. Tridimensional surface relief modulation of polymeric films 129
I. Apostol, N. Hurduc and V. Damian

9. Unconventional Layer-by-Layer Assembly for Functional Organic Thin Films 143
Guanglu Wu and Xi Zhang

10. Surface Wetting Characteristics of Rubbed Polyimide Thin Films 161
Wenjun Zheng

11. Cryochemistry of nanometals 185
Tatyana I. Shabatina and Gleb B. Sergeev

12. High Performance Organic Thin-Film Transistors and Nonvolatile Memory Devices Using High-κ Dielectric Layers 197
Albert Chin

13. High performance color conversion polymer films and their application to OLED devices 217
Masaru Nagai

14. Micro/nano scale phase front inscription on polymer thin layer for flexible beam shaping 239
Jun Ki Kim and Kyunghwan Oh

15. Organo-siloxane supramolecular polymers used in CO2 detection-Polymer thin film 253
Gabriela Telipan, Lucian Pislaru-Danescu, Mircea Ignat, Carmen Racles

16. Planar Quarter Wave Stack Reflectors Prepared from Chalcogenide Ge-Se and Polymer Polystyrene Thin Films 277
Tomas Kohoutek, Jiri Orava, Martin Hrdlicka, Jan Prikryl, Tomas Wagner and Miloslav Frumar

17. Polymeric thin film technology for neural interfaces: Review and perspectives 289
Cristina Riggio, Gianni Ciofani, Vittoria Raffa, Silvia Bossi, Silvestro Micera and Alfred Cuschieri

18. Determinations of Optical Field Induced Nonlinearities in Azo Dye Doped Polymer Film 309
Chong Hoon Kwak, Hye Ri Yang

1

Polymer dipoles relaxation and potential energy (New Simulation Model)

Abbass A. Hashim
Centre for Automation and Robotics Research (CARR)
Material and Engineering Research Institute
Sheffield Hallam University, City Campus, Pond Street, Sheffield, S1 1WB, UK

1. Introduction

Development of polymer electronic devices is one of the most interesting and required in the recent year industrial technology.

Polymeric material has various characteristics which can be controlled and monitored. These characteristics are including light-weight, mechanical flexibility, high-dielectric strength, fracture tolerance, high chemical resistance, easy processibility, and low manufacturing cost. Moreover they can be configured into almost any conceivable shape and their properties can be tailored to suit many applications.

Nowadays and near future exactingly, electronic-polymer represents an important piece of the electro-chemistry and many technological advances. Thus come from the combination of different materials in electrochemical cells.

New electro-active polymeric materials are always in the priority market requirements, with different properties, such as electroluminescence, semiconductor behaviour, electronic and ionic properties, electrochromism, etc (De Paoli & Gazotti, 2002; Noh et al., 2006).

Wide range of electronic components from micro to nano scales are in research from academics, industry, and national laboratories to present and discuss the recent research and commercial advances and needs. The researchers focus on the topics related to materials development, characterization, processing, manufacturing, analysis, device designing, implementation and applications.

The need for such nanoscale and microscale are demanding requirements for polymers as dielectrics, which have been used for insulators and charge-storage applications. Exploring polymer-based dielectrics with extremely low-k and high-k has recently become an important area of research and development. The influence of environment on electrical insulation and space-charge properties is the access topic for research on dielectric polymers (Taylor, 2006).

Recently, a significant highly competition in the markets of the tiny electronic chips technology specially and an extremely demands in the mobile industry and computer manufacturing. The future of electronic chip technology depends on the development of dielectric materials with low dielectric constant (K less than 2.2). The channel length of chip device approaches 0.1 μm, the travailed signal delay on an integrated circuit chip is dominated by the interconnect wiring. To reduce the size of the interconnect wiring; crosstalk between two adjacent lines dictates the minimum allowable spacing. This crosstalk is directly dependent upon the dielectric constant of the insulating material (Modafe et al., 2006).

Large number of capacitors employed in electronic systems, integration of capacitors is of great importance. The development of microelectronics requires decoupling capacitors with higher capacitance and shorter distance from their serving devices. In particular, the high-K materials are required for making embedded capacitors for integrated electronic devices (Müller et al., 2006; Lee et al., 2009; Popielarz & Chiang, 2007)

Microelectronic embedded capacitors are considered as a promising enabling technology. Development of an organic capacitor substrate compatible high dielectric constant material is currently unavailable (Hwang et al., 2008).

Polymer-ceramic nano-composites create potentially high-K materials. This approach could combine the low-temperature processibility of the organic polymer matrix and the high dielectric constant of the ceramic filler.

The devices working at high operating frequencies, such as fast computers, cellular phones, etc., require new high-dielectric constant (high- *K*) materials that combine good dielectric properties with both mechanical strength and ease of processing. The unique combination of dielectric and mechanical properties is hard to achieve in a one component material. Pure polymers are easy to process into mechanically robust components but generally suffer from a low dielectric constant. Thos polymers can be used for this application but the dielectric constant has to be improved to create material with high-*K* (Fan et al., 2002).

Typical high-*K* materials, such as ferroelectric ceramics, are brittle and require high-temperature processing, which is often not compatible with current circuit integration technologies. The model solution would be a high-*K* material that is mechanically robust and processable at ambient temperatures. This has raised a great interest in hybrid materials, such as ferroelectric ceramic/polymer composites, that may combine desired properties of the components (Popielarz et al., 2001).

Moreover electronic devices are transducing polymers (TPs), smart polymers (SPs), and electro-responsive polymers (ERPs) respond to an electric, magnetic, mechanical thermal, optical, chemical, and other stimulation, and respond with a chance that includes mechanical, electrical, optical, and thermal stimuli, and many others, or vice versa. The transducing nature of these polymers led to their use as sensors and actuators. The development of miniature electronics and MEMS/NEMS has attracted a great deal of attention to the development of more sophisticated transducing polymers with enhanced performance in macroscale to microscale, and even down to nanoscale (Kassiba et al., 2007).

The research subjects mentioned above are promising research field in which needs to understand the variation in the dielectric properties due to the polymer thin film internal structure modification. This study is attempting to view the tiny structural changes in polymer structure and its affect on the dielectric phase properties as a result of any kind of treatments; chemical, thermal or mechanical.

2. Polymer dielectric properties

Understanding of polymer molecular dipole influences and field interactions on dielectric characteristics is a fundamental issue for the development of electronic device applications.

Three type s of interaction of the polymer molecule with the electrical field (polarization) can be determined. These are possibly to be classified as electronic, atomic and orientation polarization. In the area of thin film, molecule orientation polarization is the most essential.
The electrical field tries to align the dipolar molecules along its own direction. This effect is hindered by thermal motion of the molecules. As a consequence orientation polarisation decreases when the temperature increases.
The electronic and atomic polarisation is independent of the temperature. The orientation polarisation is not observed when the thermal motion of the molecules is hindered, since rigidly bounded molecules or polar groups are move under the effect of the electrical field. This situation occurs at low temperatures. The orientation polarisation becomes possible when a certain, relatively high temperature is attained during the heating of the material. Increasing the temperature, the thermal energy of various polar molecules and groups successively increases, permitting orientation processes. However, the polarisation increases only in a relatively narrow range of temperature, because at higher temperature the intense thermal motion favours the disorientation of the dipoles and the polarisation decreases.

If an electric field is applied on a dielectric material, the variation of the field is followed by polarisation with a certain delay. This delay is a phenomenon common to all the three types of polarisation, and it is called "relaxation effect", characterized by the relaxation time, the time interval during which polarisation decays to the e-th fraction of its initial value. The relaxation frequencies (times) are different for the various types of polarisation. The decay of electronic polarisation is the most rapid because of the low inertia of electrons, while that of atomic polarisation is slower. The relaxation time of atomic polarisation is in the range of the period of infrared electromagnetic radiation, while the relaxation time of electronic polarisation is in the range of visible light. The relaxation time of the orientation polarisation depends on different parameters of the polar molecules (for example: size, molecular environment, etc.) and appears in the range from the radio frequencies and microwaves to several weeks.

3. Temperature dielectric relaxation

Mort and Pfister (Mort & Pfister, 1982) have shown that several distinct dielectric relaxation processes can exist in solid polymers. This is observed more clearly when the dielectric loss is studied as a function of the temperature at a given frequency.
As the temperature increases, the molecular mobility of the polymer increases leading to more dipole orientation. By convention, the dielectric relaxation processes are labelled α, β and so on, beginning at the high temperature end. The same relaxation processes are generally responsible for dispersions when mechanical properties are considered, although a particular molecular rearrangement may produce a stronger dielectric than the mechanical effect, or vice versa.
Some polymers are wholly amorphous and most of the solid materials have two phases of relaxation in solid materials. In such cases there is always a high temperature $\propto$- relaxation associated with the micro-Brownian motion of the whole chain and, at least one low-temperature (β, γ, etc) subsidiary relaxations. The relative strength of $\propto$ and β- relaxations depends on the dipole group orientation which is limited by the mobility of the β - process prior to the $\propto$-process of the higher mobility. The potential barrier U of the dipole stable

levels was studied since this leads to the values of $\varepsilon', \varepsilon''$ and consequently the mode of the relaxation (Blythe & Bloor, 1980).

4. Theoretical Model

The specimen may be regarded in terms of a series circuit. If the equivalent series components of capacitance and resistance are C_s and R_s, respectively, the total impedance will be given by (Blythe & Bloor, 1980):

$$Z = R_S + \frac{1}{jwC_S} \tag{1}$$

By comparing the out off phase and in phase currents after application of the alternating voltage, the imaginary part of the complex dielectric constant ε'', can be calculated using the measured values of R_S and C_S (Blythe & Bloor, 1980):

$$\varepsilon'' = \mathrm{R_S} \frac{\mathrm{C_S}}{\mathrm{C_o}(1+\tan^2\delta)} \tag{2}$$

C_o is the free space capacitance and $\tan\delta$ is the dissipation factor.
The temperature dependence of the dipole relaxation time constant τ often follows Arrheniu's law (Smith, 1955):

$$\tau = \tau_o \exp\left(\frac{U}{kT}\right) \tag{3}$$

Where τ_o is the free relaxation time constant at the high temperature range and τ is the time constant defined by the circuit parameters:

$$\tau = R_s\ C_s \tag{4}$$

Equations (2) can be modified using $\tan\delta = R_S\ C_S\ \omega$:

$$\varepsilon'' = \frac{\tau_o\ e^{\left(U/kT\right)}}{\mathrm{C_o}\left[1 + \tau_o^2 \omega^2 e^{\left(2U/kT\right)}\right]} \tag{5}$$

In the most cases the applied frequency makes $\tau_o^2 \omega^2 e^{\left(2U/kT\right)} << 1$, and equations (5) can be simplified as:

$$\varepsilon'' = \frac{\tau_o}{\mathrm{C_o}}\ e^{\left(U/kT\right)} \tag{6}$$

Most of the parameters are temperature independent, except the term $\exp\left(\frac{U}{KT}\right)$, which show greater temperature dependence at high temperature range.

5. The U Energy Bands Model

Experimental data have prove that the effect of the term exp $\left(\frac{U}{KT}\right)$ in the ε''-curves (equation 9) is more pronounced than in the ε'-curves (equation 8) and the small variation in U with temperature is controlled the relaxation processes. However, the $U/_T$ curves are more compatible with $\varepsilon''/_T$ curves when the polynomial fitting method was used.

Whereas, all $\boldsymbol{U}$ polynomial fitting curves are shifted on the temperature scale relative to the ε''-curves. Such a shift can add to modified equation (9) in terms of the correction shift parameter or polynomial error (P):

$$\varepsilon'' = \frac{\tau_o}{C_o} \exp\left[\frac{-1}{kT}\sum_{i=0}^{n} a_i T^i\right] \tag{10}$$

where $T = T_o \pm P$

$$\varepsilon'' = \frac{\tau_o}{C_o} \exp\left[\frac{-1}{k(T_o \pm P)}\sum_{i=0}^{n} a_i T^i\right] \tag{11}$$

Where T_o is the real temperature, T is the calculated temperature and n is the polynomial degree.

ε' and ε'' can be plotted against the temperature using equations (2) and (3). U can been deduced from $R_s\,C_s$ measurements using equations (4) and (5) as follow:

$$\ln R_s C_s = \ln \tau_o + \frac{U}{kT} \tag{12}$$

6. Implementation of the new model

6.1 Dielectric properties of a chosen example

The dielectric properties of polar polymers depend on the position, direction and the length of the dipolar group with respect to the chain. Two main relaxations were observed, namely; the high- temperature ∝ relaxation which is associated with chain backbone movement due to the rotation of the Ethers groups (C-O-C) around the main chain (crankshaft motion), and the low-temperature β-relaxation due to the hundred rotation of the acrylate groups $-COOR$ around the $C-C$ bond linking it to the chain (Blythe &Bloor, 1980; McCrum et al, 1967; Boyer, 1982). It is not known, however, how many chains are involved in the movement

Figure 1 shows the ε''/T-curves for PMA, in which an α-relaxation process can be identified clearly at a temperature of 36°C. The PMA molecules show no degree of freedom at 0^oC due to the acrylate and hydrogen side groups, but the degree of freedom increases as the temperature approaches 20°C. The molecules absorb enough energy, giving the chain backbone enough kinetic energy to vibrate with the field frequency and show the α-peak. The joint motions of the side groups and the main chains gives rise to the α-relaxation. The

β -relaxation originates from the movements of the acrylate group $(-COOCH_3)$ at 19^oC (Blythe &Bloor, 1980). This group $(-COOCH_3)$ has enough space to rotate at room temperature. It should be noted that, the acrylate group $(-COOCH_3)$ controls the dielectric properties of the PMA, making the β and α processes close to each other and overlapping in a single wide peak. The curves in Figure 1 represent the average results of two repeated dielectric spectroscopy measurements with an error of ±0.035 in the ε'' curve which are within the range of the data point resolution.
The structure of PMA is one of the most simplest and regular structure in polar polymers. PMMA and PMA have the same acrylate groups with different r groups. This fact means that PMA has more polarity than PMMA due to the effect of the hydrogen atom in its r-group while in PMMA, $r = CH_3$ is more stable.

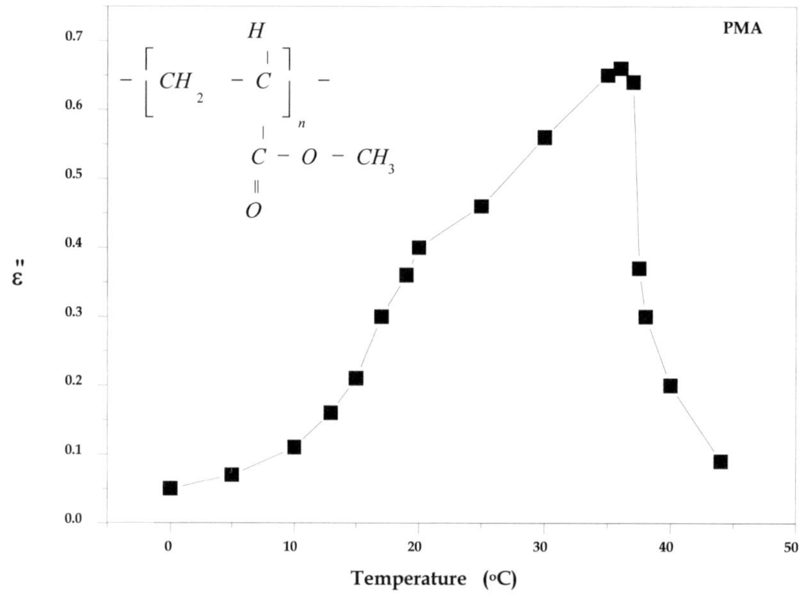

Fig. 1. Temperature dependence of $\varepsilon^{//}$ for PMA thin film.

Figure 2 shows a broad (β,α) peak in the PMMA ε''/T curve, which is observed in the temperature range of $(30-100)$°C. Since *PMMA* is a non-crystalline polymer, this peak is assigned to an α -relaxation (glass transition) as reported earlier (Bistac & Schultz, 1997) (Bistac & Schultz, 1997)(Kraise et al., 1965)(Utte et al., 1995). When this curve is compared with the ε''/T (*PMA*) curve, they are found to be very close with a shift of 28^oC towards high temperature in the PMMA data. This means that the $-COOCH_3$ group has a sufficient kinetic capability at room temperature to show a dielectric relaxation and the degree of freedom approaches its maximum value at 65^oC. The temperature shift is due to the dipole length differences between both types. Stereochemistry indicates that the *PMA* structure involves many big loop spaces when compared with *PMMA*. The small peak in Figure 2 at $120^{o}C$ is thought to originate from the space charge ρ peak (Krause et al., 1965).

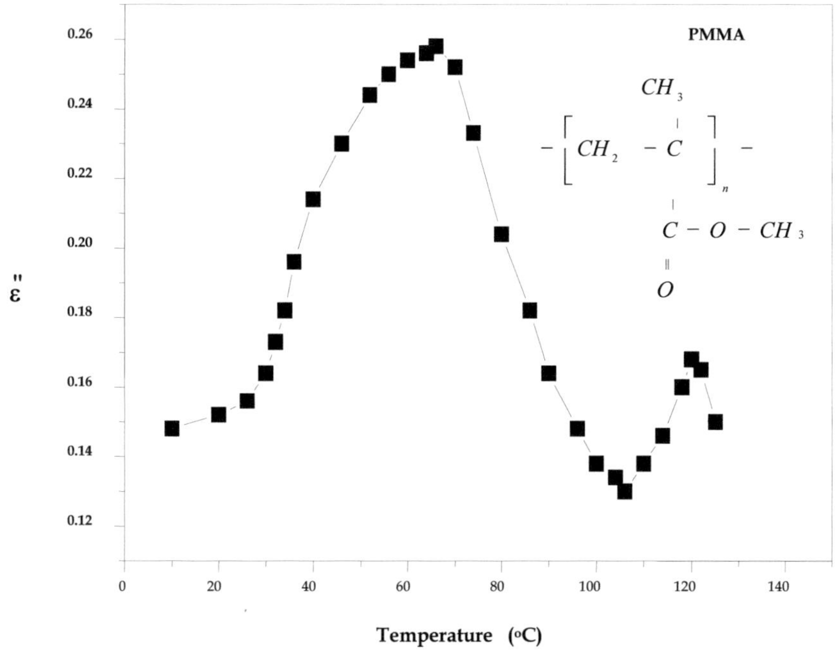

Fig. 2. Temperature dependence of $\varepsilon^{//}$ for PMMA thin film.

Figure3 shows the ε'' / T -curve of PMαClA, in which the structural differences are also in the r group, which is chlorine (Cl), associated with high polarity. The α - relaxation peak shape at $140\,^{o}C$ depends on the substitution and vibration of the chlorine loop associated with the crankshaft motion. The high polarity of the chlorine atom increases the hardness of this polymer and pushes the relaxation processes towards higher temperature range. β - relaxation occurs at temperatures around 80°C and is due to the weak rotation of the $-COOCH_3$ group and the field loops overlapping which produces strong links between the neighboured chains. The comparison of β -peaks in Figure 1, 2 and 3 indicates that there is an inverse proportional relationship between the polarity value and the strength of the β-peak.

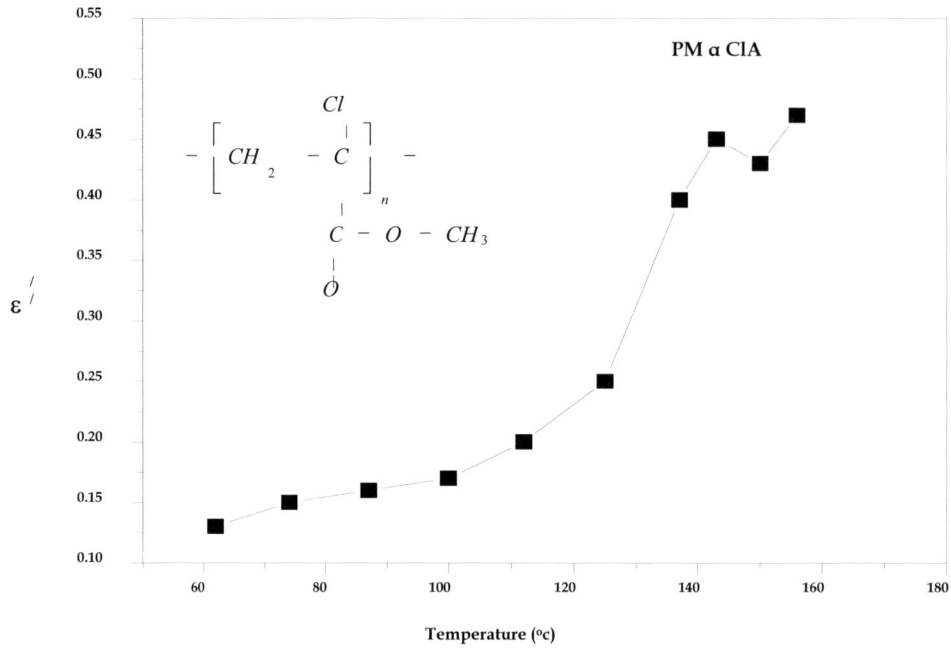

Fig. 3. Temperature dependence of ε'' for PMα ClA thin film.

The *PVAc* polarity is high as shown in Figure 4, and is increased by two fold when the temperature is doubled. The oxygen atoms substitute the carbon atoms in the acrylate group, leading to a change in the dipole length and polarity value due to the strong intermolecular interaction, which gives a sharp α - relaxation peak at $75^{o}C$.

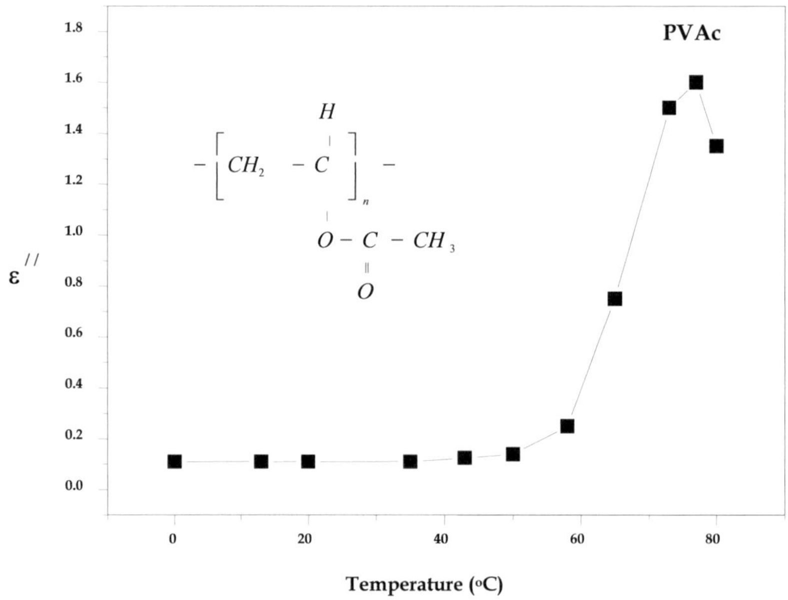

Fig. 4. Temperature dependence of $\varepsilon^{//}$ for PVAc thin film.

6.2 Simulation of the *U* Energy Bands

The simulation of the dipoles' stable energy levels U is illustrated in Figure 5. This simulation method can show the energy bands and the stable energy levels available for specific dipoles. The specific dipole group depends on the temperature range, the polymer structure and the solid state of the main polymer (Blythe & Bloor, 1980). Applying the polynomial fitting method to the data produces a smooth curve, which shows the shape of the dipole energy bands.

Figure 5 shows the $U/_T$ (PMA)-curve for the data produced from equation 9. Using the polynomial fitting approximation for $i = 4$ gives a compatible shape for $U/_T$ with $\varepsilon''/_T$-curve of Figure 1. The peak in region II flips from negative to positive signed to the dipole vector directed with or opposite to the electrical field. The simulation curve shows the density of states of the energy bands available for the main dipoles.

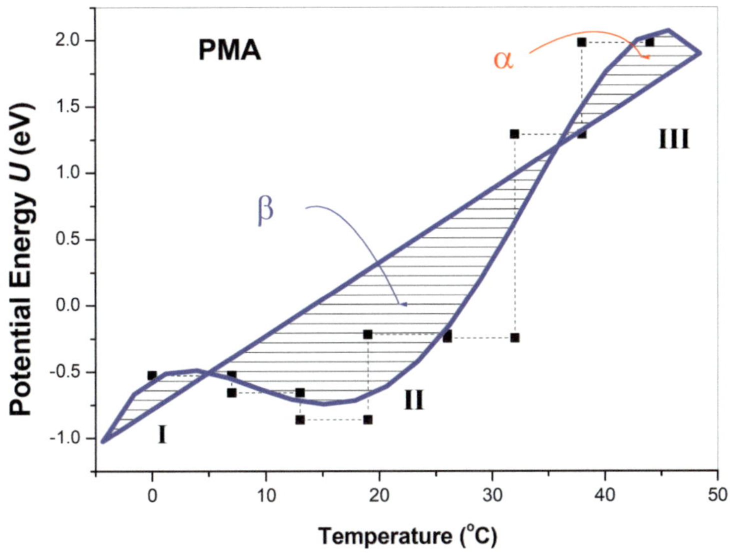

Fig. 5. The potential barrier Uversus temperature, the dot line indicates the experimental data and the solid line indicates the simulation model for the PMA thin film.

The γ -relaxation does not appear in all figures because, the r groups have, in most cases, one carbon atom. γ-relaxation is observed clearly when the side chain r is increased in length beyond 3 or 4 carbon atoms (Blythe & Bloor, 1980) and it depends also on the resolution of the measurement method. Taking this into account, the band marked I in Figure 5 (PMA), is the continuation of the second relaxation (β- relaxation). The density of states observed in region (I) is due to the temperature relaxation arising from the relaxation of the hundreds of $-COOCH_3$ groups around themselves. Region II is present in the temperature range $(5-35)$°C and related to β- relaxation indicating that there is a high density of states available for the main dipole produced from the rotation of group of $-COOCH_3$ around the $C-C$ backbones.

The positive and negative values of *U* are related to the orientation of the dipole groups to the direction of the applied field. Region III appears at 36°C indicating α- relaxation associated with the chain backbone movement (crankshaft motion).
Figure 6 shows four different density of states for *PMMA*. Comparing the first three regions I, II, III with those in Figure 1b, shows clearly the separation between the relaxation processes. Region V shows the high density of states available to the space charges.

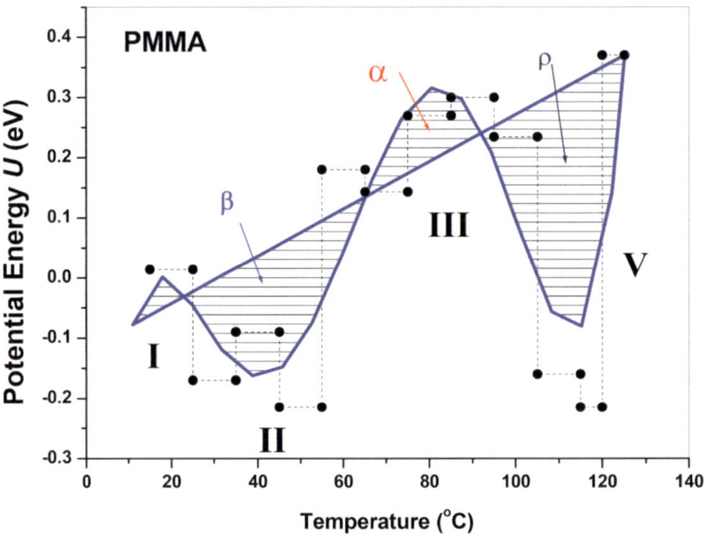

Fig. 6. The potential barrier *U* versus temperature, the dot line indicates the experimental data and the solid line indicates the simulation model for the PMMA thin film.

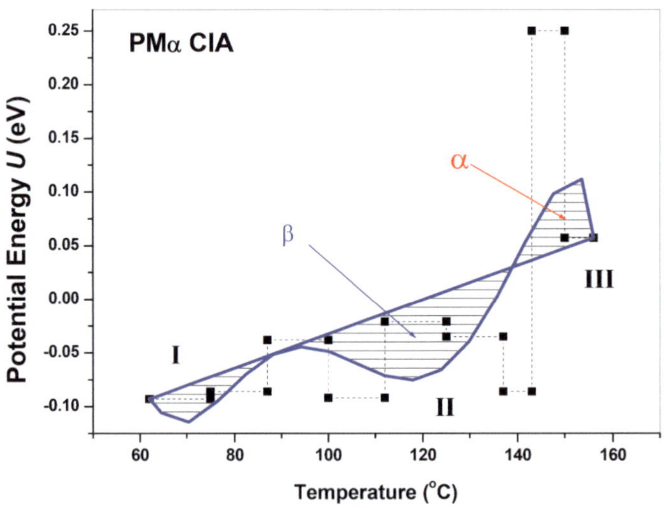

Fig. 7. The potential barrier *U* versus temperature, the dot line indicates the experimental data and the solid line indicates the simulation model for the PMα ClA thin film.

The band structure of $PM\alpha ClA$ also shows three regions as for PMA. The variation in the width and the depth of the bands is shown in Figure 7 and is due to the chlorine in the r group which replaces the $CH_3 - C$ or $H - C$ dipoles by a $Cl - C$dipole.

The high chain flexibility in the $PVAc$ dominates the relaxation response very early i.e. at low temperatures. There is a wide range of density of states for α dipole groups and this is the reason for disappearance of the other relaxation mechanisms which are seen quite clearly in $PVAc$ (Figure 8).
Figure 2, shows that the α-peak is around 65 – 70 °C and ρ is around 120 °C. However, in Figure 6, the α- peak is shifted forward to (80 – 85) °C and ρ peak is moved from 120 °C to around 110 °C, using the simulation model.

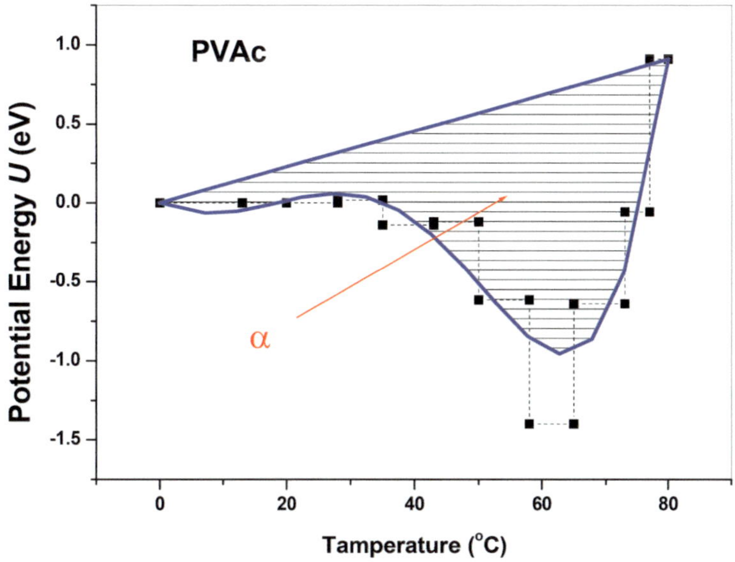

Fig. 8. The potential barrier U versus temperature, the dot line indicates the experimental data and the solid line indicates the simulation model for PVAc thin film.

This shows that $PMMA$ is most likely to be isotactic, and this is considered to be one of the advantages of our simulation model to determine the correct position of the dielectric relaxation peaks.

The low values of the potential energy U (activation energy), which were determined by this method, are due to the thick layer, the high molecular weight of the polymer, and the lower degree of freedom as well as the limited resolution of the measurement method(Kalogeras, 2003; Kalogares, 2005). For $PMMA$, the maximum potential energy value at 120 °C is found to be 0.17 eV and for $(\beta + \alpha)$activation energy peak is found around 0.4 eV. Kalogeras *et al.*, found that the activation energy for $PMMA$ at 120 °C using DRS method was about ($\approx$ 0.1 eV)and for β- relaxation it was $(0.4 - 0.9)eV$ while for α - relaxation it was ($\geq$ 0.9 eV) (Kalogares et al., 2005). For ultrathin films of $PMMA$, Wübbenhorst *et al.*, have found that α activation energy peak varied from 2 - 4 eV when the film thickness changed from 6.9 - 58.5 *nm* (Wübbenhorst et al., 2003). This means that there is a direct proportion between the film thickness and the activation energy. However, this proportion will be changed to inverse proportion when the film thickness is over the maximum limit of thickness-activation energy spectroscopy. In this case, α - relaxation is affected more than β and γ relaxations, and therefore explains the low value of $(\beta+\alpha)-$ activation energy peak in Fig.(2), which is most likely influenced by β relaxation more than α relaxation. Fig.(6), shows that β peak has a double size than the α peak. This confirms that α relaxation is less pronounced in *PMMA* dielectric spectroscopy due to the film thickness $(61\,\mu m)$.

Table 1, demonstrates the polynomial fitting coefficients and the polynomial error (P) for each polymers. The combination between these coefficients can be used to determine the dielectric relaxation before and after any kind of treatment to locate the variation in the relaxation model of the polymer under investigation. The coefficients values give a good indication and recognition for any possible structural changes.
Toluene is the basic solvent used, and interacts only weakly with the selected polymers. Its plasticizing effect is lower, due to the fact that part of the solvent molecules does not interact with the polymer chains, and is probably retained in the form of clusters (Spěváček & Schneider, 1987; Bosscher et al., 1982).

Finally using the simulation polynomial fitting technique, the U band values can be observed specifically and separately. Moreover the dielectric relaxations can be clearly detected and the temperature ranges for the relaxation processes can be obtained.

CONSTANT	PMA	PMMA	PMαClA	PVAc
a_0	-0.55749	-0.83145	21.1653	0.01961
a_1	0.05066	0.12363	-1.10272	-0.02187
a_2	-0.01037	-0.00632	0.0222	0.00152
a_3	4.99027E-4	1.32219E-4	-2.17063E-4	-1.0469E-5
a_4	-5.88414E-6	-1.17606E-6	1.0316E-6	-7.6449E-7
a_5	_____	3.71396E-9	-1.90824E-9	9.08388E-9
P	-9	-15	-10	-8

Table 1.

The energy bands encountered using the technique indicated all the density of states available for (α,β) dipoles. Comparing the different types of polymers, the effect of changing or replacing the side group on the energy band shapes is clearly demonstrated (Hashim et al., 2006].

7. References

Bistac, S. & Schultz, J. (1997). Solvent retention in solution-cast films of PMMA: study by dielectric spectroscopy, *Progress in Organic Coatings*, 31, 347-350, 0300-9440/97/$.

Bistac, S. & Schultz, J. (1997). Study of solution-cast films of PMMA by dielectric spectroscopy: Influence of the nature of the solvent on α and β relaxations, *Int. J. Adhesion and Adhesives*, 17, 197-201, 0143-7496/97/$17.00.

Blythe, A.R. & Bloor, D. (1980). Electrical properties of polymers, Cambridge University press, Cambridge, UK. ISBN-13978-0-5219-6.

Bosscher, F., Brinke, G., Challer, G. (1982). Association of stereoregular poly(methyl methacrylates). 6. Double-stranded helical structure of the stereocomplex of isotactic and syndiotactic poly(methyl methacrylate). *Macromolecules*, 15, 1442–1444. 0024-9297/82/2215-1442$01.25/0.

Boyer, R. F. (1982). *Rubb. Chem. Technol.*, 36 (1982) 1303. 00359475.

C.P. Smith, C. P. (19955). Dielectric behaviour and structure, McGram-Hill Book Company, INC, New York, USA.

De Paoli, M. and Gazotti, W. (2002). Electrochemistry, Polymers and Opto-Electronics Devices: Acombination with a Future. *J. Braz. Chem. Soc.*, Vol. 13, No. 4, 410-424, 0103-5053.

Fan, L., Rao, Y., C. Tison, Moon, K., Pothukuchi, S., Wong, C. (2002). Processability and Performance Enhancement of High K Polymer-Ceramic Nano-Composites, 8th International Symposium on Advanced Packaging Materials, IEEE, 0-7803-7434-7.

Hashim, A., Freeman, J., Hassan, A., Mohammad, M. (2006). Dipole Potential Barrier Simulation Model for studying Polar Polymers, *Materials Science and Engineering: B*, 138, 2, 161-165.

Hwang , J., Raj, P., Abothu, I., Yoon, C., Iyer, M., Jung, H., Hong, J., Tummala, R. (2008). Temperature Dependence of the Dielectric Properties of Polymer Composite Based RF Capacitors. *Microelectronic Engineering*, 85 (November 2008) 553–558, 0167-9317.

Kalogeras, I. M. (2003). Thermally Stimulated Currents of Poly(methylmethacrylate): Comments on the Molecular Origin of a Debye-Type Signal between the α and β Relaxation Modes, *Journal of Polymer Science Part B: Polymer Physics*. 42, 4 (702-713).

Kalogeras, IM. (2005). Contradicting perturbations of the segmental and secondary relaxation dynamics of polymer strands constrained in nanopores, *Acta Mater.*, 53, 1621-1630, 1359-6454.

Kassiba, A., Bouclé, J., Makowska-Janusik, M., Errien, N. (2007). Some Fundamental and Applicative Properties of [Polymer/nano-SiC] Hybrid Nanocomposites. XIII International Seminar on Physics and Chemistry. 10.1088/1742-6596/79/1/012002. *Journal of Physics*: Conference Series 79, IOP Publishing.

Krause, S., Gormley, J. J.,Roman, N., Shetter, J.A., Watanabe, W.H. (1965). Glass temperatures of some acrylic polymers, *J. Polymer Sci. A, 3,* 3573-3586, DOI: 10.1002/ pol.1965.100031020.

Lee, K. H., Lee, K., Suk Oh, M., Choi, J., Im, S., Jang, S., Kim, E. (2009). Flexible high mobility pentacene transistor with high-k/low-k double polymer dielectric layer operating at _5 V. *Organic Electronics,* 10, 194–198. 1566-1199/$.

McCrum, N. G., Read, B. E. & Williams, G. (1967). An-elastic and Dielectric effect in polymeric Solids, John Wiley (London, New York [etc.]), LCCN: 67029334.

Modafe, A., Ghalichechian, N., Frey, A., Lang, J., Ghodssi, R. (2006). Microball-bearing-supported Electrostatic Micromachines with Polymer Dielectric Films for Electromechanical Power Conversion. *J. Micromech. Microeng.* **16** (2006) S182–S190, 0960-1317.

Mort, J. & Pfister, G. (1982). Electronic properties of polymer, John Wiley and Sons, INC., ISBN-10: 0471076961, ISBN-13: 978-0471076964, UK.

Müller, K., Paloumpa, I., Henkel, K., Schmeißer, D. (2006). Organic thin film transistors with polymer high-k dielectric insulator. *Materials Science and Engineering C,* 26 (July 2006), 1028 - 1031, 0928-4931/$.

Noh, Y., Young Park, S., Seo, S., Lee, H. (2006). Root Cause of Hysteresis in Organic Thin Film Transistor with polymer dielectric. *Organic Electronics,* 7, (March 2006) 271–275, 1566-1199/$.

Popielarz, R. & Chiang, C.K. (2007). Polymer composites with the dielectric constant comparable to that of barium titanate ceramics. *Materials Science and Engineering B,* 139, 48–54. 0921-5107/$.

Popielarz, R., Chiang, C., Nozaki, R., Obrzut, J. (2001). Dielectric Properties of Polymer/Ferroelectronic Ceramic Composites from 100Hz to 10 GHz. *Macromolecules, 34 (*November 2001), 5910-5915, 10.1021/ma001576b.

Smith, C. P. (19955). Dielectric behaviour and structure, McGram-Hill Book Company, INC, New York, USA.

Spěváček, J. & Schneider, B. (1987). Aggregation of stereoregular poly(methyl methacrylates). *Adv. Colloid Interface Sci.,* 27, 81-150, 10.1016/0001-8686(87)85010-8.

Taylor, D. (2006). Space Charges and Traps in Polymer Electronics. IEEE Transactions on Dielectrics and Electrical Insulation, Vol. 13, No. 5 (October 2006), 1063-1073, 1070-9878/06/$20.00.

Utte, K., Miyatake, N., Hatada, K. (1995). Glass transition temperature and melting temperature of uniform isotactic and syndiotactic poly(methyl methacrylate)s from 13mer to 50mer, *Polymer,* 36(7), 1415-1419 , 0032-3861.

Wübbenhorst, M., Murray, CA., Dutcher, JR. (2003). Dielectric relaxations in ultrathin isotactic PMMA films and PS-PMMA-PS trilayer films, *Eur Phys J E Soft Matter.* (Nov. 2003), 12 Suppl 1,109-12. 15011027.

2

Advanced PFA thin porous membranes[1]

R. A. Minamisawa, R. L. Zimmerman, C. Muntele and D. ILA
Center for Irradiation Materials, Alabama A&M University
USA

1. Introduction

The invention of synthetic membranes in the middle of the last century was a significant development for industrial and research processes and "invaded" day-to-day life as an important technology for sustainable growth. Nowadays, nearly 50 years since the creation of synthetic polymer membranes, novel developments and refinements in membrane technology continue to be active themes of research; membrane technologies are now well accepted and cost-effective, conferring unique advantages over previous separation processes (Rogers *et al.*, 1998).

Separation membranes are broadly applied in food, chemical and pharmaceutical industries. Particularly, filtration membranes have proven to be reliable devices for water filtration (Fologea, 2005; Henriquez, 2004; Li, 2001, 2003a; Mochel, 1984; Mutoh, 1987; Schenkel, 2003). However, advances in materials and membrane processing are still a key solution to purify water at lower costs and higher flux in societies where scarce water resource is a major issue (Wiesner & Chellam, 1999).

Porous membranes are thin sheets and hollow fibers generally formed from a continuous matrix structure containing a range of open pores or conduits of small size. Porous membranes having open pores, thereby imparting permeability, are classified in nanofiltration, ultrafiltration and microfiltration membranes, depending in the pore size (Vainrot *et al.*, 2007).

Nanofiltration membranes have pores with diameter in the range of 3 nm and are used for treatment of slightly polluted water and for pretreatment in desalination processes. Commonly, an electrostatic charge is applied in the NF membrane in order to enhance salt rejection.

Ultrafiltration membranes and microfiltration membranes have, respectively, pore diameters in the range of 10-100 nm and up to 1 μm. Combined, these membranes are extensively used in wastewater treatment equipment for removing virus and bacteria, organic molecules and suspended matter. Separation capacity in these membranes is based on simple filtration, therefore, depending on the contaminant size in solution and on the diameter of the pores.

Ideally, porous membranes require high permeability, high selectivity, enhanced resistance to biofouling, and resistance against solvents, high- and low-pH environments, and

---

[1] U.S. Patent pending.

oxidizers. In other words, the material precursor for the membranes must be chemically resistant and the pores are required to have a homogeneous distribution in the pore size, fulfilling the high selectivity requirement, and a homogeneous spatial distribution of the pores, leading to enhanced mechanical resistance. Microbial fouling or biofouling has been the most complex challenge to eliminate (Girones *et al.*, 2005, Vainrot *et al.*, 2007). The solution for these problems lies in the development of innovative processes for fabrication of porous membranes as well as in the availability of new polymer precursors (Vainrot *et al.*, 2007).

Up to now, several materials and methods have been proposed to enhance properties in porous membranes, exploring from the polymer to the microelectronic technologies. However, currently there is no membrane available that fulfills all cost, quality and performance requirements, suggesting that the membrane technology is still in its early stage of development. The challenge lies in developing new fabrication methods able to process resistant materials into high flux porous membranes structures.

In this chapter a review is given of the main issues related to the fabrication of high performance porous thin film membranes and how this technology has been developed to keep the bottom-line of cost-benefit. We later introduce our recent results in the development of Perfluoroalkoxyethylene (PFA) fluorpolymer based thin porous membranes with enhanced separation capacity as well as being resistant to biofouling and harsh chemicals, using an ion beam nanofabrication technique. In addition, we describe the development of a feedback ion beam controlled system able to fabricate well shaped and well distributed micro and nanopores, and to monitor in real-time the pore formation.

2. Advances in porous membranes

Firstly, we briefly provide an overview in the current status and the advances in porous membrane fabrication. Membranes in separation modules are usually fluorpolymer based membranes due to their cost-effectiveness as well as their thermal stability and chemically inert properties, attributes that give excellent resistance to the devices. While these polymer properties are desirable for porous membranes, they also render the polymer unamenable to casting into well-shaped membranes by conventional processes. Because it is difficult to chemically etch this material, it is impractical to fabricate membranes with high pore quality regarding spatial and size distribution in fluorpolymer films; consequently, this type of membrane has low selectivity as well as low mechanical stability (Caplan *et al.*, 1997). Figure 1 displays an example of this type of tortuous path membrane.

Track-etched membranes (TEMs) are typically used for high-specification filtration in many laboratory applications. The fabrication process consists in the ion bombardment of membranes, commonly PET, at high energy and low fluencies and in a post-chemical etching of the damaged material along the ion track. This ion beam technique creates energetic particles that are nearly identical and have almost the same energy; consequently the tracks produced by each particle are almost identical. The etching process involves passing the tracked film through a number of chemical baths, creating a clean, well-controlled membrane with good precision in terms of pore size (Ferain & Legras, 1997, 2001a, Quinn *et al.*, 1997). This etching process determines the size of the pores, with typical pore sizes ranging from 20 nm to 14 μm. Although the shape of the pores is significantly better than the tortuous path membranes, the spatial distribution is inhomogeneous. As can be seen in the TEMs shown in figure 1, there are undamaged areas in the membrane as well as regions where two or more etched tracks combine. These broad pores are propitious points for mechanical fracture and decrease the filtration selectivity in respect to the majority of smaller pores.

So far, the membranes with highest flux performance were introduced by the Dutch company Aquamarijn, using the well established semiconductor technology (van Rijn *et al.*, 1999). These membranes, called Microsieves, are fabricated using optical lithography and chemical etching of a silicon nitride thin film grown on a silicon substrate. After defining the membrane in the silicon nitride film, the silicon substrate is back etched (Girones *et al.*, 2005). The final membranes have pores with excellent pores size and spatial distribution (Figure 1). The drawback of the Microsieve technology is the difficult control of fouling and, mainly, the high cost of the substrates. Whereas 200 mm diameter silicon wafers cost some hundreds of dollars, few kilometers of fluorpolymers films can be obtained at similar expense. Additionally, although silicon nitride is chemically resistant, it is not as chemically inert as fluorpolymer materials, which decreases its applicability. Similarly to the TEMs, the fabrication process of Microsieves is relatively time consuming and expensive.

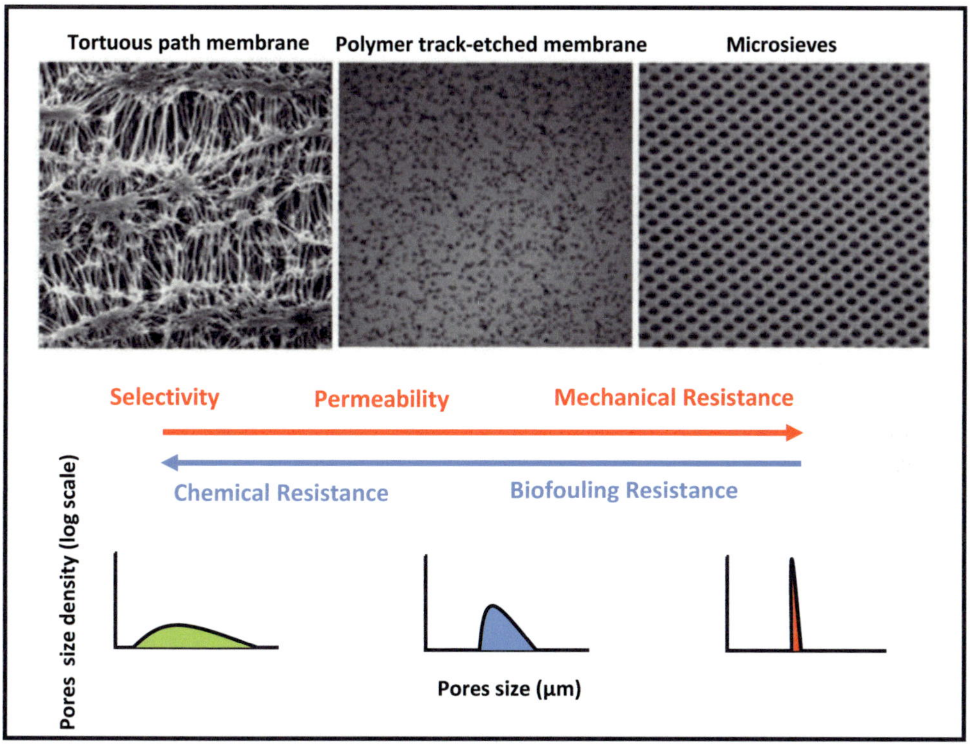

Fig. 1. Comparison of performance for different types of porous membranes. The direction of the arrows indicates improvement of the described properties. The bottom graphs schematically compare the pores size distribution for the membranes shown above.

3. Ion beam processing of PFA

Perfluoroalkoxyethylene is a fluoropolymer that has a carbon chain structure fully fluorinated in radicals and with a small amount of oxygen atoms. The chains are cross linked and are expressed by the molecular formula $[(CF_2CF_2)_nCF_2C(OR)F]_m$. PFA thin films have a broad range of applications in the packing and coating industry due to its thermal stability (melting point of approximately 304ºC), low adhesion, biological suitability and low frictional resistance (DuPont, 1996). PFA is solvent resistant to virtually all chemicals, which makes wet etch processing of these materials difficult or even impossible (Caplan *et al.*, 1997).

In this section, we evaluate the use of ion bombardment as an alternative tool for the processing of fluorpolymers, specifically Perfluoroalkoxyethylene irradiated with 5 MeV Au^+ ions. When ion beam irradiation is applied to process polymers, some parameters must be taken into consideration such as the surface modification, the polymer mobility and destruction, the charge-up effect in insulators, heat dissipation and recombination with molecules in the post bombardment environment. (Bachman *et al.*, 1988; Balik *et al.*, 2003; Evelyn *et al.*, 1997; Parada *et al.*, 2004, 2007a; Minamisawa *et al.*, 2007, 2007a).

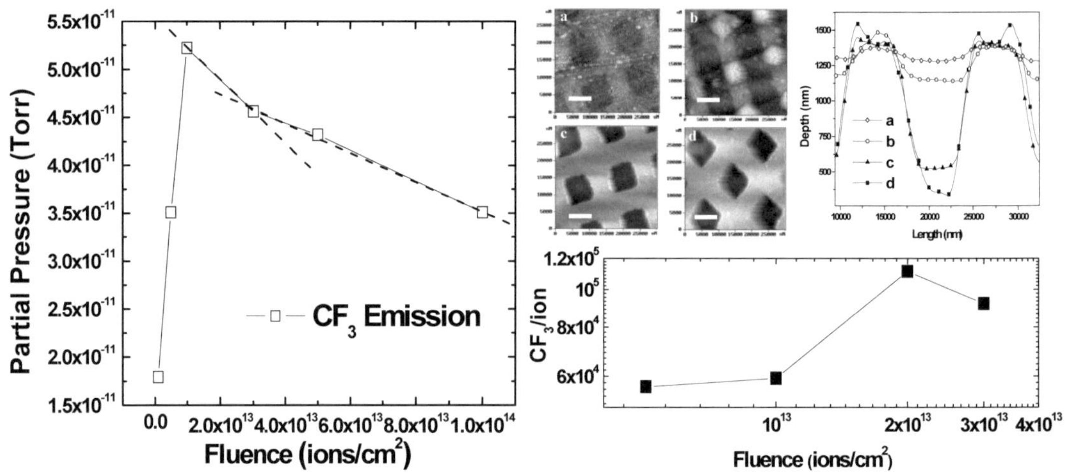

Fig. 2. Gas emission and mass loss of bombarded PFA thin films. The left plot shows the RGA profile of PFA films bombarded at different accumulated fluence. Atomic force microscopy images and the depth profiles (top-right) of the patterned films bombarded respectively with: a) 5×10^{12}, b) 1×10^{13}, c) 2×10^{13} and d) 3×10^{13} Au^+/cm^2. The scale in the AFM images represents 5 μm. The calculated physical etching yield for different implantation fluencies is shown in the bottom-right graph.

Data concerning mass loss of the ion bombarded PFA have been provided by two kinds of experiments: Measurment of the released gaseous species during bombardment and the the physical etching yield by surface analysis after irradiation. The PFA film thickness was 12.5 μm in all experiments.

In-situ Residual Gas Analyser (RGA) monitored a substantial emission of CF_3 molecules species from the PFA polymer film while bombarded at a maximum accumulated fluence of 1×10^{14} Au^+/cm^2 (Figure 2). The weak bonds between conjugated carbon when compared with F-C bonds and the relative higher mobility of the small radicals compared with the carbonic chains justify the higher emission of CF_3 gases during the ion beam modification. The idea is that CF_2 radicals are broken from the carbonic chains and recombined with adjacent fluorine atoms. For fluences up to 1×10^{13} Au^+/cm^2, the gas emission increases and after this value decreases due to the high level of fluorine loss and induced carbon crosslinking to form a more stable graphitelike material.

Figure 2 shows the atomic force microscopy AFM image of samples stenciled while bombarded at different accumulated fluences. The calculated physical etching yield extracted from the topographic AFM images of PFA films is about 9.0×10^4 CF_3 molecules emitted per incident ion. This value is more than 10^3 times higher than the sputtering yield simulated by TRIM06 software (Ziegler *et al.*, 1985). This deviation is attributed to thermal evaporation of the polymer. At low ion beam currents, low physical etching yield was observed, supporting the influence of thermal sublimation.

Figure 3a displays the Raman spectra of a thin PFA film and one bombarded at 1×10^{13} Au^+/cm^2 fluence, showing the presence of CF and CO bonds with peaks around 731.0 and 1381.1 cm^{-1}, respectively. At 1×10^{14} Au^+/cm^2 fluence the accumulated yield increased by a factor of ten for the same acquisition time while conserving the original bonds. This effect is attributed to enhanced fluorescence due to the influence of the implanted Au particles impurities on the PFA surface that formed nanometer sized metal clusters or surface grains. The PFA characteristic CF and CO bonds signals disappear in the sample bombarded at 1×10^{15} Au^+/cm^2 fluence, giving place to the D and G vibrational modes from amorphous carbon with peaks around 1329 and 1585 cm^{-1}, respectively. The D band is assigned to zone centers phonons of the E_{2g} symmetry and the G band to K-point phonons of the A_{1g} symmetry (Ferrari & Robertson, 1999).

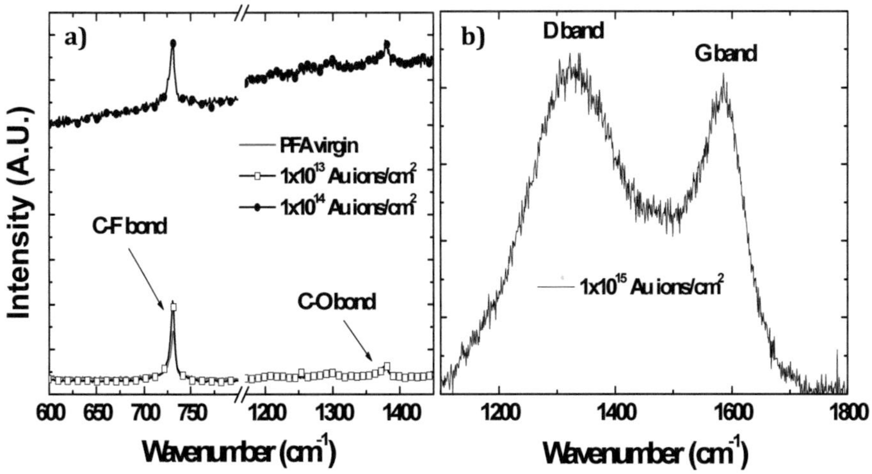

Fig. 3. Raman analysis of PFA thin films bombarded at different accumulated fluencies. At fluencies lower than 1×10^{14} Au^+/cm^2 (a), no significant change is observed in the PFA chemical bonds. At 1×10^{15} Au^+/cm^2 (b) accumulated fluence, the polymeric chains are modified to a graphite-like chemical structure due to substantial fluorine emission.

4. Probing pore formation

The fabrication of pores in freestanding PFA thin membranes by direct ion bombardment was controlled by a feedback system. The apparatus monitors the nanopore diameter when the ion beam impinges the polymer membrane defining a hole through which He gas is released and detected in an *in-situ* RGA (figure 4). PFA films were stenciled by a 2000 sq/inch mesh (5 × 5 μm^2 square shape openings) while bombarded by a 5 MeV Au^{+3} ion beam.

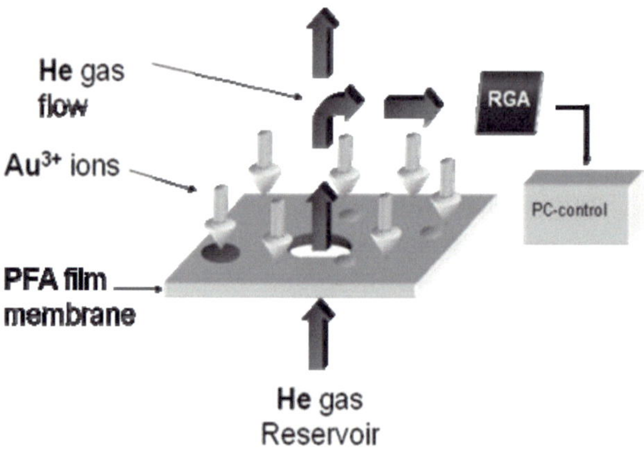

Fig. 4. Core idea of the feedback system built to monitor the pore formation. The pore formation in the PFA thin membrane, created by ion-induced physical etching, releases He gas from the reservoir, which is detected by the RGA and monitored in the PC control.

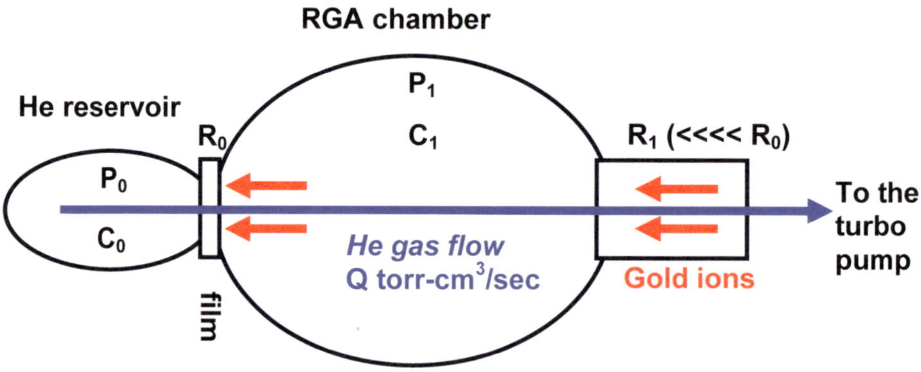

Fig. 5. Schematic of the experimental set-up for pore fabrication. The finite He gas supply contained in the reservoir has an initial pressure P_0 in the order of 10^3 Torr, while the RGA reads the He partial pressure P_1 of the order of 10^{-7} to 10^{-10} Torr. Helium partial pressure near the turbo pumps is in the order of 10^{-12} Torr. C_0 is the He reservoir volume (about 3 mm^3), C_1 is the RGA chamber volume (about 10^3 cm^3) and R_0 is the impedance (sec/cm^3) of the membrane that is many orders larger than the impedance of the 3 mm orifice R_1 to the beam line vacuum pumps. The red arrows symbolize the 5 MeV gold ion beam.

Figure 5 shows a schematic of the feedback system, where ***P*** and ***C*** denote, respectively, pressure and volume of the compartments and ***R*** the impedance of the gas channels. Specifically, the behavior of the system can be described as a gas flow through a channel with a difference of pressure, which defines the conductance ***1/R*** as the rate flow per unity difference of pressure. Because of the light atomic mass, He diffusion through the PFA membrane is observed even before the pore formation. Therefore, the gas flow through the membrane has one dynamic before the opening of the pore ($t<t_0$) and another after the pore formation ($t>t_0$). The approximate solutions of the pressure behavior inside the RGA chamber (P_1) for $t<t_0$ and $t>t_0$ are given, respectively, by equations 1 and 2 (Dushman, 1962):

$$P_1 = P_0 \frac{R_1}{R_0}\left[1 - e^{-\frac{t}{R_1 C_1}}\right] e^{-\frac{t}{R_0 C_0}} \qquad t<t_0 \tag{1}$$

$$P_1 = P_0 e^{-\frac{t_0}{R_0 C_0}}\left[e^{-(t-t_0)\left(\frac{1}{R_p}+\frac{1}{R_0}\right)\frac{1}{C_0}}\right] R_1\left(\frac{1}{R_0}+\frac{1}{R_p}\right) \qquad t>t_0 \tag{2}$$

$$\tau_2 = R_{eff} C_0 \qquad 1/R_{eff} = 1/R_p + 1/R_0$$

When pores are formed, i.e., two or more gas conductances are connected in parallel, the total impedance is determined by the reciprocal of the sum of the inverse of ***R*** for each channel. The observance of the time constants before $\tau_0 = R_0 C_0$ and after $\tau_2 = R_{eff} C_0$ pore formation determines the conductance $1/R_p$ of the pores with the formula below equation 2. The value of the conductance $1/R_p$ enables the calculation of the pore dimensions.

Figure 6 displays a logarithm representation of the pressure in the RGA chamber $P_1(t)$, observed for two samples. The experimental results are fit using equations 1 and 2. The ***1/e*** characteristic time constant $R_0 C_0$ is about 2.65 minutes for both samples. Using the value of the volume C_0 of 2.5 x 10^{-3} cm^3, we have an accurate determination of the gas diffusion impedance of the film $R_0 = 7.9 \times 10^4$ sec/cm^3. A transient rise in pressure observed during pores formation is a "relaxation" effect, easily understood as the transition to a higher pressure in the RGA chamber volume. The observance of the transient rise signal is used as the initial time t_0 for triggering the feedback system, since it corresponds to the opening of pores. The ion beam is then blocked after a final time t_f, so the ion fluence accumulated during $\Delta t = t_f - t_0$ is used to control the final pores diameter. Notice that the ion beam current was optimized and kept constant during the system calibration. After Δt, the time constant $R_{eff} C_0$ is lowered in respect to $R_0 C_0$ as an indication of the additional conductance of the created pores. From figure 6, the time constants $R_{eff} C_0$ extracted for Δt_1 = 1 and Δt_2 = 1.5 minutes are, respectively, 1.88 minutes and 1.0 minutes. Considering that 1360 pores were simultaneously fabricated, the average of conductance per pore calculated using the formula below equation 2, are $\boldsymbol{1/R_p(\Delta t_1)}$ = 5.7 × 10^{-9} sec/cm^3 and $\boldsymbol{1/R_p(\Delta t_2)}$ = 2.3 × 10^{-8} sec/cm^3.

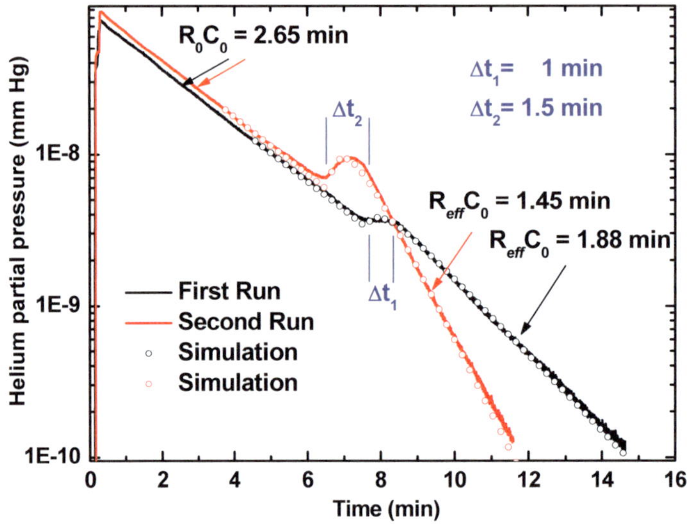

Fig. 6. RGA monitoring signal measured for two PFA samples bombarded during different times $\boldsymbol{\Delta t}$. After blocking the ion beam, the time constant $\boldsymbol{R_{eff}C_0}$ is lowered with respect to $\boldsymbol{R_0C_0}$ as an indication of the additional conductance of the created pores. The off-set between $\boldsymbol{t_0}$ in $\boldsymbol{\Delta t_1}$ and $\boldsymbol{\Delta t_2}$ is attributed to a small variation on the film thickness.

Considering that the mean free path of He in the experimental conditions is $\boldsymbol{\lambda}$ = 8.8 × 10^{-4} cm or 8800 nm, larger than the 50 nm to 2 μm pores produced by ion bombardment, the He gas flow through the pores is in the free molecular flow regime, where the atoms do not collide with each other while passing through a pore. The following equation gives a convenient numerical version of the conductance of a cylindrical tube with length $\boldsymbol{L}$ and radius $\boldsymbol{a}$ at 300 K temperature (Dushman):

$$1/R = 0.032\frac{a^3}{L} \quad \text{liters / sec} \tag{3}$$

Substituting the conductance per pore extracted from figure 6 in equation 3 for a 12.5 μm channel, pore diameters measured by the RGA are $\boldsymbol{D_{RGA}(\Delta t_1)}$=260 nm and $\boldsymbol{D_{RGA}(\Delta t_2)}$=415 nm. These results are higher than the ones measured by AFM images of $\boldsymbol{D_{AFM}(\Delta t_1)}\approx$100 nm and $\boldsymbol{D_{AFM}(\Delta t_2)}\approx$300 nm, which suggests that the effective channel lengths are smaller and that the conduits are not perfect cylindrical tubes.

5. PFA thin porous membranes

Optical microscopy inspection of the fabricated membranes reveals different dimensions of the pores in the bombarded and in the opposite face of the film. Whereas the pore sizes in the bombarded face is constant and equivalent to the stencil mask shape, the pore dimensions in the non-bombarded face can be controlled by tuning the accumulated ion fluence. Therefore, the membrane conduits have conical-like shapes and the pore size in the non-bombarded face defines the effective filtration area. This result confirms the variation in the pore diameters extracted from the RGA measurements.

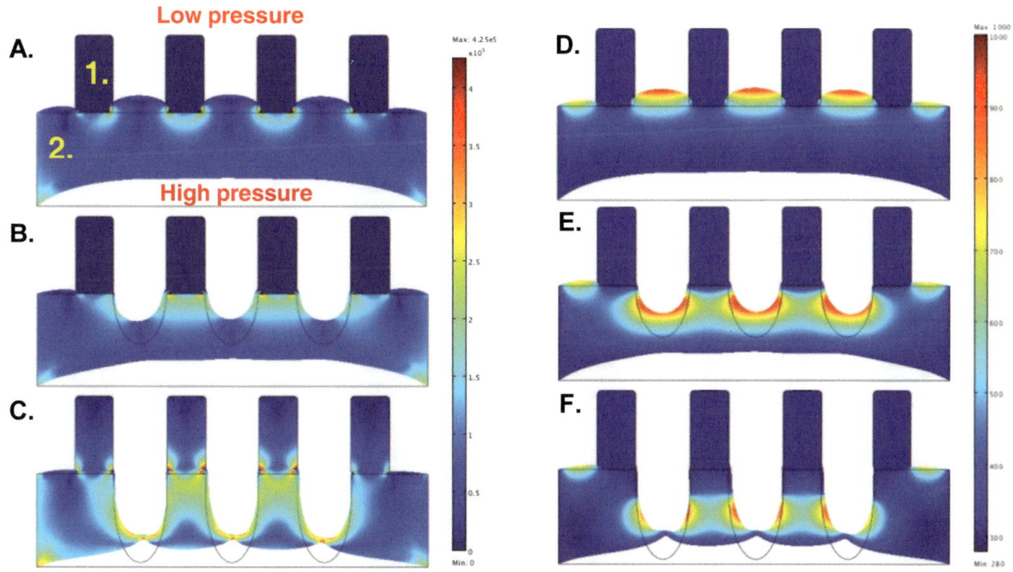

Fig. 7. Finite element simulations of strain and temperature superposed by deformation of the film during bombardment of the stencil masked PFA films. The black line represents the shape of the film without pressure-induced deformation. The PFA film is indicated by 1 while the metal mask is indicated by 2. The pressure across the film leads to deformation of the soft film, constrained by the metal mask. The strain (A, B and C) in the unmasked areas is lower than in the masked ones, which creates low density regions that allow higher ion penetration. Simultaneously, the metal mask acts as a heat sink for the power delivered by the ion beam, which focus the temperature in the center of the unmasked areas (D, E and F).

The pore shape and the high physical etching yield is explained in terms of thermal and strain effects that act in the polymer during irradiation as shown in the COMSOL simulations in figure 7. The strain acting in the unmasked areas decreases the density of polymer chains in the center of these regions, allowing higher penetration of gold ions, and consequently concentrated bond scissoring (figure 7A and B). In the instant of the pore formation, the strain effect is directly responsible for the pores opening (figure 7C). Simultaneously, the metal mask acts as a heat sink for the power delivered by the ion beam, leading to high temperature concentration (up to 1000°C) in the center of the unmasked areas of the polymer film (figure 7D and E). Although a complete phase diagram for PFA is not readily available in the literature, assuming the PFA melting point of around 310°C, sublimation at 1000°C may be a possible explanation for the high physical etching yield during bombardment. Combined, both effects lead to a concentration of physical etching in the center of the unmasked areas, and consequently, to the conical-like shape formation of the conduits.

Figure 8 shows the Raman scattering spectra extracted inside and in a non-bombarded adjacent area of a pore fabricated at 1 × 10^{13} Au^{+}/cm^{2} fluence. The C-F and C-O bonds inside the pore are conserved when compared with the masked area. This is an evidence of the fluorpolymer property to decompose under ion bombardment leaving relatively undamaged material.

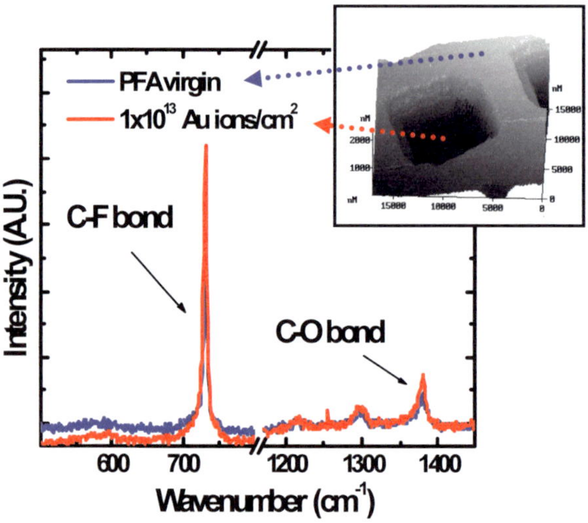

Fig. 8. Damage analysis in one processed pore evaluated by Raman spectroscopy. The graph compares the Raman spectra measured inside one pore and in an adjacent masked area (inset AFM image). The chemical structure of the pores is unchanged after 1 × 10^{14} Au^{+}/cm^{2} implantation, consequently, keeping the polymer material properties.

At 1 × 10^{14} Au^{+}/cm^{2}, circular micropores with ~2 μm diameter and uniform distribution in space were fabricated in the non-bombarded face of the PFA thin film membrane as shown in the optical microscopy image on figure 9. The impression that some pores are closed is attributed to focus artifact, however, the inset AFM image confirms that all the pores are effectively opened. Distances between adjacent pores have an average of approximately 12 μm, which matches with the center of the collimation squares in the stencil mask.

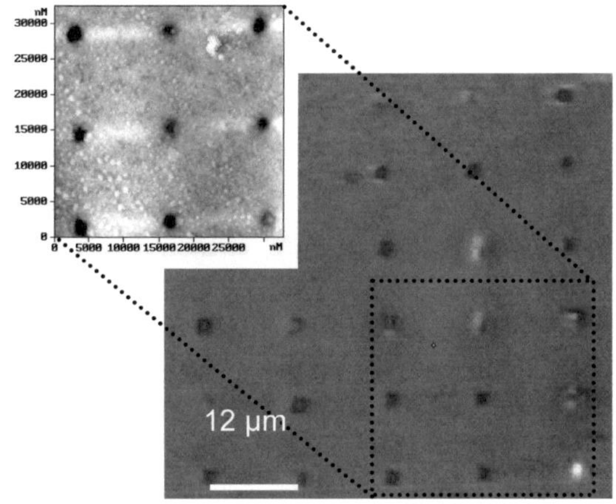

Fig. 9. Optical microscopy and AFM images of a PFA microporous membrane. Although the optical microscopy images gives the impression that some pores are closed due to artifacts of measurement, the AFM topography image (inset) confirms that all pores are opened.

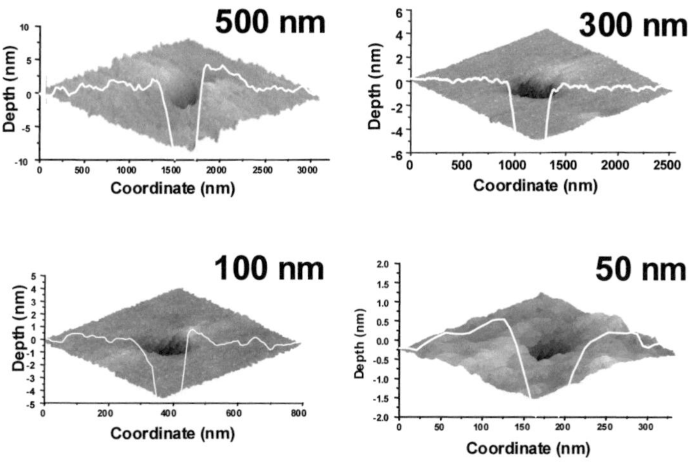

Fig. 10. Atomic force microscopy images superposed by topography profiles of nanopores with diameters ranging from 50 to 500 nm.

The PFA porous membranes fabricated by direct ion-induced physical etching have better pore size and spatial distribution compared to the tortuous path and polymer track-etched membranes, while maintaining the fluorpolymer properties. Simultaneously, the PFA porous membrane filtration capacity is almost comparable to the high-flux microsieve silicon nitride membranes. Figure 10 shows the AFM images superposed on the topography profiles of nanopores with 50, 100, 300 and 500 nm diameters fabricated at different accumulated fluence. Having pores at scales smaller than 500 nm, the PFA porous membranes are strong, chemically resistant membranes for air monitoring and sampling in aggressive environments. At this scale, bacteria or other microorganisms can be filtered in water or air treatment.

6. Conclusions

In this chapter we firstly assessed the current stage of development of porous thin film membranes for filtration applications. As discussed in the introduction, porous membrane performance requires high permeability, high selectivity, enhanced resistance to biofouling, and resistance against solvents, high- and low-pH environments, and oxidizers. Additionally, the cost-effectiveness of this technology is a major concern for financially disadvantaged societies with scarce water resources. ***Innovative processes for fabrication of porous membranes as well as the development of new polymer precursors is the key solution.***

Finally, we report our recent development in the fabrication of PFA fluorpolymer porous thin film membranes by ion-induced physical etching using a stencil mask technique. PFA fluorpolymers are candidates for advanced filtration membranes for being chemically inert, potentially resistant to biofouling because of its lowest adhesion coefficient known. We show that Au+ ion bombardment at specific conditions is able to induce a high level of physical etching through mechanical sputtering, enhanced by thermal sublimation of the PFA film. In our process, porous membranes with homogeneously distributed pores down to 50 nm diameter were fabricated. Chemical structure analysis demonstrate that the post-processed pore channels are relatively undamaged, consequently, maintaining their chemical resistant properties. Lastly, we presented the development of a feedback ion beam controlled system able monitor in real time the pore formation, by monitoring the gas flow through the bombarded films.

PFA porous membranes potentially ***offer a combination of the high performance of the microsieve membranes and the cost-effectiveness and resistant material performance of fluorpolymer membranes***. At the present stage, our PFA porous thin film membranes are a realistic alternative to substitute the low-flux performance fluorpolymer tortuous path membranes. Further research must be focused in different ion beam parameters, possibly in high current ion beam bombardment, and in the application of the process in similar fluorpolymers materials such as ETFE and FEP.

7. References

Bachman B. J., & Vasile M. J. (1988). *Proceedings of the IEEE 38th Electronics Components Conference*. Los Angeles, CA.

Balik, C. M., Said, M. A., & Carlson, J. D. (2003). High-energy ion implantation of polymers: Poly(ethylene terephthalate). *Journal of Polymer Science, Part B: Polymer Physics*, 25(4), 817-827.

Caplan, M. R., Chiang, C. Y., Lloyd, D. R., & Yen, L. Y. (1997). Formation of microporous Teflon® PFA membranes via thermally induced phase separation. *Journal of membrane science*, 130, 219-237.

Dushman, S. (1962). *Scientific Foundations of Vacuum Technique*. New York, NY: John Wiley & Sons.

DuPont. (1996). PFA Films bulletin: http://www2.dupont.com/Teflon_Industrial/en_US/assets/downloads/h04321.pdf

Evelyn, A. L., Ila, D., Zimmerman, R. L., Bhat, K., Poker, D. B., & Hensley, D. K. (1997). Resolving the electronic and nuclear effects of MeV ions in polymers. *Nuclear Instruments and Meth*ods B, 127/128, 694.

Ferrari, A. C., & Robertson, J. (1999). Interpretation of Raman spectra of disordered and amorphous carbon. *Physics Review B*, 61, 14095-14107.

Ferain, E., & Legras, R. (1997). Characterization of nanoporous particle track etched membrane. *Nuclear Intruments and Methods in Physics Research Section B*, 131, 97-102.

Ferain, E., & Legras, R. (2001). Characterization of nanoporous particle track etched membrane. *Nuclear Intruments and Methods in Physics Research Section B*, 174, 116-122.

Fologea, D., Gershow, M., Ledden, B., McNabb, D. S., Golovchenko, J. A., & Li, J. (2005). Detecting single stranded DNA with a solid state nanopore. *Nanoletters,* 5, 1905-1909.

Girones, M., Lammertink, R. G. H., & Wessling M. (2005). Protein aggregate deposition and fouling reduction strategies with high-flux silicon nitride microsieves. *Engineering with membranes: Medical and Biological Applications*, 273(1-2), 68-76.

Henriquez, R. R., Ito, T., Sun, L., & Crooks, R. M. (2004). The resurgence of Coulter counting for analyzing nanoscale objects. *Analyst*, 129, 478-482.

Li, J., Gershow, M., Stein, D., Brandin, E., & Golovchenko, J. A. (2003). DNA molecules and configurations in a solid-state nanopore microscope. *Nature Materials*, 2, 611–615.

Li, J., Stein, D., McMullan, C., Branton, D., Aziz, M. J., & Golovchenko, J. A. (2001). Ion-beam sculpting at nanometer length scales. *Nature*, 412, 166-169.

Minamisawa, R. A., Almeida, A., Abidzina, V., Parada, M. A., Muntele, I., & Ila, D. (2007). Effects of low and high energy ion bombardment on ETFE polymer. *Nuclear Instruments and Methods in Physics Research B*, 257, 568-571.

Minamisawa, R. A., Almeida, A., Budak, S., Abidzina, V., & Ila, D. (2007). Surface damage studies of ETFE polymer bombarded with low energy Si ions (<100 keV). *Nuclear Instruments and Methods in Physics Research B*, 261, 1159-1161.

Mochel, M. E., Eades, J. A., Metzger, M., Meyer, J. I., & Mochel, J. M. (1984). Electron beam cutting in amorphous alumina sheets. *Applied Physics Letters*, 44(5), 502-505.

Mutoh, Y. (1987). Porous fluorine resin membrane and process for preparing the same. U.S. Pat. N° 4,702,836. Fujisawa, JP.

Parada, M. A., Minamisawa, R. A., Almeida, A., Muntele, C., Zimmerman, R. L., Muntele, I., & Ila, D. (2004). Fluoropolymer studies for radiation dosimetry. *Brazilian Journal of Physics*, 34, 948-950.

Parada, M. A., Minamisawa, R. A., Moreira, M. V., Almeida, A., Muntele, I., & Ila, D. (2007). Damage effects of gamma and X-rays in polymer film electrets. *Surface and Coatings Technology*, 201, 8246-8249.

Quinn, J., Anderson, J. L., Ho, W. S., & Petzny, W. J. (1972). Model pores of molecular dimension: the preparation and characterization of track-etched membranes. *Biophysical Journal*, 12 (8), 990-1007.

Rogers, R. (1998). Membrane Materials. *Chemical Engineering News*, 76(34), 38-39.

Schenkel, T., Radimilovic, V., Stach, E. A., Park, S. J., & Persaud, A. (2003). Formation of a few nanometer wide holes in membranes with a dual beam focused ion beam system. *Journal of Vacuum Technology B,* 21(6), 2720-2723.

Vainrot, N., Eisen, M. S., & Semiat, R. (2008). Membranes in Desalination and Water Treatment. *MRS Bulletin*, 33, 16-20.

Van Rijn, C. J. M., Nijdam, W., Kuiper, S., Veldhuis, G. J., van Wolferen, H., & Elwenspoek, M. (1999). Microsieves made with laser interference lithography for micro-filtration applications. *Journal of Micromechanical Microengineering*, 9, 170-172.

Ziegler, J. F., Biersack, J. P., & Littmark, U. (1985). *The Stopping and Range of Ions in Solids*, New York, NY: Pergamon Press.

3

Ion Transfer in Layer-by-Layer Films

Bin Wang and Ritesh N. Vyas

Abstract
The need for miniaturized devices has driven the exponential development of nanotechnology during the past two decades. One special field of paramount practical applications is electrochemical systems at the nano-scale. For successful development in such fields, an in-depth analysis of mass and/or charge transfer mechanism is highly desired. In such a system, mass transfer often takes the form of ion transfer, which can also be viewed as charge transfer. For example, in fuel cell development, the catalyst layer deposited on a polyelectrolyte membrane would demand higher ionic (protonic) conductivity. In the electrochemical sensors, the overall performance depends largely upon sensitivity of the thin films to recognize the analyte and the speed to communicate the resultant signals with the underlying electrodes. These phenomena are closely related to the (ionic) mass transfer within such films. A clear understanding of mass transfer in such an electrochemical system is the prerequisite for any significant progress in devices.
Fundamentally, three different types of mass-transfer phenomena exists for ionic species in electrolytes at electrodes: (1) diffusional transport under concentration gradient, (2) migration transport of oppositely charged ions under electric field of the electrode, and (3) convection transport due to physical stirring of the electrolyte. For electrodes modified with electroactive or redox films, the redox behavior is much more complicated. When the applied potential reaches the oxidation potential of the redox-active species in the film, the electron transfer from the electrode surface to the film is coupled with the simultaneous ionic transfer from electrolyte to the film for maintaining electro-neutrality. Thus, we observe two simultaneous mass-transfer processes at the same time and each one needs to be characterized individually.
A detailed study of mass-transfer within such films would first require conceptual understanding of the main characterization technique used. We will conduct a literature survey on various models used to characterize such mass transfer in layer-by-layer (LbL) films, and then propose a model for characterizing mass-transfer in LbL films that contain nanoparticles.

1. Introduction

Polyelectrolyte multilayers built through the layer-by-layer (LbL) method has been one of the most promising systems in the field of materials science.[1] Layered structures can be constructed by adsorbing various polyelectrolyte species onto the surface of solid or liquid material by means of electrostatic adsorption. The thickness of the adsorbed layers can be tuned precisely in the nanometer range. Materials of all forms besides polyelectrolytes, such as organic and inorganic particles and crystals, biomolecules, lipids, and viruses, can be used to form diverse shapes with complicated stratified structures. Many factors influence the formation process of polyelectrolyte LbL films, including polymer type, molecular weight, concentration, deposition time, salt type and concentration, pH, and solvent composition.[2-5] Among them, salt concentration has shown to have predominant effect on the resultant film structure. Polyelectrolytes are the most thoroughly investigated class of substances in LbL studies; thus, the preparation parameters are well correlated with the film structures made of polyelectrolytes. Within an LbL film, the anionic and cationic polymer chains complex to each other to such a high degree that small salt counterions are usually absent. The intrinsic structure of the polyelectrolyte LbL films provides them unique properties in membrane separations.[6-11] In general, LbL films pertain no internal structures because of heavy interpenetration between adjacent layers, a result of the intrinsic nature of the charge compensation within the layers.[12,13] The LbL films can be viewed as amorphous and the parameters are not excessive when modeling the structures. Schlenoff *et al.* have employed electrochemical methods to evaluate the transport of redox-active ions across LbL films, addressing variables such as number of layers, ion charge, total ion concentration, and temperature.[14,15] The critical importance of the mechanism of internal charge balance within the layered structures in separating ions in solutions has been revealed by a quantitative treatment of equilibria.

The mass transfer phenomenon in polyelectrolyte LbL films has been well documented. Schlenoff *et al.* have employed electrochemical methods to evaluate the transport of redox-active ions across LbL films, addressing variables such as number of layers, ion charge, total ion concentration, and temperature.[14-17] On the basis of the observed increases of limiting current with increases in the external salt concentration, they proposed that the diffusion of the probe ion in the film is described as hopping between "exchange sites" created within the film by the supporting electrolyte. The strong attenuation of the current is explained as a combination of a low diffusion coefficient and a low concentration of the electroactive ion in the film. The decrease in permeability with the number of layers is attributed to a decrease in the concentration gradient in the film due to an increase in film thickness. However, this and many other mass-transport studies in LbL films have been accomplished through different techniques like cyclic voltammetry and chronoamperometry.[10,18-21] In this study, we plan to systematically investigate the mass transfer phenomena in LbL thin films through the electrochemical impedance spectroscopy (EIS). EIS has been widely used as a versatile technique to characterize interfacial and transport properties of polymer films adsorbed on surface of an electrode.[22] The estimation of parameters such as diffusion coefficient and electron transfer rate constant can explain the kinetics at interfaces, thus this technique becomes relatively easy and accurate as compared to other techniques.[23-25] In applying EIS for electrochemical system analysis, an important consideration is to fit experimental data to a suitable model. The use of EIS to study LbL films has become more common once LbL films find potential future in the field of bio-sensors and electrochemical sensors.

An early study shows the use of EIS in studying membrane resistivity and capacitance where phospholipid bilayers were deposited on LbL polyelectrolyte surface.[26] EIS has been used to characterize the stability and insulating properties[27-30] of LbL films that are stabilized by different approaches like cross-linking and thermal conversion. The charge transfer resistance and film resistance are obtained directly from the fitting of experimental impedance spectrum to Randle's equivalent circuits (vide infra) by simulation software. The films were exposed to different conditions of *p*H and temperature and insulating or corrosion resistant properties of the films were observed by variation in the calculated resistances. Self-assembled monolayers[31] and multilayers of polyelectrolytes have been widely studied for analyzing permeability towards ionic diffusion at various stages of LbL deposition.[32-34] Zhao *et al.* have demonstrated potential applications of LbL assembly in preparing hybrid compounds containing Prussian-blue (PB) redox-mediators for future applications in bio-sensors.[35] The authors used PB nanoparticles protected by PDDA in alternating fashion with negatively charged glucose oxidase. The modified electrodes were able to catalyze the electro-reduction of hydrogen peroxide formed from the enzymatic reaction at a lower potential than bare electrodes. EIS has been one of the primary techniques to characterize the electronic communication of the analyte with the modified electrodes.[36-41] Recently, Calvo *et al.* demonstrated the use of EIS in measuring ionic diffusion coefficients in pyridine derivatized (PAH | PSS) multilayer films.[42] Other advances also include EIS studies for explaining catalytic and charge storage efficiencies of LbL assembled nano-scale species such as gold,[43-45] titanium dioxide,[46] vanadium pentoxide,[40] polyoxometalates,[47] and carbon nanotubes.[48] Thus, EIS has been widely applied in extracting absolute values of the electrochemical parameters governing electronic and ionic transfer in LbL films. However, in most of the cases studied above, the experimental data has been directly fitted to an equivalent circuit proposed by the researchers to explain the electrochemical phenomena in LbL films. In other words, the conceptual modeling of LbL films that can directly relate the microstructure of the film with the experimental EIS spectrum has been omitted. Silva *et al.* have performed such EIS modeling in LbL films of polyelectrolytes[49,50] as well as gold nanoparticles.[38,51] Here, we first review the proposed models for gaining further insight in EIS modeling of LbL films. Followed, we will develop an EIS model to describe the mass transfer in functional LbL films. We then will analyze experimental results by using such modeling.

Capillary Membrane Model (CMM)

The CMM model is developed to consider progressive reduction of the active area of the electrode with the increase in deposition of the films. The diffusion impedance of partially blocked electrodes is often modeled by assuming that the ensemble of active sites behaves as an ideal array of microelectrodes (Figure 1).[52] These models can be adapted for polyelectrolyte multilayers having capillaries (or preferential transport paths) and areas through which transport is hindered or very unfavorable. At the initial stages of the deposition effective coverage of the electrode for these films is small (Figure 2.1). A formation of pin-holes resulting from the disordered arrangement of polyelectrolyte chains is assumed such that these pinholes also persist during the sequential assembly of the polyelectrolyte layers, although their area is progressively reduced Eventually, at larger number of layers, the remaining electroactivity may still be associated with the previously formed pin-holes because these continue to be active-sites were the film is less dense but now redox-species has to diffuse through external layers before reaching the capillary (Figure 2.2). Let us assume that species **Ox** and **R** are electroactive species residing in the film. For such a non-linear diffusion Finklea *et al.*[52] have compared the problem with an electron transfer step of species **Ox** and **R** within the film with preceding and succeeding chemical steps of redox couple **A** and **B** as follows:

$$\mathbf{A} \underset{k_2}{\overset{k_1}{\longleftrightarrow}} \mathbf{O_x} \tag{1}$$

$$\mathbf{O_x} + e^- \leftrightarrow \mathbf{R} \tag{2}$$

$$\mathbf{R} \underset{k_1}{\overset{k_2}{\longleftrightarrow}} \mathbf{B} \tag{3}$$

For a blocking layer with coverage of θ, (1- θ) represents the area fraction of microelectrodes. When (1- θ) is less than 0.1, then coverage θ and the radius r_b of inactive area surrounding the active site in above figures are related by the following relation

$$1-\theta = \frac{r_a^{\,2}}{r_b^{\,2}} \tag{4}$$

Thus for disk shaped pin-holes, the rate constants have been expressed:

$$\frac{k_1}{k_2} = \frac{(1-\theta)}{\theta} \tag{5}$$

$$k_1 + k_2 = \frac{\left[\frac{D}{4r_b^{\,2}}\right]\left[\frac{\theta}{(1-\theta)^2}\right]}{\left(0.3(1-\theta)^{-1/2}\right)^2} \tag{6}$$

Where D is the diffusion coefficient of redox-species.

The real and imaginary components of the faradaic impedance for coupled reaction schemes have been obtained as:

$$Z_{re} = R_{ct}\frac{\theta}{(1-\theta)} + \frac{\sigma}{\sqrt{\omega}} + \left(\frac{\sigma\theta}{(1-\theta)}\right) \tag{7}$$

$$Z_{img} = \frac{\sigma}{\sqrt{\omega}} + \left(\frac{\sigma\theta}{(1-\theta)}\right) \tag{8}$$

where q is the average time taken for diffusion of the oxidized and reduced species. The last two terms in above equations depict the diffusion process in the system. Based on the above equations Silva *et al.*[49] have extracted an expression solely for diffusion impedance in polyelectrolyte films as under:

$$Z_d^{(c)} = \sum_k \frac{\sigma_k^{(c)}}{\sqrt{\omega}}\left(1 + \frac{\theta}{(1-\theta)} + j + \frac{j\theta}{(1-\theta)}^{1/2}\right) \tag{9}$$

The Warburg parameter σ and diffusion coefficient D can be obtained according to the following equation:[53]

$$\sigma = \frac{RT}{n^2 F^2 A\sqrt{2}}\left(\frac{1}{c_{\mathrm{Ox}}{}^{*}\sqrt{D_{\mathrm{Ox}}}} + \frac{1}{c_{\mathrm{R}}{}^{*}\sqrt{D_{\mathrm{R}}}}\right) \tag{10}$$

If we consider the values for concentrations and diffusion coefficients of oxidized (c_o^*) and reduced species (c_R^*) to be same as those in the bulk then we can assume $c_{\mathrm{Ox}}{}^{*} = c_{\mathrm{R}}{}^{*} = c$ and $D_{\mathrm{Ox}} = D_{\mathrm{R}} = D$ and we get

$$\sigma_{\mathrm{k}}{}^{(c)} = \frac{\sqrt{2}RT}{n^2 F^2 A D_k^{f\,1/2} c_k^f} \tag{11}$$

where, the concentration of species k (k = 1 for $[Fe(CN)_6]^{4-}$ and k = 2 for $[Fe(CN)_6]^{3-}$) in the vicinity of the film surface. c_k^f is equal to its concentration in the bulk solution, c_k^S. The equivalent circuit for such a system is shown in Figure 3.

Homogeneous Membrane Model

Silva and co-workers studied the above model for the effects of supporting electrolyte and temperature on impedance response.[49,50] PSS | PAH multilayers have been characterized with EIS, however, a third situation is added where the number of layers become so large that the polyelectrolyte multilayers become practically homogeneous as shown in Figure 4. The film has also been modeled as a homogeneous membrane where the increase in the number of layers leads to an increase in the thickness L of the film, a decrease in the concentration gradient, and then a decrease in the current. This tendency, for the decrease in the permeability with the increasing layers may not occur when the last layer deposited and the electroactive species have opposite charge. In this case the decrease in concentration gradient associated with the increase in film thickness may be compensated by an enhanced inclusion of the species in the film. The film-solution interface is considered to be in electrochemical equilibrium. The diffusion coefficient D_k^f and the partition coefficient K_k of species k are allowed to vary with the number of layers. For a thicker homogeneous film, the diffusion is predicted by assuming that diffusion layer thickness l is finite and in the case of homogeneous polymer films is equal to the film thickness L. The diffusion impedance in this case has been derived as

$$Z_{\mathrm{k,d}}{}^{(m)} = \frac{\sigma_{\mathrm{k}}{}^{(m)}}{\sqrt{j\omega}}\,\frac{\tanh(a_{\mathrm{k}}\sqrt{\omega}) + K_{\mathrm{k}}\sqrt{D_{\mathrm{R}}}}{1 + b_{\mathrm{k}}\tanh(a_{\mathrm{k}}\sqrt{\omega})} \tag{12}$$

$$\sigma_{\mathrm{k}}{}^{(m)} = \frac{\sqrt{2}RT}{n^2 F^2 A D_k^{f\,1/2} c_k^f(0)} \tag{13}$$

where, $$a_{\mathrm{k}} = \frac{L}{\sqrt{2D_{\mathrm{k}}^{\ f}}}(1+j) \tag{14}$$

LbL Assembled Films of (PMo_{12} | PDDA)

It is clear from the above study that modeling of ionic and electronic conductivity in LbL films by EIS would be a significant research contribution. Recently, Hammond *et al.* separately measured ionic and electronic conductivities in LbL assembled conducting polymer films by de-doping the conducting polymer to minimize the contribution from electronic conductivity.[54] However, EIS modeling of LbL films to measure ionic and electronic conductivites individually has not been accomplished till date. Moreover, LbL films containing nanoparticles (NPs) are much less studied in this respect. Nanoparticles with dimension in the 1-100 nm range can be assembled with polyelectrolytes through the LbL method.[55] Both ionic and van der Waals interactions have significant contributions to depositing NPs, most water-based and highly charged, into LbL films. A combination of strong electrostatic attraction and firm binding to the surface polyelectrolyte layer makes the adsorbed NP layer thermodynamically more preferable than the dissolved state. Nanoparticles such as CdSe, TiO_2, SiO_2, and gold are among the well studied. These NPs are highly interesting for applications ranging from light emitting diodes to photovoltaics to sensors and to semiconductors. One particular field is electrochemical reactions in which NPs have widely been employed as catalysts due to the hugely enhanced surface areas and hence higher catalytic efficiencies.[56-58] The combination of the ultrathin feature of LbL films and the high surface area of NPs seems a highly desirable architecture for developing new catalysis systems. A fundamentally critical aspect in understanding electrochemical reactions within the NP-embedded LbL films is the mass transfer phenomenon through the membrane. The proposed study will deal with the ion diffusion in a thin film that is regulated not only by electrostatic forces between the composing polyelectrolyte chains and NPs, but also by the geometric constraints brought by the particulate matter. The knowledge obtained from such a study has both very broad industrial application potential and is the key to understanding other thin membrane systems immensely encountered in biology.

To successfully establish the preparation parameters in constructing NP-containing LbL films, we can maneuver two important parameters in such a process, NP concentration and solution ionic strength. We choose polyoxometalate (POM) clusters as the NP component to probe the LbL deposition parameters. Polyoxometalates are a well-known class of ångström-scale anionic clusters with much diversity in size, composition, and function. Their interesting properties include the high stability of most of their redox states, the possibility to tune their redox potentials by changing the heteroions and/or the addenda ions without affecting their structure, the variability of the transition metal cations which can be incorporated into the hetero-polyoxometalate structure, and the possibility of multiple electron transfer.[59] Because most POM clusters are water soluble, they are ideal candidates for the LbL assembly technique. There has been an increasing interest in constructing POM thin films through the LbL process.

We attempt to find a generic fabrication rule that applies to most POM clusters.[60] In the study, we used Keggin type POM clusters to study their layer-by-layer deposition behavior. We found that POMs can be adsorbed onto LbL assemblies from concentration range 0.1 to 5 mM. We use two situations, without added salt and with added salt (0.1 M). When the deposition was performed without added salt, the ionic strength of a solution was at a few mM scale originated from the counter ions and HCl added to adjust the pH. Under this condition, POM clusters adsorb at sub-monolayer coverage. The cluster concentration shows limited influence on the deposition process. When the cluster concentration was increased from 0.1 mM to 5 mM, a 50-fold increase, the coverage merely increased by ~15-30% for both POM species. When NaCl is added to the deposition solutions, the results are more complex. For depositions from a low POM concentration, the adsorption is still at the about-monolayer level. But when POM concentration is above 1 mM, multilayer structures start to form.

Here, we study the ionic diffusion phenomena of a redox couple in LbL assembled PMo_{12} films. It is our interest to observe the effects of varying porosity and loadings of the film on the diffusion process of redox-active ions. We first construct an EIS model that can be applied to POM films with varying loading concentrations. We find diffusion coefficients for the ferrocyanide redox couple to understand the effect of electrostatic attraction/repulsion between redox-probe ion and the microstructure of the LbL films. The microstructure of the films as interpreted with EIS has been compared with that obtained from cyclic voltammetry (CV) measurements.

2. Methodology

As stated in the Introduction, EIS has been widely used as a versatile technique to characterize interfacial and transport properties of polymer films, organic-inorganic coatings, and self-assembled monolayers adsorbed on surface of an electrode. The estimation of parameters such as diffusion coefficient and electron transfer rate constant that can explain the kinetics at interfaces becomes relatively easy and accurate as compared to other techniques. The typical Nyquist plots for polyelectrolyte films exhibit a characteristic semicircle at higher frequencies corresponding to kinetic control, and a straight line at lower frequencies (slope ~ 1) corresponds to mass-transfer control. A Randle's equivalent circuit model can be used in case of self-assembled films with or without modifications.[28] Briefly, the circuit consists of solution resistance R_s in series with the parallel combination of double-layer capacitance (C_{dl}) and charge-transfer resistance (R_{ct}) with Warburg impedance (Z_w) in series (Fig. 1). The equations for a simple Randle's circuit can be directly used for calculations if Warburg line on Nyquist plot has a slope close to 1. In our case, diffusion coefficient for $Fe(CN)_6^{3-/4-}$ redox species were calculated. The real and imaginary values for impedance for an ideal Randle's circuit (Figure 5) can be given as:[39]

$$(Z_{re})_1 = R_s + \frac{R_{ct} + \sigma\omega^{1/2}}{\left(C_{dl}\sigma\omega^{1/2} + 1\right)^2 + \omega^2 C_{dl}^2\left(R_{ct} + \sigma\omega^{-1/2}\right)^2} \tag{15}$$

$$(Z_{im})_1 = \frac{\omega C_{dl}\left(R_{ct} + \sigma\omega^{-1/2}\right)^2 + \sigma\omega^{-1/2}\left(C_{dl}\sigma\omega^{1/2} + 1\right)}{\left(C_{dl}\sigma\omega^{1/2} + 1\right)^2 + \omega^2 C_{dl}^2\left(R_{ct} + \sigma\omega^{-1/2}\right)^2} \tag{16}$$

where ω is the radial frequency and σ is the Warburg parameter. If the deposited film behaves as an ideal capacitor on the electrode, then we always observe a vertical line with a unity slope in low-frequency region.[9] For the high frequency region, the intercept with real impedance axis would also give the accurate values for solution resistance R_s of Figure 5. However, the charge transfer resistance R_{ct} and double layer capacitance C_f only describe the resistance and capacitance provided by the electrochemical double layer at the interface, an ideal semi-infinite diffusion. In our study of POM films, the experimental data show depressed semi-circles in high frequency region with a slight deviation from unity slope in lower frequency region. Thus, a resistance to the movements of redox ions inside and out of the deposited film needs to be addressed to the circuit along with the charge transfer resistance. Adding a film resistance (R_f) in parallel to double layer capacitance in the circuit has been employed for self-assembled monolayer. Silva and coworkers dealt with this problem differently and employed two models in the case of polyelectrolyte films with increasing thickness.[49,50] Here, the Randle's circuit was modified by addition of two more resistances, namely film resistance R_f offered by multilayers and R_m due to ohmic conduction within the film. The films in their studies exhibited non-linear diffusion; in contrast, our films demonstrate semi-infinite diffusion patterns. Here, we adapt the same Randle's circuit as in Figure 2.1.3.5.3, but add two more elements to the circuit to define the properties of film: film resistance (R_f) and film capacitance (C_f). Equations (15) and (16) are thus modified as

$$(Z_{re})_2 = R_s + \frac{\dfrac{R_f + (Z_{re})_1}{(\omega C_f)^2}}{[R_f + (Z_{re})_1]^2 + \left[(Z_{img})_1 - \dfrac{1}{\omega C_f}\right]^2} \tag{17}$$

$$(Z_{img})_2 = R_s + \frac{\dfrac{(Z_{img})_1}{(\omega C_f)^2} + \dfrac{((Z_{img})_1)^2 - ((Z_{re})_1)^2 - R_f (Z_{re})_1}{\omega C_f}}{[R_f + (Z_{re})_1]^2 + \left[(Z_{img})_1 - \dfrac{1}{\omega C_f}\right]^2} \tag{18}$$

where, $(Z_{re})_2 = Z'$ is the real component of the modified Randle's circuit, and $(Z_{re})_2 = Z''$ is the imaginary component of the circuit. After modifying the circuit, we can develop equations to calculate the diffusion coefficients for thin film samples. At low frequencies ($\omega \rightarrow 0$), Eq. 17 and 18 become:

$$Z' = R_s + R_f + R_{ct} + \sigma\omega^{-1/2} \tag{19}$$

$$Z'' = \sigma\omega^{-1/2} + 2\sigma^2 C_{dl} + 4\sigma^4 C_{dl}^2 C_f - R_f^2 C_f - R_{ct}^2 C_f \tag{20}$$

From Eq. 19, the plot of Z' vs. $\omega^{-1/2}$ gives slope = σ and intercept = (R_s + R_f + R_{ct}). The intersection of a Nyquist plot with x-axis gives the values for solution resistance R_s that can be used to calculate R_{ct} from the intercept values. The Warburg parameter σ and diffusion coefficient D can be obtained according to Ref. 61:

$$\sigma = \frac{RT}{\sqrt{2}n^2F^2A}\left(\frac{1}{D_{Ox}^{1/2}c_{Ox}} + \frac{1}{D_{red}^{1/2}c_{red}}\right) \tag{21}$$

where n is the number of electrons transferred (in this case 1), F is Faraday's constant (96485 C mol^{-1}), A is the electrode area (1 cm^2), R is gas constant (8.314 J mol^{-1} K^{-1}) and T is room temperature (298 K). Assuming diffusion coefficients $D_{ox} = D_{red} = D$ and concentrations $c_{ox} = c_{Red} = c_{bulk}$ we get:

$$D = \left(\frac{\sqrt{2}RT}{n^2F^2A\sigma c_{bulk}}\right)^2 \tag{22}$$

This equation resembles the one obtained by Bobacka and coworkers.[40] From Eq. 2, the calculated values for diffusion coefficients become approximation values and are called apparent diffusion coefficients.

3. Experimental Section

Chemicals. 3-aminopropyltriethoxysilane (APTES), phosphomolybdic acid hydrate (PMO_{12}, reagent grade) and poly(diallyl dimethyl ammonium chloride) (PDDA; M_W 250,000) were purchased from Sigma-Aldrich and their aqueous solutions were filtered immediately before use. Potassium ferricyanide, potassium ferrocyanide and sodium phosphate were purchased from Malinckrodt. The nanopure water used for all experiments was purified by Barnsted Nanopure II purification system. The resistivity was about 18 MΩ/cm. The PDDA solution (10 mM, pH 2.5) was used with or without the addition of 0.1 M NaCl while POM solutions are explained in Results sections.

Preparation of ITO electrode. Indium tin oxide (ITO) coated glass slides (Sigma-Aldrich, 80–100 Ω) were cleaned and coated with a monolayer of APTES according to published procedures.[62,63] Briefly, freshly cut ITO slides (25 × 10 × 1 mm) were consecutively sonicated in acetone, base-bath (10% KOH in isopropyl alcohol) and nanopure water for 20 minutes each. Afterwards, the slides were given electrochemical cycling (-200-700 mV) in 0.1 M HCl solution until the characteristic ITO curve was obtained. The slides were then dipped in APTES (2 v %, methanol) solution overnight to form a cationic monolayer. The slides were sonicated for 5 minutes in methanol solution before the deposition of multilayers.

Preparation of the multilayer. PMO_{12}|PDDA multilayer films were deposited according to our recently published procedure.[60] APTES modified ITO slides were dipped in PMO_{12} solution (10 min) and PDDA solution (10 min) in an alternating fashion for 10 consecutive times, between each dipping the slides were exposed to three consecutive water baths for 5 seconds each.

Electrochemical cell and the measurements. Electrochemical measurements were conducted on a Voltalab 10 PGZ 100 Potentiostat equipped with VoltaMaster 4 Electrochemical Software version 2.10. A conventional three-electrode setup was used. The working electrode was the ITO slide modified with self-assembled films, while a platinum wire was used as the counter electrode. An Ag/AgCl (3 M KCl) reference electrode was used for all measurements. EIS measurements were performed within the frequency range 1 Hz–100 kHz with amplitude of the applied sine wave potential as 10 mV and applied dc potential as

220 mV. A mixture of 5 mM concentration of $K_3[Fe(CN)_6]$ and $K_4[Fe(CN)_6]$ (1:1) in 25 mM sodium phosphate solution was used as redox probes.

4. Results and Discussion

Preparation and characterization of $(PMo_{12}/PDDA)_{10}$ films. $(PMo_{12}|PDDA)_{10}$ films (samples **1-4**) were prepared according to the conditions in Table 1. Cyclic voltammetry (CV) was used to characterize the electrochemical behavior of the films. A recent study has shown that electrostatic attraction/repulsion between terminating layer in multilayer assembly and solubilized charged redox species play an important role in diffusion as well as electron transfer at the interface.[43] A multilayer assembly with a layer of negatively charged nanoparticles on the top has proved to be barrier for redox processes of negatively charged couples such as $[Fe(CN)_6]^{3-/4-}$.[38,45,46] However, we observe a different behavior when the samples in this study are coated with PMo_{12}^{3-} clusters. The CV curves for $(PMo_{12}|PDDA)_{10}$ multilayers, in which a PMo_{12} layer was the outmost layer, were obtained when $[Fe(CN)_6]^{3-/4-}$ was used as the redox couple. CV curves (Figure 7) for sample **1** at increasing scan rates (10-70 mV·s^{-1}) show a linear increase in redox peak currents with the square roots of their corresponding scan rates. Similar behavior has been observed with samples **2**, **3**, and **4**. The CV shapes are not broad or having a plateau, indicating that the current is not originated from an array of microelectrodes nor there is a slow diffusion through films.[64,65] Combined with Figure 8, there seems like a semi-infinite linear diffusion in the film for each sample. Figure 8 shows the comparison of redox curves for samples **1–4** when the scan rate was 100 mV·s^{-1}. We observe anodic peak at ~300 mV and cathodic peak at ~-50 mV for all samples, with peak-to-peak separation ΔE_p = 350±50 mV indicating quasireversible voltammograms. The difference in peak-to-peak separation ΔE_p suggests variation in film structure or charge.[23] The increase in the peak currents for films made with higher ionic strengths (samples **2** and **4**) may suggest an increase in the permeability of the films due to increased porosity. The high permeability may be originated from a moderate delamination of films and less-stratified microstructure. Compared to the PMo_{12} surface coverage trend,[60] the diffusion of $[Fe(CN)_6]^{3-/4-}$ couple into the multilayer assembly seems independent of the PMo_{12} loading.

Mass transport analysis of $(PMo_{12}/PDDA)_{10}$ films. Figure 9 illustrates the fits obtained with the modified Randle's circuit to impedance data for samples **1–4**. The fitting of the data to the equivalent circuit (Figure 6) was performed by using Zview software. The parameter values, given in Table 2, agree well with the EIS experimental data. In Figure 9, the Nyquist plots for samples **1–4** exhibits a characteristic semicircle at higher frequencies corresponding to kinetic control and a straight line at lower frequencies corresponding to mass-transfer control. The impedance data for each sample do confirm the trends obtained by cyclic voltammetry. The slope of the straight line at lower frequency in the Nyquist plots remains close to one, reaffirming semi-infinite planar diffusion. Compared to other samples, the decrease in the diameter of semi-circle for sample **4** can be explained by decrease in film resistance, which further confirms increase in its permeability as explained by cyclic voltammetry. When mercaptoundecanoic acid (MUA) stabilized gold nanoparticles are adsorbed layer-by-layer with poly(L-arginine), the films have shown a higher permeability for $[Fe(CN)_6]^{3-/4-}$ couple at increasing layers.[51] The authors describe that the negative COO^- charge becomes further away from nanoparticle surface which in turn helps reduce the repulsion between the redox species and negatively charged NPs. However, in (PMo_{12} | PDDA)$_{10}$ samples, the change in the microstructure of the films causes enhanced diffusion. The double layer capacitance C_d depends upon the dielectric and insulating features at the interface of electrolyte and electrode. A decrease in C_d for films at higher surface coverage (sample **4**) is observed. An apparent increase in the diffusion coefficient for sample **4** is observed as compared with other samples (Table 2). Thus, high ionic strengths of dipping solutions may induce high porosity in the films that tend to demonstrate enhanced diffusion of redox species.

Microstructure interpretation of ion transfer in LbL films. In previous work we have predicted two different micro-structures for POM films prepared with dipping solutions of different ionic strengths and concentrations. At lower ionic strengths and concentrations, the multilayer films are predicted to observe a stratified structure owing to the flat, train configuration adapted by PDDA chains (Figure 11.1). At higher ionic strengths and concentrations, PDDA chains adapt a loop and tail configuration that allows the formation of a more porous structure into which larger amount of POM clusters could occupy if available (Figure 11.2). However, along with the variation in porosity of the films we have also varied the charge on the terminating layer. Thus, we need to consider the microstructure of the films as well as the electrostatic forces at the interface tandem while explaining ionic diffusion. For the films with stratified structure, the ionic diffusion would largely depend upon the thickness of the overall film,[66] loading of POMs and finally the electrostatic attraction/repulsion at film-electrolyte interface.[38] Due to lesser porosity, ionic diffusion in such films would also depend upon the surface coverage of the film itself.[58] However, for porous microstructure, each pore inside the film can be imagined as an empty hole surrounded by a cluster of negatively charged POMs rendering a highly negative electric field on the outer edge of the hole. Thus, it would be increasingly difficult for a negatively charged redox ion to diffuse through a multilayer assembly by overcoming the repulsive effect on each pore. Meanwhile, it would be easier for a positively charged redox-probe to diffuse through such a hole. The amount of electrostatic attraction/repulsion inside the film would largely depend on the amount of POM loaded and available porosity of the film. For example, while comparing samples **2** and **4** one can imagine a microstructure with higher loadings of POM in sample **4** with a high porosity as compared to sample **2**. Thus, the electrostatic attraction/repulsion between POM clusters in the film and redox ions for sample **4** should be higher than sample **2**. Overall, the electrostatic forces at the interface as well as within the films would play a role in diffusion of porous films.

5. Conclusion

Results obtained from this study contain information for mass transfer through thin membranes. A model developed for membrane with particle components has broader applications. The conditions considered in this study can be easily applied to many situations in both industry and basic science. Here, we are able to develop a modified Randle's circuit equivalent to calculate redox species diffusion coefficients through layer-by-layer thin films deposited on electrodes. The films generally show Nyquist plots with a Warburg line slope ~ 1. From the diffusion coefficient calculations, it appears that using high ionic strength solutions would not help greatly in achieving higher ionic diffusion in the case of POM films. However, the ionic strength and concentrations of the dipping solutions also influence the ionic diffusion of the films. For electronic diffusion, higher ionic strength solutions provide a better loading of POMs, which in turn help in enhancing the electronic conduction.

6. References

1. G. Decher, Science **277**, 1232–1237 (1997).
2. J. B. Schlenoff, H. Ly, and M. Li, J. Am. Chem. Soc. **120**, 7626–7634 (1998).
3. S. T. Dubas and J. B. Schlenoff, Macromolecules **32**, 8153–8160 (1999).
4. J. B. Schlenoff and S. T. Dubas, Macromolecules **34**, 592–598 (2001).
5. H. W. Jomaa and J. B. Schlenoff, Macromolecules **38**, 8473–8480 (2005).
6. G. B. Sukhorukov, A. A. Antipov, A. Voigt, E. Donath, and H. Möhwald, Macromol. Rapid Commun. **22**, 44–46 (2001).
7. J. H. Dai, A. M. Balachandra, J. I. Lee, and M. L. Bruening, Macromolecules **35**, 3164–3170 (2002).
8. A. M. Balachandra, J. H. Dai, and M. L. Bruening, Macromolecules **35**, 3171–3178 (2002).
9. M. D. Sullivan and M. L. Bruening, J. Am. Chem. Soc. **123**, 11805–11806 (2001).
10. L. Krasemann and B. Tieke, Langmuir **16**, 287–290 (2000).
11. W. Jin, A. Toutianoush, and B. Tieke, Langmuir **19**, 2550–2553 (2003).
12. J. Schmitt, T. Grünewald, G. Decher, P. S. Pershan, K. Kjaer, and M. Lösche, Macromolecules **26**, 7058-7063 (1993).
13. M. Lösche, J. Schmitt, G. Decher, W. G. Bouwman, and K. Kjaer, Macromolecules **31**, 8893–8906 (1998).
14. T. R. Farhat and J. B. Schlenoff, Langmuir **17**, 1184–1192 (2001).
15. H. H. Rmaile, T. R. Farhat, and J. B. Schlenoff, J. Phys. Chem. B **107**, 14401–14406 (2003).
16. T. R. Farhat and J. B. Schlenoff, J. Am. Chem. Soc. **125**, 4627–4636 (2003).
17. S. T. Dubas, T. R. Farhat, and J. B. Schlenoff, J. Am. Chem. Soc. **123**, 5368–5369 (2001).
18. J. Pozuelo, E. Riande, E. Saiz, and V. Compañ, Macromolecules **39**, 8862–8866 (2006).
19. A. A. Antipov, G. B. Sukhorukov, and H. Möhwald, Langmuir **19**, 2444–2448 (2003).
20. A. Fery, B. Schöler, T. Cassagneau, and F. Caruso, Langmuir **17**, 3779-3783 (2001).
21. I. Rubinstein and I. Rubinstein. J. Phys. Chem. **91**, 235–241 (1987).
22. B. Lindholm-Sethson, Langmuir **12**, 3305–3314 (1996).
23. J. J. Harris, P. M. DeRose, and M. L. Bruening, J. Am. Chem. Soc. **121**, 1978–1979 (1999)..
24. M. Zhao, Y. Zhou, M. L. Bruening, D. E. Bergbreiter, and R. M. Crooks, Langmuir **13**, 1388–1391 (1997).
25. S. A. Merchant, D. T. Glatzhofer, and D. W. Schmidtke, Langmuir **23**, 11295–11302 (2007).
26. R. N. Vyas and B. Wang, Electrochem. Commun. **10**, 416–419 (2008).
27. R. P. Janek, W. R. Fawcett, and A. Ulman, Langmuir **14**, 3011–3018 (1998).
28. J. J. Harris and M. L. Bruening, Langmuir **16**, 2006–2013 (2000).
29. S. Han and B. Lindholm-Sethson, Electrochim. Acta **45**, 845–853 (1999).
30. V. Pardo-Yissar, E. Katz, O., Lioubashevski, and I. Willner, Langmuir **17**, 1110–1118 (2001).
31. W. Zhao, J. J. Xu, C. G. Shi, and H. Y. Chen, Langmuir **21**, 9630–9634 (2005).
32. R. Pei, X. Cui, X. Yang, and E. Wang, Biomacromolecules **2**, 463–468 (2001).

33. N. Kohli, D. Srivastava, J. Sun, R. J. Richardson, I. Lee, and R. M. Worden, Anal. Chem. **79**, 5196–5203 (2007).
34. W. Sun, R. Gao, and K. Jiao, J. Phys. Chem. B. **111**, 4560–4567 (2007).
35. F. Battaglini, E. J. Calvo, C. Danilowicz, and A. Wolosiuk, Anal. Chem. **71**, 1062–1067 (1999).
36. F. Huguenin, M. Ferreira, V. Zucolotto, F. C. Nart, R. M. Torresi, O. N. Oliveira, Jr. Chem. Mater. **16**, 2293–2299 (2004).
37. M. G. Friedrich, V. U. Kirste, J. Zhu, R. B. Gennis, W. Knoll, R. L. C. Naumann, J. Phys. Chem. B. **112**, 3193–3201 (2008).
38. M. Chirea, V. García-Morales, J. A. Manzanares, C. Pereira, R. Gulaboski, and F. Silva, J. Phys. Chem. B. **109**, 21808–21817 (2005).
39. J. O'M. Bockris, B. E. Conway, and R. E. White, Modern Aspects of Electrochemistry No 14; Plenum Press: New York, 1982; Chapter 2.
40. F. Sundfors, J. Bobacka, A. Ivaska, and A. Lewenstam, Electrochim. Acta **47**, 2245–2251 (2002).
41. P. Suresh, and V. Lakshminarayanan, Langmuir **23**, 1548–1554 (2007).
42. M. Tagliazucchi, D. Grumelli, and E. J. Calvo, Phys. Chem. Chem. Phys. **8**, 5086–5095 (2006).
43. S. Tian, J. Liu, T. Zhu, and W. Knoll, Chem. Mater. **16**, 4103–4108 (2004).
44. J. Zhao, C. R. Bradbury, and D. J. Fermín, J. Phys. Chem. C. **112**, 6832–6841 (2008).
45. H. Zhang, H. Lu, and N. Hu, J. Phys. Chem. B. **110**, 2171–2179 (2006).
46. N. A. Galiote, A. J. F. Carbvalho, and F. Huguenin, J. Phys. Chem. B. **110**, 24612–24620 (2006).
47. N. Gu., D. Wei, L. Niu and A. Ivaska, Electrochim. Acta **51**, 6038–6044 (2006).
48. L. Su, F. Gao, and L. Mao, Anal. Chem. **78**, 2651–2657 (2006).
49. S. V. P. Barreira, V. García-Morales, C. M. Pereira, J. A. Manzanares, and F. Silva, J. Phys. Chem. B. **108**, 17973–17982 (2004).
50. T. H. Silva, V. García-Morales, C. Moura, J. A. Manzanares, and F. Silva, Langmuir **21**, 7461–7467 (2005).
51. M. Chirea, C. M. Pereira, and F. Silva, J. Phys. Chem. C. **111**, 9255–9266 (2007).
52. H. O. Finklea, D. A. Snider, J. Fedyk, E. Sabatani, Y. Gafni, and I. Rubinstein, Langmuir **9**, 3660–3667 (1993).
53. A. J. Bard and L. R. Faulkner, Electrochemical Methods; Wiley: New York, 2000; Chapter 10.
54. T. R. Farhat and P. T. Hammond, Chem. Mater. **18**, 41–49 (2006).
55. N. A. Kotov, in Multilayer Thin Films: Sequential Assembly of Nanocomposite Materials; G. Decher and J. B. Schlenoff, Eds.; Wiley-VCH: Weinheim, Germany, 2003; Chapter 8.
56. N. Toshima and T. Yonezawa, New J. Chem. 1179–1201 (1998).
57. K. Y. Chan, J. Ding, J. Ren, S. Cheng, and K. Y. Tsang, J. Mater. Chem. **14**, 505–516 (2004).
58. A. S. K. Hashmi and G. J. Hutchings, Angew. Chem. Int. Ed. **45**, 7896–7936 (2006).
59. M. Sadakane and E. Steckhan, Chem. Rev. **98** 219–37 (1998).
60. B. Wang, R. N. Vyas, and S. Shaik, Langmuir **23**, 11120–11126 (2007).
61. E. Barsoukov and J. R. Macdonald, Impedance Spectroscopy; John Wiley and Sons: New York, 2005; Chapter 1.
62. W. Cheng, S. Dong, and E. Wang, Chem. Mater. **15** 2495–2501 (2003).

63. J. Qiu, H. Peng, R. Liang, J. Li, and X. Xia, Langmuir **23** 2133–2137 (2007).
64. V. P. Menon and C. R. Martin, Anal. Chem. **67**, 1920–1928 (1995).
65. O. Chailapakul and R. M. Crooks, Langmuir **11**, 1329–1340 (1995).
66. M. K. Park, D. C. Lee, Y. Liang, G. Lin, and L. Yu, Langmuir **23**, 4367–4372 (2007).

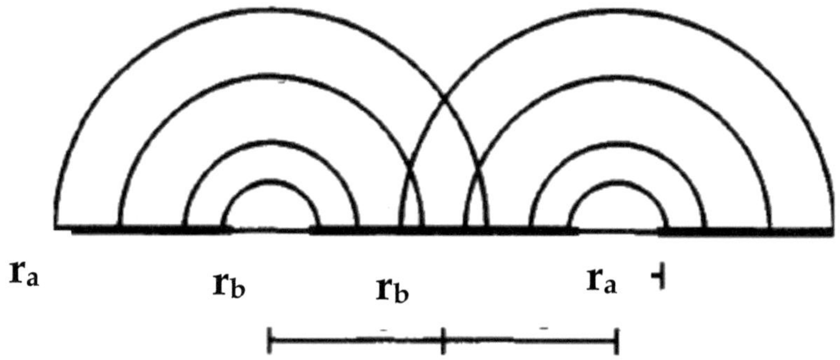

Fig. 1. Microarray parameters and diffusion profiles. Note: $\mathbf{r_a}$ is the radius of the microelectrode site and $\mathbf{r_b}$ is the radius of the inactive area surrounding the microelectrode site. Diffusion layers indicated by semicircles are isolated at short times (high frequencies) and overlapped at long times (low frequencies).

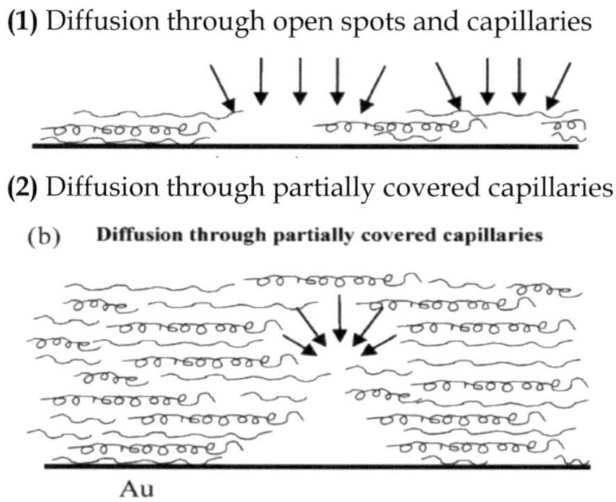

Fig. 2. Electrochemically active site configuration at two stages of film growth and their associated diffusion profiles.

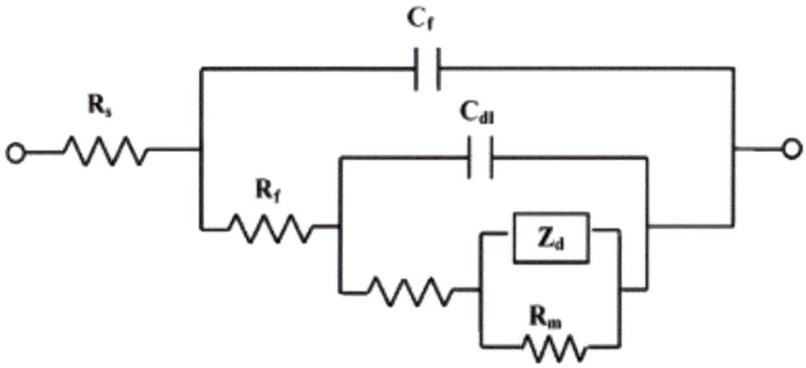

Fig. 3. Equivalent circuit for the PEM-modified electrode. Note: R_s is the solution resistance, C_f is the film capacitance, R_f is the film resistance, C_{dl} is the double layer capacitance associated with metal surface, R_{ct} is the apparent charge-transfer resistance, R_m is the resistance representing Ohmic conduction in the film, and Z_d is the diffusion impedance.

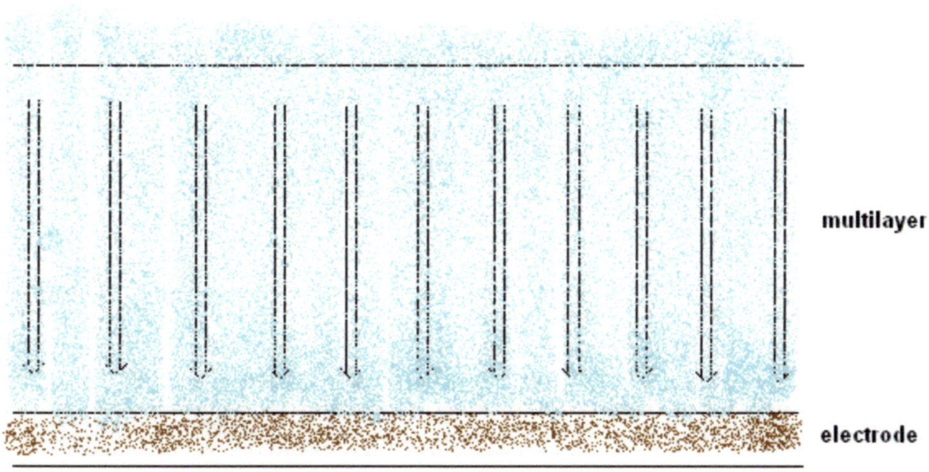

Fig. 4. Diffusion paths across the homogeneous membrane when the number of layers is large.

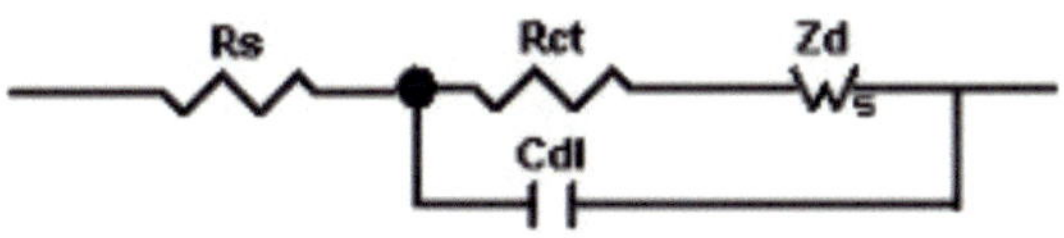

Fig. 5. A conventional Randle's circuit. R_s is solution resistance, R_{ct} is charge transfer resistance, Z_d is the Warburg impedance, and C_{dl} is the double layer capacitance.

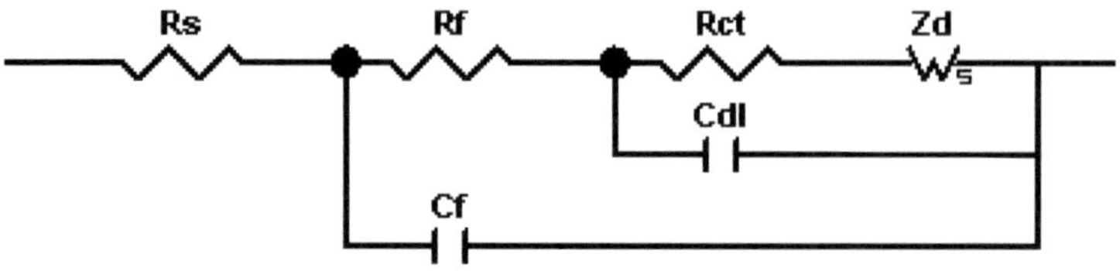

Fig. 6. A modified Randle's equivalent circuit. R_s is solution resistance, R_{ct} is charge transfer resistance, R_f is the film resistance, Z_d is the Warburg impedance, C_f is the film capacitance, and C_{dl} is the double layer capacitance.

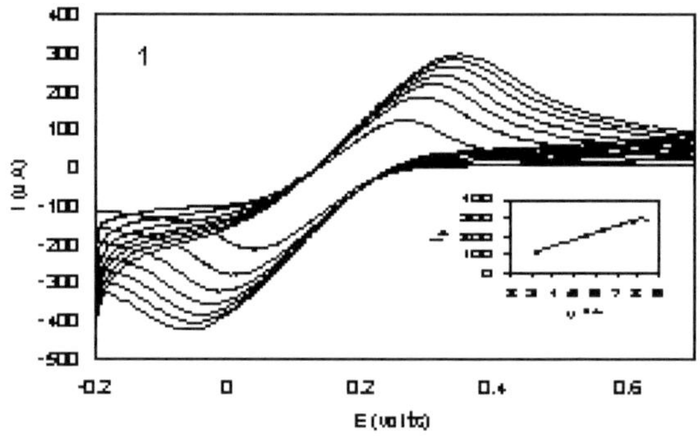

Fig. 7. Cyclic voltammograms for $(PMo_{12}|PDDA)_{10}$ sample **1** at scan rates 10, 20, 30, 40, 50, 60, 70 mV·s^{-1}. Inset shows the anodic peak current vs square root of scan rate.

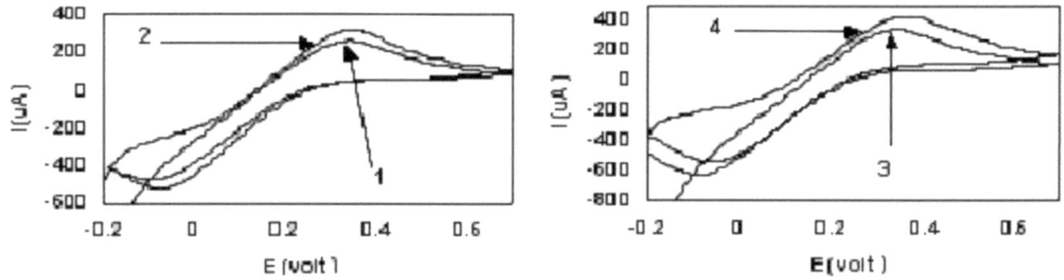

Fig. 8. Cyclic voltammograms obtained at 100 mV·s^{-1} in presence of (0.005 M $[Fe(CN)_6]^{3-/4-}$ in 0.025 M Na_2HPO_4, pH 6.3) for $(PMo_{12}|PDDA)_{10}$ films: a) samples **1** and **2**; b) samples **3** and **4** in Table 1.

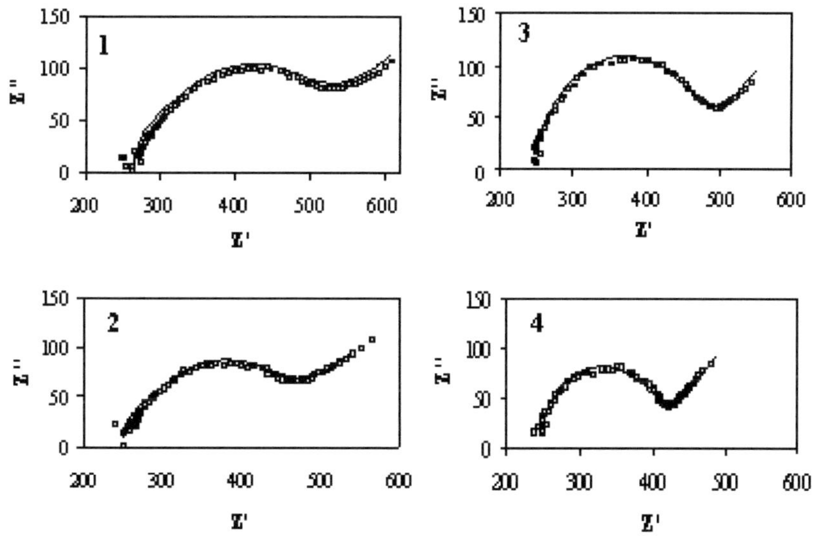

Fig. 9. Electrochemical impedance spectra (0.005 M $[Fe(CN)_6]^{3-/4-}$ in 0.025 M Na_2HPO_4, pH 6.3) for $(PMo_{12} | PDDA)_{10}$ films. Frequency range: 1-10^5 Hz. Sinusoidal Voltage: 10 mV. The dc potential: 220 mV. The electrode was immersed 5 min prior to data acquisition. (□) Experimental curve; (—) Fitting curve.

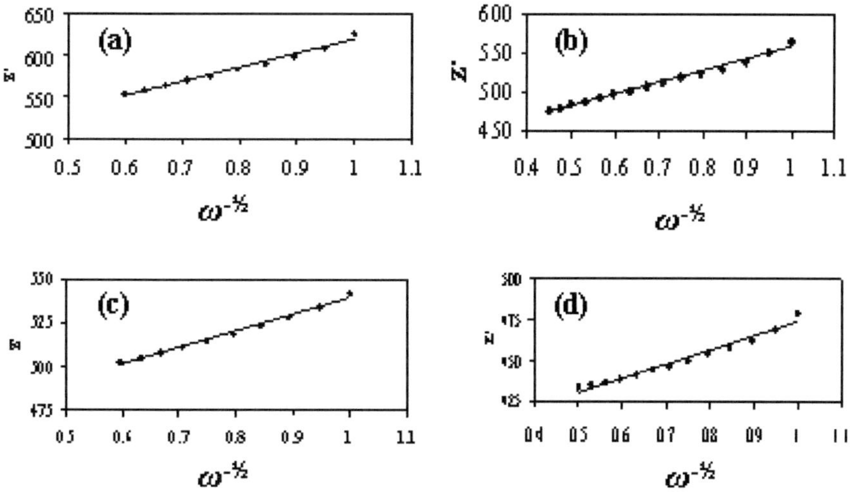

Fig. 10. Z' vs. $\omega^{-½}$ plots for samples: (a) **1**, (b) **2**, (c) **3**, and (d) **4**. Low frequency range selected for Warburg line from the Nyquist plot.

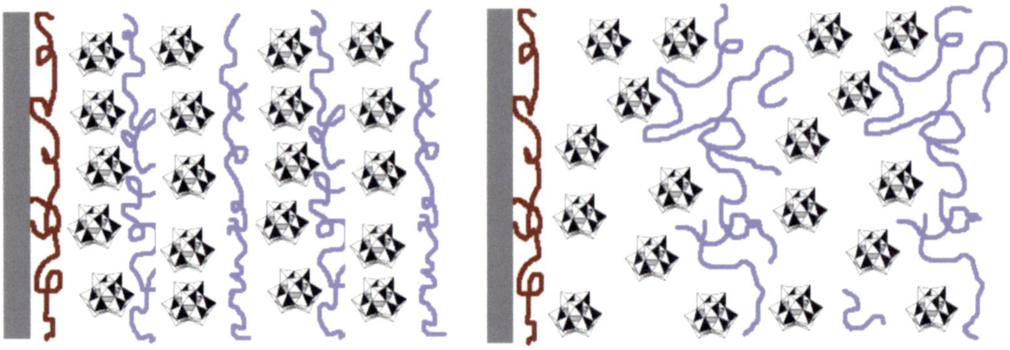

Fig. 11. Proposed (POM | PDDA)$_{10}$ multilayer microstructures: **(1)** under low ionic strength, **(2)** under high ionic strength. Legends: gray slab, substrate; brown chains, PEI; polyhedrons, POM; blue chains, PDDA.

sample	PMo_{12} (mM)	pH of solutions	NaCl (M)
1	0.1	3.3	0
2	0.1	3.3	0.1
3	10.0	2.0	0
4	10.0	2.0	0.1

Table 1. Preparation Parameters used in Layer-by-Layer Construction of (PMo_{12} | PDDA)$_{10}$ Films.

sample	R_s $\Omega \cdot cm^2$	R_{ct} $\Omega \cdot cm^2$	R_f $\Omega \cdot cm^2$	C_f $\mu F \cdot cm^{-2}$	C_{dl} $\mu F \cdot cm^{-2}$	D 10^{-7} $cm^2 \cdot s^{-1}$
1	265	105	115	7.8	2.4	0.8
2	250	100	90	8.0	2.5	1.3
3	245	100	120	8.0	10.0	2.9
4	245	95	60	8.0	3.0	4.2

Table 2. Parameter Values Obtained by Fitting the Impedance Data of (PMo_{12} | PDDA)$_{10}$ Films (Samples **1-4**) from Table 1 to Modified Randle's Equivalent Circuit in Figure 6.

Non-equilibrium charge transport in disordered organic films

Vladimir Nikitenko[1] and Alexey Tameev[2]
[1]National Research Nuclear University (MEPhI),
[2]A.N. Frumkin Institute of Physical Chemistry and Electrochemistry of RAS
Russia

1. Introduction

Charge transport in disordered organic layers has been intensively investigated in recent years both experimentally and theoretically. In these investigations, basic studies are needed for practical and technological developments in modern organic electronics and photonics. The concept of Gaussian disorder model (GDM) i. e. the temperature- and field- assisted tunneling (hopping) of charge carriers between localized states (LSs), forms the background for understanding of the physical nature of transport (Bässler, 1993; Novikov et al., 1998). Energetic disorder is described by Gaussian distribution of energies of LSs. The ubiquitous experimental option for investigating charge transport is the time- of- flight (TOF) experiment, the observable being the transient current in the organic layer.

Excess charge carriers are generated in an organic film by light in course of TOF experiments. These carriers are not yet in quasi- equilibrium shortly after their generation, notably if there is an excess energy during excitation (Bässler, 1993). This circumstance together with a strong variance of transition rate of carriers between LSs in disordered materials causes a decrease of the average mobility with time, while the spatial dispersion of carriers relative to their mean position is anomalously large, i.e. the transport occurs in non-equilibrium conditions. Usually this case is referred to dispersive transport (Bässler, 1993; Arkhipov & Bässler, 1993b), whereas at long time transport is characterized by time-independent mobility and diffusion coefficients. The latter transport mode, referred to quasi- equilibrium, or Gaussian transport, is the topic of recent works (Arkhipov et al., 2001a; Schmechel, 2002; Fishchuk et al., 2002; Pasveer et al., 2005). It is often realized in materials with moderate energetic disorder. Indeed, the TOF transients of $\geq 1 \mu m$ thick samples at room temperature bear out a well-developed plateau. This circumstance, however, does not always imply that transport is completely quasi- equilibrium. An unambiguous signature of the deviation is the anomalously large dispersion of formally non- dispersive TOF signals and the concomitant scaling of the tails of TOF signal as a function of sample thickness and electric field strength. Moreover, quasi- equilibrium transport is questionable for the case of thin (<100 nm) organic films, suitable for organic light- emitting diodes (OLEDs), when the transit time is short enough (Blom & Vissenberg, 1998).

Transport of charge carriers from electrodes to the zone of most efficient recombination is an essential part of operation of OLEDs. Transient electroluminescence (TrEL) is the important experimental method of investigations of charge transport in single- layer OLEDs (Pinner et al., 1999), where application of TOF method is difficult. Initial rise of TrEL in single- layer OLEDs based on organic material with strong asymmetry of mobilities is controlled by incoming flux of carriers with higher mobility to the recombination zone. The present methodology of TrEL analyses is usually based, however, on the questionable assumption that transport of charge carriers in OLEDs can be described in terms of time- independent mobility and diffusion coefficients (see, for example, Pinner et al., 1999).
The main computational method in GDM is the Monte- Carlo numerical simulation. However, simulating the effect of positional disorder is a notoriously difficult and time-consuming task, especially at low field strengths and temperatures (Bässler 1993). This problem can be overcome by analytic modelling. Theoretical description of hopping transport can be greatly simplified by means of the concept of effective transport energy (see, for example, Monroe, 1985; Baranovskii et al., 2000; Arkhipov et al., 2001a). The effective transport energy is analogous to the mobility edge in the multiple- trapping (MT) model, while deeper LSs can be considered as traps. This concept allows adapting the earlier results of MT model (Arkhipov & Nikitenko, 1989) for the hopping.
In the recent work (Nikitenko et al., 2007) an analytic theory of non- equilibrium hopping charge transport in disordered organic materials, based on the concept of effective transport energy (see section 2 of this chapter), is proposed. Previously this concept has not been applied for the general case of non- equilibrium hopping transport, only the asymptotic regimes of quasi- equilibrium (Arkhipov & Bässler, 1993a) and extremely non- equilibrium (dispersive) (Arkhipov & Bässler, 1993b) hopping (long and short times after generation, respectively) have been considered analytically. In this chapter this theory is described briefly (section 3) with special emphasis on transport in thin samples, e. g. used in OLEDs. Results of the theory are in good agreement with both TOF experiments for molecularly doped polymers and Monte- Carlo simulations of GDM (Bässler, 1993; Borsenberger et al., 1993; Borsenberger & Bässler, 1994), see section 4. In the section 5 the theory applied to the modelling of initial rise of TrEL from a single- layer OLED (Nikitenko & von Seggern, 2007; Nikitenko et al., 2008) both in injection- limited (IL) and space- charge limited (SCL) transport regimes. A method for determination of TrEL mobility is proposed.

2. Effective transport energy

An inherent feature of disordered materials is a broad distribution of LSs in energy and mutual distances. The energy distribution of states (DOS) is described by the distribution function $g(E)$. The appropriate choice for organic materials is the Gaussian DOS,

$$g(E) = \exp\left(-E^2/2\sigma^2\right)\Big/\sqrt{2\pi\sigma^2} \tag{1}$$

Analyses of transport in disordered materials can be greatly simplified by the concept of transport energy. It rest on the notion that (i) for the typical case $\sigma/kT >> 1$ charge transport is controlled by thermally activated jumps from LSs in a deep tail of *g(E)* and (ii) for these jumps the energy of target state E_{trans} does not depend on the energy of initial state *E*, if $E<<E_{trans}$ since *g(E)* decreases drastically to lower energies. The states around E_{trans} (the latter is usually called 'transport energy'), contributes mostly to the transport process. The deep

states, $E<E_{trans}$, can be considered as traps, while E_{trans} and LSs with $E>E_{trans}$ are mobility edge and 'conductive states', respectively. The method to calculate E_{trans} should be introduced here.

The transition rate of charge carrier from LS of the energy E to the LS of energy E', which are separated by the distance r, is described in this work in terms of Miller- Abraham expression, i. e. $\nu(r,E,E')=\nu_0\exp\left[-\eta(E'-E)(E'-E)/kT-2\gamma r\right]$, where $\eta(x)$ is the unit step function, ν_0 is the frequency prefactor, γ is the inverse localization length, k is the Boltzmann constant and T is the temperature. If transport is dominated by energetic rather than by positional disorder one can assume that the typical release frequency of a carrier from LS with the energy E, $\omega(E)$, depends only on the energy of this LS, E. This assumption will be justified by comparing of theoretical results with data of experiments and Monte- Carlo simulations in GDM (see below). The steep dependence of the transition rate on energy difference and separation in space together with positional disorder implies that one of neighbor target states is strongly preferable. Following the work (Arkhipov et al., 2001a), one can define the typical release frequency $\omega(E)$ by the condition, that the mean number of neighbors for the LS of energy E, which are available during the time shorter than ω^{-1}, is equal to unity. This number is a sum of 2 terms, $n(E, \omega)= n_{\downarrow}+ n_{\uparrow}$, the latter and former accounting for upward and downward (in energy) jumps, respectively. The result is (Nikitenko et al. 2007)

$$\omega(E)=\nu_0\exp\left[-(E_{trans}-E)/kT-2\gamma a\right], \qquad E<<E_{trans}, \tag{2}$$

where $a=(E_{tr}-E_{trans})/2\gamma kT$ is the typical hopping distance, and the maximal energy of the target state, E_{tr}, is defined by the conditions

$$\omega(E)\to\nu_0\exp\left[-(E_{tr}-E)/kT\right],\ E\to-\infty, \tag{3}$$

and $n\approx n_{\uparrow}(E,\omega)=1$, if release frequency from deep LSs is described by Eq. (3). One should remember that the energy E_{tr} cannot be considered as the upper energetic limit of sites that can act as traps because Eq. (3) does not contain the tunneling term $2\gamma a$, see Eq. (2).

Obviously, for $n_{\uparrow}>1/2$ jumps are preferably upward in energy and, hence, the upper limit of traps, i. e. the transport energy E_{trans}, can be defined by the condition

$$n_{\uparrow}\left[E_{trans},\ \omega(E_{trans})\right]=1/2. \tag{4}$$

The calculation of $n_{\uparrow}\left[E,\omega(E)\right]$ is described in the works Arkhipov et al., 2001a; Nikitenko et al. 2007. Neighbours from which a carrier preferably returns to the initial state are not included to the number $n_{\uparrow}$ because these roundtrip jumps do not contribute to transport.

Since jumps from the LSs with energies $E\geq E_{trans}$ occur preferably downwards in energy, $\omega(E)\approx\nu_0\exp(-2\gamma a)\equiv\tau_0^{-1}$. In other words, τ_0 is the lifetime of carriers in 'conductive' LSs, i. e. in the states with energies above E_{trans}. It should be noted that the condition $2\gamma N^{-1/3}kT/\sigma>>1$, where N is the spatial density of LSs, is usually fulfilled in organic materials even at moderately low temperatures. For this case, Eqs. (2), (4) yields $E_{trans}\approx 0$ and $a\approx E_{tr}/2\gamma kT\approx 0.745N^{-1/3}$ (Nikitenko et al. 2007). This is plausible because the jump rate to the neighbor LS is defined by the spatial distance rather then by the energy difference.

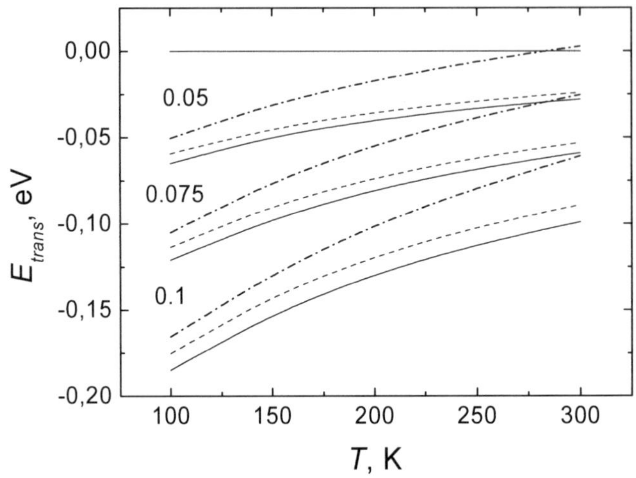

Fig. 1. Temperature dependence of the transport energy level for several values of the energy width of the Gaussian trap distribution, as shown in the figure.

One has to note that transport energy was introduced previously by different ways (see Baranovskii et al., 2000; Schmechel, 2002). The results of the calculation of the temperature dependence of E_{trans} from Eqs. (3) and (4), using for several values of σ are shown in Fig. 1 (solid lines), providing that $2\gamma N^{-1/3}=10$. For comparison, Fig. 1 shows the temperature dependences of the transport energy, obtained using the method of Baranovskii et al., 2000 (dash–dotted lines), and the results of the calculations under the assumption that returns to the initial states are absent, as in the cited work (dashed lines). The difference between the two latter results at not too high temperatures, T < 200 K, does not exceed kT. The same conclusion (Nikitenko et al., 2007) results from the comparison with results of Schmechel, 2002 (in that work the transport energy defined as the maximum of energy dependent "differential conductivity", that describes the contribution of states with given energy E to the total conductivity). This supports the notion that the definition of E_{trans} by Eq. (4) is in good agreement with the statement that LSs with energies around E_{trans} contribute mostly to the transport and hence E_{trans} is in analogy with 'mobility edge' of MT model. Thus, both release and capture of charge carriers for LSs with energies $E<E_{trans}$, which are characterized by the typical frequency $\omega(E)$, see Eq. (2), and by energy- independent capture time $\tau_0=exp(-2\gamma a)$, respectively, can be described in complete analogy with MT model, $E<E_{trans}$ being the mobility edge. The states with energies $E<E_{trans}$ are referred below as traps.

3. Analytic theory of non-equilibrium transport

3.1 Equation of non- equilibrium transport

One has to remember, that in a TOF experiment, charge carriers are generated in the upper portion of the DOS, relative to quasi- equilibrium state. That implies subsequent energetic relaxation (thermalization) of charge carriers. It occurs simultaneously with transport, and should be considered as a two- step process (Monroe, 1995; Arkhipov & Bässler, 1993b; Nikitenko, 1992). The first step is a sequence of fast jumps downwards in energy right after a carrier was started at $t=0$. The characteristic time of this process cannot exceed considerably the time τ_0 under the presumed condition $2\gamma N^{-1/3} kT/\sigma > 1$, because (i) it is the time of the slowest jump in this sequence and (ii) the number of downward jumps is small. Upward jumps to the transport energy, which are described above, are the rate- limiting step for transport at $t >> \tau_0$, hence under this condition the concept of transport energy is valid. The moment $t = \tau_0$ will be referred as $t = 0$ below, because τ_0 is much shorter than all other characteristic times.

One has to remind also the approximation of demarcation energy, which simplifies greatly the analytic modeling of energetic relaxation of charge carriers [Arkhipov & Rudenko 1982b; Arkhipov & Nikitenko 1992; Arkhipov & Bässler 1993b]. In this approximation all the traps (i.e. LSs with energies $E < E_{trans}$) are considered as being either currently deep or either currently shallow. The trap is currently deep, if release of a previously captured carrier is not probable up to time t, and vice versa. The demarcation energy $E_d(t)$ is defined from the condition $\omega\left[E_d(t)\right]t = 1$ and Eq. (2) as

$$E_d(t) = E_{tr} - kT\ln(\nu_0 t) = E_{trans} - kT\ln(t/\tau_0), \qquad t > \tau_0 . \tag{5}$$

In the simplest approximation, all traps are deep, if $E < E_d(t)$, and vice versa.

On the basis of the formal analogy between MT and hopping at $t >> \tau_0$, as discussed above, one can describe the time- and coordinate- dependent density of charge carriers $p(x,t)$ by the following differential equation, which was derived previously in the framework of MT model (Arkhipov & Nikitenko, 1989),

$$\frac{\partial p(x,t)}{\partial t} + \mu(t)F_0\frac{\partial p(x,t)}{\partial x} - D_F(t)\frac{\partial^2 p(x,t)}{\partial x^2} = -\lambda(t)\left[p(x,t) - p(x,0)\right], \tag{6}$$

where F_0 is the strength of applied electric field, $\lambda(t)$, $\mu(t)$, and $D_F(t)$ are the trapping frequency, time- dependent mobility and coefficient of field- assisted diffusion (FAD), respectively. These values are defined as

$$\mu(t) = \mu_0\theta_1(t), \quad \lambda(t) = \theta_1(t)/\tau(t), \quad D_F(t) = \tau_0^{-1}(\mu_0 F_0\tau_0)^2\theta_1^3(t)/\theta_2(t), \tag{7}$$

where the mobility in «conductive» states is introduced as $\mu_0 = (1/6)(e/kT)a^2/\tau_0$, and

$$1/\theta_m(t) = \int_{E_d(t)}^{E_{trans}} dEg(E)\left[\tau_0\omega(E)\right]^{-m}, \quad m = 1, 2, \tag{8}$$

$$1/\tau(t) = \int_{-\infty}^{E_d(t)} dE\, g(E)/\tau_0 . \tag{9}$$

Obviously, $\theta_m(t)$ decreases with time, while $\tau(t)$ increases.

In a typical TOF experiment the condition $eF_0L >> kT$ is fulfilled and hence diffusion could be neglected in comparison to drift. However, in a disordered system there is an additional spreading of an initially narrow sheet of charge carriers that gives rise to the diffusion- like term in Eq. (6). This kind of spreading was analyzed analytically for quasi- equilibrium conditions (Arkhipov & Bässler, 1993a) and called as field diffusion. It appears due to (i) small deviations from equilibrium in the shallow trap population, caused by electric field and (ii) dispersion of release times of carriers. The mean distance of drift in «conductive states», $\mu_0 F_0 \tau_0$, is a «mean free path» in coefficient D_F, see Eq. (7), and hence $D_F \sim F_0^2$. Details of thermalization of charge carriers and the way of derivation of Eq. (6) are described in the paper Nikitenko et al. 2007.

Initial condition for conducting a TOF experiment is to generate a narrow sheet of carriers at $x=0$, hence $p(x,0)=A_0\delta(x)$ is proportional to Dirac delta function, and A_0 is the area density of generated carriers. Fourier transformation with respect to the variable x yields the following solution of Eq. (6) in the infinite sample

$$p(x,t)=G(x,t,0)+\int_0^t dt'\lambda(t')G(x,t,t'), \tag{10a}$$

$$G(x,t,t')=A_0\exp\left\{-\Lambda(t,t')-\left[x-F_0M(t,t')\right]^2/4S_F(t,t')\right\}\Big/\sqrt{4\pi S_F(t,t')} \tag{10b}$$

where $S_F(t,t')=\int_{t'}^{t} d\tau D_F(\tau)$, $\Lambda(t,t')=\int_{t'}^{t} d\tau\lambda(\tau)$, $M(t,t')=\int_{t'}^{t} d\tau\mu(\tau)$. Remember that conventional diffusion is neglected and hence charge carriers cannot penetrate to the region $x<0$, i.e. $p(x,t)\sim\eta(x)$. Disagreement of this statement with Eq. (10) is, however, not principal. One should remember that diffusion- like term in Eq. (8) is only an approximate way to describe field- assisted dispersion of a carrier packet, and this approximation is not valid at short time domain. The condition of applicability of field- diffusion approximation, namely $t\theta_2(t)/\theta_1(t)>>1$ (Arkhipov & Rudenko 1982a), is fulfilled, however, in all calculations that are performed in this work. It means that shift of the mean position of charge carriers exceeds their dispersion considerably, and unphysical «tail» of $p(x,t)$ at $x<0$, is unimportant.

The transit time t_{tr} is defined by the condition that the mean position of charge carriers is equal to the film thickness L. Eq. (6) yields

$$F_0\int_0^{t_{tr}} dt'\exp\left[-\Lambda(t,0)\right]\mu(t')=L. \tag{11}$$

Obviously, $E_d(t)=E_{trans}$ at $t=\tau_0$ i.e. all traps are deep and occupation of all traps is off-equilibrium. In course of thermalization, the lower energetic boundary $E_d(t)$ of shallow states, that are in quasi- equilibrium, moves towards deeper states. Shortly after charge carriers were generated, however, transport is exclusively dominated by deep traps i.e. it is entirely dispersive. Time duration of this period is defined by the condition $\theta_1(t)^{-1}d\tau(t)/dt<1$ (Arkhipov & Rudenko 1982b). Under this condition one can neglect the term $\partial p/\partial t$ as well as field- diffusion term can be neglected in comparison with right- hand side of Eq. (6). Since eq. (6) transforms to the ordinary dispersive transport equation (Arkhipov & Bässler 1993b),

$$\mu_0\tau(t)F_0\frac{\partial p(x,t)}{\partial x}=-\left[p(x,t)-p(x,0)\right]. \tag{12}$$

The solution of Eq. (12) is

$$p_{disp}(x,t)=A_0\eta(x)\exp\left[-x/\mu_0F_0\tau(t)\right]/\mu_0F_0\tau(t). \tag{13}$$

Spatial profiles of charge carrier density, as calculated from Eqs. (10), are shown in the Fig. 2

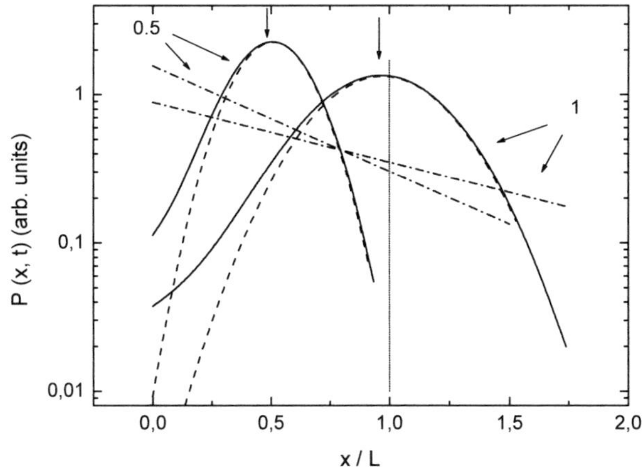

Fig. 2. Spatial profiles of the charge carrier density $p(x,t)$ (solid lines). Dashed lines shows the respective Gaussians. Dash- dotted lines shows $p(x,t)$ under the assumption that transport is dispersive, see Eq. (13). $F_0 = 2 \cdot 10^5\, V/cm$, $\sigma/kT = 3.5$, $2\gamma N^{-1/3} = 10$, $L = 5\mu m$.

for $t = 0.5t_{tr}$ and $t = t_{tr}$ (see solid lines) considering $t_{eq\_\mu} << t_{tr}$. Obviously, maximum of a carrier distribution is practically in the position $x = L$ at $t = t_{tr}$. The front of $p(x,t)$ profile is practically Gaussian while the carrier density near $x = 0$ greatly exceed the values that are expected from Gaussian function. Consequently, the mean position of carriers is slightly behind the maximum, see arrows in the figure, while both values increases practically linear with time. A non- Gaussian tail of the distribution appears due to capture of carriers on deep traps with release times greater than transit time. Meanwhile these spatial profiles differ significantly from an exponential function $p_{disp}(x,t)$, see Eq. (13).

Vice versa, $\lambda(t) \to 0$, $\mu(t) \to \mu_{eq}$, $D_t(t) \to D_{eq}$ in the limit of long time and hence Eq. (6) reduces to the usual drift- diffusion equation, which describes quasi- equilibrium transport with time- independent mobility μ_{eq} and field- dependent diffusion coefficient D_{Feq}. These constants are defined by Eqs. (7), (8) with $E_d(t) \to -\infty$.

Thus, the system of Eqs. (6)- (9) provide an analytic description of hopping transport of charge carriers in dispersive transport regime and allow also to analyze subsequent relaxation towards quasi- equilibrium. Coefficients and time dependencies from Eq. (8) will be analyzed extensively in the following section.

3.2 Non- equilibrium drift and field- stimulated diffusion

Using Eqs. (1), (2) and (8) one obtains

$$\frac{\theta_m(t)^{-1}}{(\nu_0\tau_0)^m} = \frac{1}{2}\exp\left[\frac{m}{2}\left(\frac{\sigma}{kT}\right)^2 + \frac{mE_{tr}}{kT}\right]\left\{erfc\left[\frac{m\sigma}{\sqrt{2}kT} + \frac{E_d(t)}{\sqrt{2}\sigma}\right] - erfc\left[\frac{m\sigma}{\sqrt{2}kT} + \frac{E_{trans}}{\sqrt{2}\sigma}\right]\right\}, \quad (14)$$

where $m = 1$, 2, and $erfc(x) = \left(2/\sqrt{\pi}\right)\int_x^\infty dt \exp\left(-t^2\right)$ is the complementary error function. Eqs. (2), (7) and (14) for the limit $t \to \infty$ yields

$$\mu(t) \approx \mu_{eq} = \frac{e}{kT}\chi_\mu \frac{\nu_0 a^2}{6}\exp\left[-\frac{1}{2}\left(\frac{\sigma}{kT}\right)^2 - \frac{E_{trans}}{kT} - 2\gamma a\right], \quad t \ge t_{eq\_\mu}, \tag{15}$$

$$D_F(t) \approx D_{Feq} = \chi_D \nu_0 a^2 \left(\frac{eF_0 a}{6kT}\right)^2 \exp\left[\frac{1}{2}\left(\frac{\sigma}{kT}\right)^2 - \frac{E_{trans}}{kT} - 2\gamma a\right], \quad t \ge t_{eq\_D}, \tag{16}$$

where $\chi_\mu = 2/erfc\left[-\sigma/\sqrt{2}kT - E_{trans}/\sqrt{2}\sigma\right]$, $\chi_D = \chi_\mu^{\ 3} erfc\left[-\sqrt{2}\,\sigma/kT - E_{trans}/\sqrt{2}\sigma\right]$. Both factors χ_μ, χ_D are nearly 1 for the practically relevant case $\sigma/kT >> 1$. For the limit of dilute systems, $2\gamma N^{-1/3} kT/\sigma >> 1$, one obtains $E_{trans} \approx 0$ and $a \approx 0.745 N^{-1/3}$. Thus, $\mu_{eq} \sim \exp\left(-C 2\gamma N^{-1/3}\right)$, and the value $C = 0.745$ is close to the well- known result of percolation theory, $C_{perc} = 0.865$ (Shklovskii & Efros, 1984).

In general, the temperature dependence of D_{Feq} is determined by the factor $\exp\left[0.5(\sigma/kT)^2\right]$, as in the work Arkhipov & Bässler, 1993a. However, value of D_{Feq}, see eq. (16), exceeds considerably the respective result of this work, although the latter does not include the exponential dependence of hoping distance. Comparable values of D_{Feq} can be obtained only for the case of relative „dense" systems, $2\gamma a \le 5$. This is not surprising because only in this limit the method of the work Arkhipov & Bässler, 1993a, i. e. the method of configurational averaging of hopping rates in respect to hopping distances and energies of starting and target states, is applicable.

One has to note that the field dependence of $\mu(t)$, has not been analyzed in this paper because it is restricted to the low- field limit, i. e. $eF_0 a/kT < 1$. Therefore the effect of field strength on mobility is of minor importance. On the other hand, at room temperature and $a \approx 0.6 nm$, one obtains $F_0 < 4 \cdot 10^5\, V/cm$, that is strong enough to neglect conventional diffusion.

Analyzing Eq. (14) yields the following relaxation times t_{eq_μ} and t_{eq_D} of mobility and field diffusion, respectively

$$\nu_0 t_{eq\_\mu} = \exp\left[(\sigma/kT)^2 + E_{tr}/kT\right], \quad \nu_0 t_{eq\_D} = \exp\left[2(\sigma/kT)^2 + E_{tr}/kT\right]. \tag{17}$$

From eqs. (7) and (14) one obtains the following analogue of the Einstein relation of time dependent mobility and field diffusion coefficient,

$$\frac{e}{kT}\frac{D_F(t)}{\mu(t)} \approx \frac{1}{12}\left(\frac{eF_0 a}{kT}\right)^2 erfc\left[\frac{\sqrt{2}\sigma}{kT} + \frac{E_d(t)}{\sqrt{2}\sigma}\right]\exp\left[\left(\frac{\sigma}{kT}\right)^2\right]. \tag{18}$$

In the long time limit, $t >> t_{eq\_D}$, this ratio reaches a maximal value, $\approx (1/6)(eF_0 a/kT)^2 \exp[(\sigma/kT)^2] >> 1$, $\sigma/kT >> 1$. In Fig. 4 time dependencies $eD_F(t)/\mu_{eq} kT$ and and $\mu(t)/\mu_{eq}$ are calculated from Eqs. (7), (14), and plotted vs normalized time t/t_{eq_μ} for several values of energetic disorder parameter σ/kT. It demonstrates enhanced field-assisted diffusion in the time domain $t_{eq\_\mu} << t < t_{eq\_D}$, if $\sigma/kT \geq 2.5$. Obviously, the FAD coefficient increases at long time domain, although mobility remains practically constant, see Eq. (17).

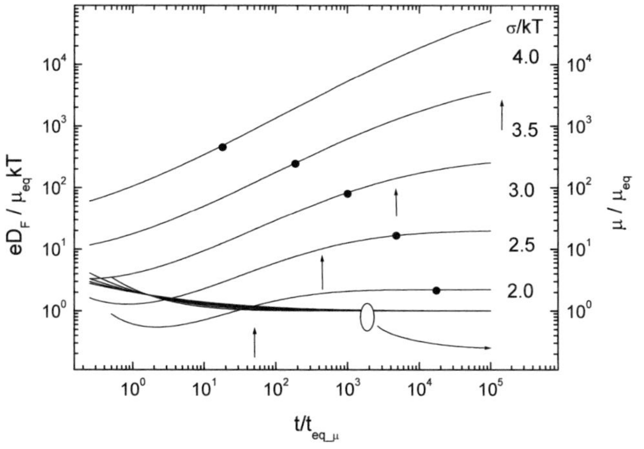

Fig. 3. Time dependences of the field- assisted diffusion coefficient, normalized by equilibrium value of the coefficient of usual diffusion, $\mu_{eq}kT/e$, parametric in σ/kT values. Respective dependences $\mu(t)/\mu_{eq}$ are also shown in the figure. Arrows mark relaxation time t_{eq_D}. Full circles show the values of $eD_F(t_{tr})/\mu_{eq}kT$ at $t = t_{tr}$, providing that $L = 5\mu m$. Other parameters: $F_0 = 2 \cdot 10^5\, V/cm$, $N = 4.6 \cdot 10^{21} cm^{-3}$, $\gamma^{-1} = 0.12nm$, $T = 295K$.

Simulation data in the work Pautmeier et al., 1991 on a system with pure energy disorder ($\sigma/kT = 3.0$) also delineated the different time scales for relaxation of mobility and diffusivity. Time dependence of the diffusivity is taken from Fig. 1 of the work Pautmeier et al., 1991 and compared with the function $D_F(t)$ in the Fig. 4. The latter is calculated from Eqs. (7), (14) and is normalized by the minimal value of GDM diffusivity. The time is normalized by the typical hopping time $t_0 = (1/6)\exp(2\gamma N^{-1/3})$ (Bässler, 1993). Both dependences are in good agreement. They bear out a minimum at $t \cong t_{eq\_\mu}$. For GDM data the latter is defined by the condition $\mu(t_{eq\_\mu})/\mu_{eq} \approx 2$, in accordance with Fig. 3. At shorter times the transport is dispersive. Meanwhile the time t_{eq_μ} is practically the same as the time when the averaged energy of localized carrier approaches to the equilibrium when the averaged energy of localized carrier approaches to the equilibrium value σ^2/kT, i. e. the difference becomes less than kT, while both the maximum of distribution of occupied states (DOOS) and demarcation energy $E_d(t)$ reaches σ^2/kT (see inset to the Fig. 4). The time t_{eq_D} is of the same order of magnitude as the time when the dispersion of energies approaches to the equilibrium value σ (Pautmeier et al., 1991). Slow relaxation of the latter reflects slow relaxation of carriers towards the very tail states. This is shown in the inset to the Fig. 4, where time evolution of spatially averaged DOOS, is calculated, see details in the

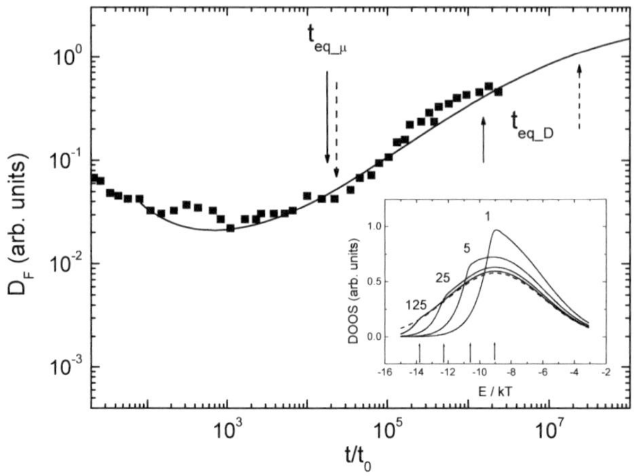

Fig. 4. Comparison of time dependences $D_F(t)$ from this work (solid line) and from GDM (points, see the work Pautmeier et al., 1991). Relaxation times of μ and D_F as defined from GDM and this model are marked by solid and dashed arrows, respectively. Inset shows the time evolution of the density of occupied states. Ratios t/t_{eq_μ} are shown in the figure, respective positions of demarcation energy $E_d(t)$ are shown by arrows. Steady- state distribution is denoted by dotted line.

work Nikitenko et al., 2007. The coefficient of FAD is controlled by entire DOOS because it determines the variance of dwell times for carriers. On the other hand, shallow LSs contribute preferably to the current, and their equilibration takes much less time.

4. Transient current in time- of- flight experiment

Time-of-flight is the conventional technique for studies of electron-hole transport behaviors in materials with long dielectric relaxation times. By this method, one measures the time required for a sheet of carriers generated by a short flash of radiation to transit a sample of known thickness. A polymer layer, typically 1-30 μm in thickness is placed between two blocking electrodes, at least one of which must be semitransparent. The sample is used as a capacitor in an *RC* circuit. The applied voltage charges the capacitor to a potential *V* and then sample is illuminated by a highly absorbed flash of radiation. The duration of the light impulse is short compared to the transit time. The flash generates a sheet of carriers that then drift across the sample. *R* is selected such that *RC* is much less than the transit time. This arrangement is described as the small current mode. Under these conditions, the voltage across *R* is proportional to the current flowing in the sample, *j(t)*. When the carriers exit the sample, a sharp decrease in the current is observed. The time corresponding to the decrease in current is usually defined as the transit time, the time required for the sheet of charge to transit a thickness *L*. If transport is quasi- equilibrium, the transit time is related to the mobility μ as

$$t_{tr} = L/\mu F_0 = L^2/\mu V \tag{19}$$

For polymers, it can usually be assumed that prior to charge injection, the field in the sample is uniform and given as *V/L*. In order that the field in the sample is not perturbed by the

injected charge, the experiments are performed so that the injected charge is small compared to the charge on the electrodes. The analysis of non- equilibrium transport is highly complicated, see below.
Eqs. (10) together with the following equation (Arkhipov & Rudenko, 1982b)

$$j(t) = (e/L)\partial/\partial t \int_0^L dx(x-L)p(x,t) \tag{20}$$

solve the problem to calculate the transient current under TOF conditions. Accounting the condition $t\theta_2(t)/\theta_1(t) >> 1$, Eqs. (10), (20) yield

$$j(t) \approx j_{drift}(t) = (eA_0F_0/L)\left[\mu(t) - \lambda(t)\int_0^t dt'\mu(t')\exp[-\Lambda(t,t')]\right], \quad t << t_{tr}, \tag{21}$$

Obviously, $j_{drift}(t)$ is also the asymptotic solution of Eqs. (10), (20) for the limit of pure drift, i. e. $D_F \to 0$, $t < t_{tr}$ (Arkhipov & Rudenko, 1982b).

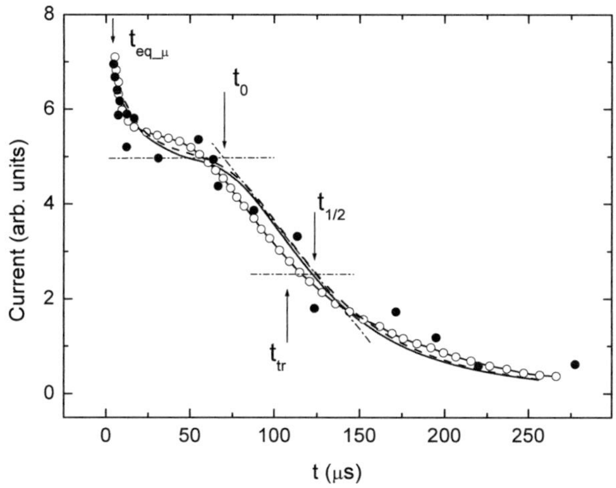

Fig. 5. Time dependence of the TOF current. Solid and dashed lines are the results of calculations from Eqs. (10), (20) and (22), respectively. Results of TOF experiment (line with circles) and Monte- Carlo simulations of GDM (full circles). $T = 312K$, $\sigma/kT = 3.5$, $N = 4.6\cdot10^{21}cm^{-3}$, $\gamma^{-1} = 0.12nm$.

One can simplify the highly complicated time dependence $j(t)$ as defined by Eqs. (10), (20) for the relevant case of moderately non- equilibrium transport, $t \geq t_{eq\_\mu}$, or $t\lambda(t) < 1$. Estimating integrals in Eqs. (10), (20) for this case yields $p(x,t) \approx G(x,t,0)$ and

$$j(t) \approx (1/2)(eA_0F_0/L)\mu(t)\exp[-t\lambda(t)]erfc\left[(F_0M(t,0)-L)/2\sqrt{S_F(t,0)}\right]. \tag{22}$$

Fig. 5 shows good quantitative agreement with the TOF signal as calculated from the approximate Eq. (22) and from Eqs. (10), (20), see dashed and solid lines, respectively, and qualitative agreement both with experimental, see line with circles, and simulated in GDM data, see full circles. Data for polycarbonate (PC), doped by 1,1-bis(di-4-tolylaminophenyl) cyclohexane (TAPC) are taken from Fig. 22 of the paper Bässler, 1993. Although t_{eq_μ} is much less than transit time, see arrows on the figure, the current is not strictly constant at any time, because of strong spatial dispersion of carriers. The „plateau" level j_0 is defined operationally by the time of minimal tangent and then characteristic values $j_{1/2} = j_0/2 = j(t_{1/2})$, t_0 and $t_{1/2}$ are defined by dot- dashed lines in Fig. 5. Obviously, the transit time t_{tr} from Eq. (11) is good approximation to the experimentally determined time $t_{1/2}$. Eq. (18) explains why in TOF experiments the dispersion of the carrier arrival times, $W = (t_{1/2} - t_0)/t_{1/2}$, greatly exceeds the value predicted by conventional diffusion (Bässler, 1993; Borsenberger et al., 1993b). Using Eq. (22) one obtains

$$W \approx L^{-1}\left[\pi \int_0^{t_{tr}} dt D_F(t)\right]^{1/2} \tag{23}$$

In the limit $t_{tr} > t_{eq\_D}$, one can reduce Eq. (23) to the well- known form (Bässler 1993)

$$W \approx \sqrt{\pi D_F(t_{tr})/\mu_{eq} F_0 L}\ , \tag{24}$$

implying $t_{tr} \approx L/\mu_{eq} F_0$, see Eq. (19), and $D_F(t) \approx D_F(t_{tr})$. The ratio $eD_F(t_{tr})/kT\mu_{eq}$, that can be derived from the measured dispersion *W*, differs considerably from that predicted under premise of quasi- equilibrium, $eD_{Feq}/kT\mu_{eq}$, if $t_{tr} < t_{eq\_D}$ (Fig. 3).

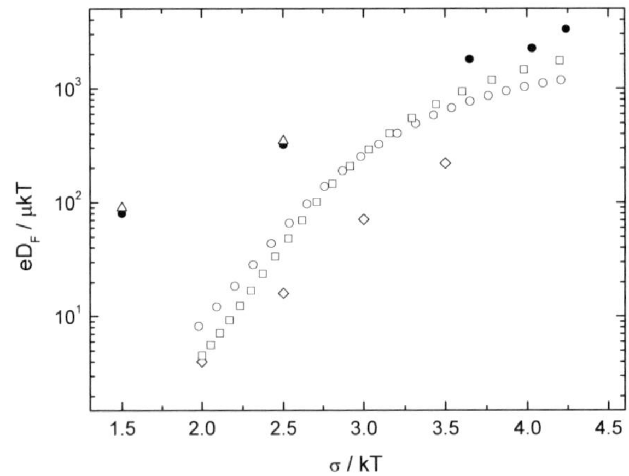

Fig. 6. Normalized field- assisted diffusion coefficient as a function of σ/kT. The values of $eD_F(t_{tr})/\mu(t_{tr})kT$ are shown by circles (the variable is σ) and by squares (the variable is kT). Data of GDM at $\Sigma = 0$ (diamonds), $\Sigma = 3.25$ (triangles), and experimental results (full circles) are shown for comparison.

The dependences of the ratio $eD_F(t_{tr})/\mu(t_{tr})kT$ on the energy disorder parameter σ/kT are shown on the Fig. 6, providing that $F_0 = 3.6 \cdot 10^5\, V/cm$, $L = 6.75\mu m$, $N = 4.6 \cdot 10^{21} cm^{-3}$, $\gamma^{-1} = 0.12 nm$. Experimental and simulation data are taken from Fig. 8 of the paper (Borsenberger et al., 1993a). Experimental TOF data for TAPC, doped by bisphenol-A-polycarbonate (BPPC), are indicated by full circles on the Fig. 6. Diamonds and triangles are show results of GDM simulations for the case of vanishing and large positional disorder, respectively. Results of the present theory, as calculated from Eqs. (14)-(16), (11), are denoted on Fig. 6 by squares (variation of σ, $T = 295K$) and open circles (variations of T, $\sigma = 0.075eV$). These results are in qualitative agreement with experiment for the system with large positional disorder, if $\sigma/kT \geq 3$, that means $\sigma/kT \geq \Sigma$, where $\Sigma = 3.25$ is the parameter of positional disorder in GDM (Bässler, 1993). This confirms the validity of the present approach to the materials with predominance of energetic disorder. In Fig. 7 calculated *W(L)* dependences are compared with experimental data at several values of σ/kT, providing that $F_0 = 2 \cdot 10^5 V/cm$, $N = 4.6 \cdot 10^{21} cm^{-3}$, $\gamma^{-1} = 0.15nm$. Data on TAPC-doped polystyrene are taken from the work Borsenberger & Bässler, 1994. A $W \sim L^{-0.5}$ law is indicated by dashed straights. This is confirmed by the $W(L)$ dependence for $\sigma/kT = 2.6$. For $\sigma/kT = 3.0$ $W(L)$ become weaker ($L < 3\mu m$), and for $\sigma/kT = 4.4$ the dispersion is practically independent of sample thickness. Values of W, as calculated from Eqs. (23) and (24), are in qualitative agreement with experimental data (see solid and dash- dotted lines).

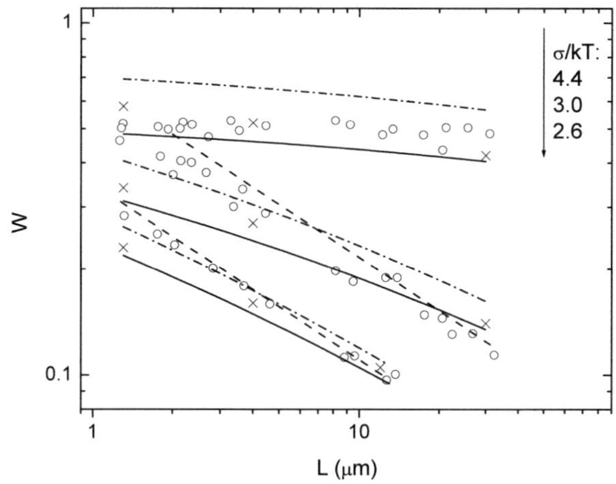

Fig. 7. Dependences of the relative dispersion of the transient current W on the sample thickness L for several values of σ/kT. Experimental data are shown by circles. Solid and dash- dotted lines are calculated from approximate Eqs. (23) and (24), respectively. The dashed lines are the dependences $W \sim L^{-1/2}$. Crosses are results of straightforward determination of $W = \left(t_{1/2} - t_0\right)/t_{1/2}$ from calculated $j(t)$ curves.

The peculiarities of $W\left(L, F_0, \sigma/kT\right)$ dependences, mentioned above, are a signature of transport being not completely in quasi- equilibrium although the mobility has equilibrated already. The coefficient of field- assisted diffusion continues to increase during several orders of magnitude in time even at moderate energy disorder i. e. $\sigma/kT = 3.0$ (Fig. 4). This transport regime will therefore be referred to as quasi-dispersive.
One should expect that in the quasi- dispersive regime the normalized time dependences $j(t/t_{tr})/j_0$ for different L and F_0 are universal, since W does not depend on these parameters. This kind of scaling is a well- known signature of dispersive transport. However, in

accordance with previous GDM results (Bässler, 1993; Pautmeier et al., 1991), this work indicates, that there is also scaling in non- dispersive TOF transients provided that there is

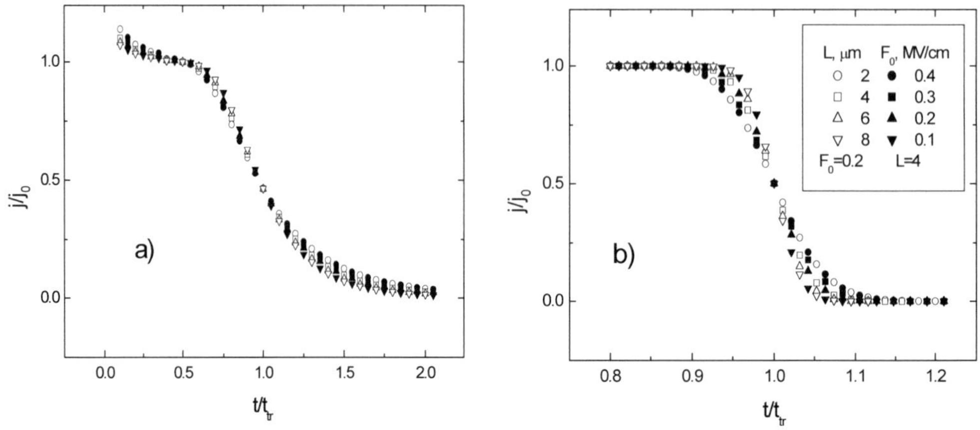

Fig. 8. Time dependences of transient current at several values of F_0 and L, $\sigma/kT = 3.5$ (a) and $\sigma/kT = 2.0$ (b). Other parameters are: $N = 4.6 \cdot 10^{21} cm^{-3}$, $\gamma^{-1} = 0.12nm$, $T = 295K$.

also scaling in non- dispersive TOF transients provided that disorder is sufficiently strong ($t_{eq\_D} > t_{tr}$) but below the critical value at which transit time dispersion commences ($t_{eq\_\mu} < t_{tr}$). Figs. 8a and 8b illustrate both occurrence and violation of scaling in TOF signals.

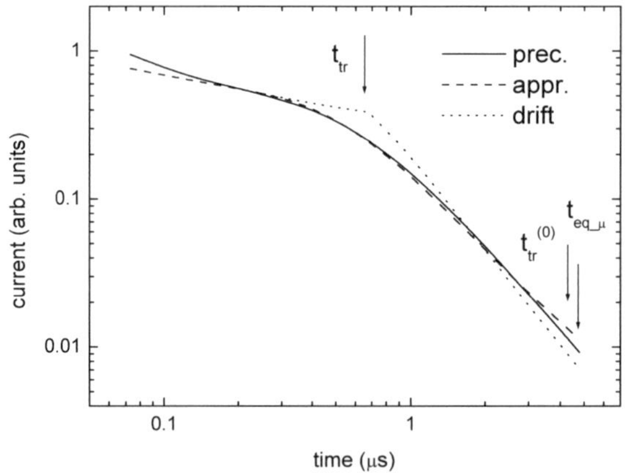

Fig. 9. TOF current in thin film, $L = 100nm$. Solid line is the calculation from Eqs. (10), (20) while the dotted line is the same but field diffusion is neglected. Dashed line is calculated from the approximate Eq. (22).

Current and time are normalized by the plateau value $j_0 = e\mu_{eq} A_0 F_0 / L$ and by the transit time t_{tr}. Obviously, charge transport must become dispersive in thin samples when the transit time is short enough (Bässler, 1993; Pautmeier et al., 1991). Fig. 9 shows the calculated occurrence of dispersive transport in a thin film, $L = 100nm$. Other parameters are the same as for Fig. 5. Neglecting the field diffusion one obtains the break of TOF current at $t = t_{tr}$ (see dotted line). Fig. 9 shows that the time t_{tr}, see Eq. (11), is a good estimate of the transient time even for dispersive regime. Obviously, $t_{tr} << t_{tr}^{(0)}$, where $t_{tr}^{(0)} = L/\mu_{eq}F$ is the time of flight of carriers calculated under the (violated) assumption that the mobility has reached its quasi- equilibrium value already. Therefore the apparent mobility, defined as $\mu_{app} = L/F_0 t_{tr}$, increases considerably at low temperatures or strong energy disorder upon decreasing the sample thickness. Fig 10 shows the calculated dependence of the ratio μ_{app}/μ_{eq} on L at several temperatures, providing that $N = 4.6 \cdot 10^{21} cm^{-3}$, $\gamma^{-1} = 0.12nm$ $F_0 = 2 \cdot 10^5 V/cm$, $\sigma = 0.075eV$. One should remember that $L \leq 100$ nm are typical values for organic light- emitting diodes. In single- layer diodes (Pinner et al., 1999) the transit time of charge carriers determine the characteristic time of onset of electroluminescence.

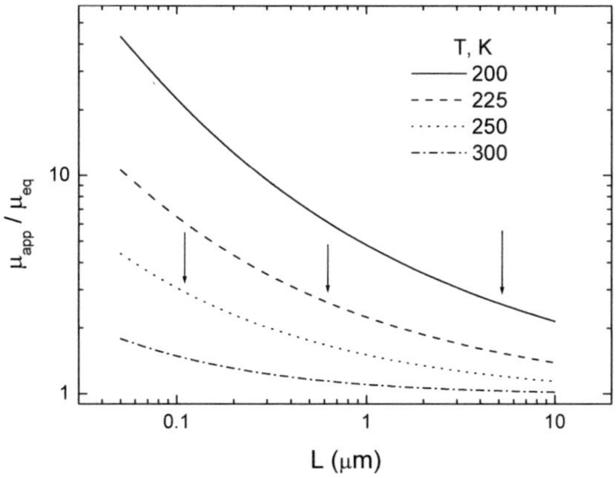

Fig. 10. Ratio μ_{app}/μ_{eq} is calculated as a function of film thickness L for several temperatures. Arrows shows the values of L, which are satisfy to the condition $t_{tr} = t_{eq\_\mu}$.

It is obvious that the temperature and field dependences of mobility as defined from the latter time, see the next section, should differ from values derived from TOF experiments at $L >> 1\mu m$ if dispersive character of transport is disregarded.

5. Transient electroluminescence from light- emitting diodes

Transient electroluminescence (TrEL) is widely considered as a general technique for determination of charge carrier mobility in thin organic films ($L \leq 100$ nm). A typical single-layer OLED is similar to a sample for TOF measurements. TrEL measurements have the advantage that they provide information directly from the light emitting device. The delay time, i.e., the time lag between applying a rectangular voltage pulse to the device and the first appearance of electroluminescence (EL), is identified as the time until the two leading fronts of injected carriers—holes and electrons—meet in the device. Zone of the most intensive EL is typically a narrow sheet (several nanometers in thickness) in proximity to one of the electrodes because of strong asymmetry of hole and electron mobilities (Crone et al., 1999; Friend et al.. 1999). Consequently, the transport of the fast carriers across almost whole layer is the key process for the EL onset.

5.1 The injection- limited regime

Injection of holes is limited by a sufficiently high energetic barrier in this regime. Space charge density of holes is small, hence electric field is approximately uniform, except of a thin layer near the cathode. It is instructive to examine the possibility of non- equilibrium transport for this case. At the first glance, initial energy distribution of injected carriers should be rather "cold" in order to preclude their subsequent energetic relaxation. However, one has to remember that (i) the DOS is shifted to lower energies in the proximity of the electrode by the image- force Coulomb potential and (ii) energetic disorder provide the possibility of downward jumps. One can estimate the initial density of occupied states (DOOS) as a product of the DOS and the rate of initial jumps (Gartstein & Conwell, 1996), namely $\rho_1(E) \propto g(E)\exp\left[-\eta(E+U(a))(E+U(a))/kT\right]$ (the zero point of energy is the peak of Gaussian DOS), where $U(a) = H_h - e^2/16\pi\varepsilon\varepsilon_0 a - eF_0 a$, H_h is the injection barrier. Thus, the energetic dependence of initial DOOS resembles the quasi- equilibrium ("cold") distribution, namely $\rho_{eq}(E) \propto \exp(-(E-E_{eq})^2/2\sigma^2)$, $E_{eq} = -\sigma^2/kT$, if $E > -U(a)$. Otherwise, the energetic dependence of DOOS is "hot", $\rho_1(E) \propto g(E)$. Considerable energetic relaxation (and therefore the dispersive transport) is possible in course of subsequent motion, if $E_{eq} < -U(a)$, i.e. $H_h < H_*$, where $H_* = \sigma^2/kT + e^2/16\pi\varepsilon\varepsilon_0 a + eF_0 a$. Indeed, the simple calculation yield for the average energy of initial DOOS $\langle E\rangle \ge E_{eq} + kT$, if $H_h \le H_*$ (the same result is obtained by the use of the method of the work (Arkhipov et al., 1999), i.e. including dispersion of lengths of initial jumps). One obtains $H_* = 0.5$ eV ($\sigma = 0.075$ eV) and $H_* = 0.7$ eV ($\sigma = 0.1$ eV), providing that $a = 0.6$ nm, $\varepsilon = 2.5$, and $F_0 = 3\cdot 10^5\ V/cm$ (Nikitenko & von Seggern, 2007). One has to note that the equality $\langle E\rangle \ge E_{eq} + kT$ is equivalent to $t \le t_{eq\_\mu}$ (Nikitenko et al., 2007), providing that the DOOS is "hot" at $t = 0$, as in the previous section (respectively, $t \le t_{eq\_D}$, if $H_h \le H_* + \sigma^2/kT$). Therefore the *quasi-dispersive* (although not *dispersive*) regime of charge transport can be realized, even if $H_h \ge H_*$. The difference in initial conditions, mentioned above, should be unimportant, if the time interval $t >> t_{eq\_\mu}$ is considered.

One has to remember that the injection is a multi- step hopping motion of a charge carrier (Gartstein & Conwell, 1996; Arkhipov et al., 1999). It include two competitive processes: (i) "cooling" by energetic relaxation and (ii) "heating" in course of their motion across the energetic barrier, which is formed by the potential energy $U(x)$, because the probability to overcome this barrier is smaller for a carrier with lower energy. One obtains practically the same DOOS of *eventually* injected carriers, as $\rho_1(E)$, following the ref. (Arkhipov et al., 1999). The arguments mentioned above approve the application of the simple model of this section to the description (at least qualitative) of the initial TrEL.

TrEL is considered here providing that holes moves much faster than electrons, hence the recombination zone is located next to the cathode at the initial (after the switching of voltage pulse at $t=0$) time period. The anode and the cathode are located at $x=0$ and $x=L$, respectively, L is the film thickness. The recombination current density, $J_R(t)$, which is proportional to TrEL intensity, is the product of the conduction current density of holes, $J(L,t)$, incoming to the recombination zone, and the probability of radiative recombination, $\phi(t)$,

$$J_R(t)=J(L,t)\ \phi(t). \tag{25}$$

The latter function increases with time (adiabatically slow relative to the first one) due to (i) slow increase of electron density and (ii) slow increase of mean lifetime of singlet excitons because of slow (dispersive) transport of electrons apart from the cathode. Therefore, the function $\phi(t)$ is determined by rather complicated physical processes. It can be introduced, however, in a simple phenomenological way in order to reduce the number of model parameters. It is known that TrEL kinetics shows often two different rise times, which are connected with transport of electrons and holes, respectively (Pinner et al., 1999). Thus, one can write the function $\phi(t)$ in this case as follows:

$$\phi(t)\approx J_R(t)/J_R^{\ st}=1-(1-\phi_0)\exp(-t/\tau_e), \quad t>>t_{tr}, \tag{26}$$

where t_{tr} is the transit time of holes, $J_R^{\ st}$ is the long- time limit of $J_R(t)$, and $\phi(t)$ is normalized by the condition $\phi(\infty)=1$, providing that long-time TrEL kinetics can be described by time τ_e, $\tau_e>>t_{tr}$. One can obtain the parameter $\phi_0=\phi(0)<1$ by extrapolation of Eq. (26) to the zero time.

The time dependence of $J(L,t)$ is discussed at first, providing rather high energy barrier for injection of holes, hence the electric field is uniform. One obtains [Nikitenko & von Seggern, 2007]

$$J(L,t)=J_h\left(1-\int_0^L dx p(x,t)\right), \tag{27}$$

where J_h is the injection current density of holes and $p(x,t)$ is the distribution function of holes, being injected by short pulse at $t=0$ see Eq. (10). It can be approximated by the Gaussian function, if $t>>t_{eq\_\mu}$:

$$p(x,t)\approx\exp\left\{-\left(x-\mu_{eq}F_0t\right)^2/4S_F(t,F_0)\right\}\Big/\sqrt{4\pi S_F(t,F_0)}. \tag{28}$$

This function is characterized by time- dependent coefficients of FAD, $D_F(t,F_0)$, and mobility $\mu(t)\approx\mu_{eq}$, $F_0=(V-V_{bi})/L$ is the strength of applied electric field, V and V_{bi} are applied and built-in voltages, respectively. Eqs. (27) and (28) yields

$$J(L,t)\approx J_h(1/2)erfc\left[\left(L-\mu_{eq}F_0t\right)/2\sqrt{S_F(t,F_0)}\right], \quad t>>t_{eq\_\mu}. \tag{29}$$

Quasi-equilibrium transport regime is established at the long- time limit, $t>>t_{eq\_D}$: $D_F(t,F_0)\approx D_{Feq}(F_0)$ and $S_F(t)\approx D_{Feq}t$, although $D_{Feq}/\mu_{eq}>>kT/e$ $(\sigma/kT>>1)$. The transit time of holes can be estimated as $t_{tr}\approx L/\mu_{eq}F_0$ in quasi-equilibrium or quasi-dispersive regime. Obviously, $J(L, L/\mu_{eq}F_0)=0.5J_h$, see Eq. (29). Therefore, $t_{tr} \approx t_{1/2}$, where the half-rise time of TrEL, $t_{1/2}$, is defined as $J_R(t_{1/2}) = 0.5J_R(\infty)$.

5.2 The space charge- limited regime

If the space- charge limited (SCL) regime of hole's transport is realized, an electric field strength, $F(x,t)$, is determined by space charge of holes (except of several nanometers next to the cathode, where electron's charge is important), in accord with the Poisson's equation. If the dispersive transport is finished, $t > t_{eq\_\mu}$, one can connect the current with the density of holes, $p_h(x,t)$:

$$J(x,t) \approx \mu(t)\exp\left[-\lambda(t)t\right]F(x,t)p_h(x,t) . \tag{30}$$

Following to the work Many & Rakavy, 1962, hence neglecting any diffusion, the density of holes next to the cathode can be written as the step- like function,

$$p_h^{(0)}(x,t) = \left[\varepsilon\varepsilon_0/eM(t)\right]\int_0^{x_1(t)} dx_*\delta(x-x_*), \tag{31}$$

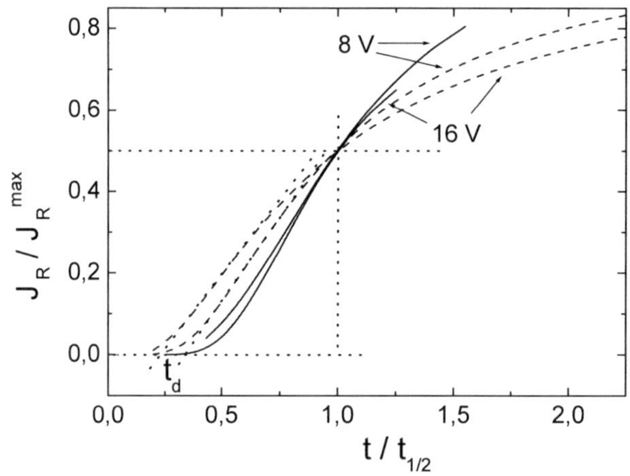

Fig. 11. Time dependencies of the initial rise of recombination current, calculated in SCL regime (solid lines) and in IL regime (dashed lines). Dotted lines show how the delay time t_d is defined. The values of applied voltage are shown in the figure.

where, ε is the relative dielectric constant, and $x_1(t) = 2L\ln\left\{1/\left[1-M(t)F_0/2L\right]\right\}$ is the leading front position. One can include approximately the field- stimulated broadening of step- like leading front of the distribution (31), replacing the Dirak delta- function in this equation by Gaussian function (28), with x_* and $S_F\left[t,F(x_*,t)\right]$ instead of $\mu_{eq}F_0t$ and $S_F(t,F_0)$, respectively. By the use of Eq. (30), one obtains

$$J(L,t) \approx \frac{\varepsilon\varepsilon_0 \mu(t)}{2eM(t)} F(L,t) \exp[-\lambda(t)t] \; erfc\left[\frac{L - x_1(t)}{2\sqrt{S_F[t, F(L,t)]}}\right], \quad t < t_1, \quad (32)$$

$F(L,t) = F^{(0)}(L,t) - (e/\varepsilon\varepsilon_0)\int_L^\infty [p_h(x,t) - p_h^{(0)}(x,t)]dx$, where $F^{(0)}(L,t) = F_0/[1 - M(t)F_0/2L]$.

The time t_1 is the transit time of holes, it is defined as $x_1(t_1) = L$, i.e.

$$F_0 M(t_1)/L \approx 0.787 . \quad (33)$$

Obviously, $t_1 \approx 0.787 L/\mu_{eq} F_0$ (Many & Rakavy, 1962), because $\mu(t) \approx \mu_{eq}$, if $t_1 >> t_{eq\_\mu}$. Eqs. (25), (26) and (32) (or Eq. (29), in the injection- limited regime), yields approximate analytic description of initial rise of TrEL. One has to note, that the effects of filling of deep states and field dependence of non- equilibrium mobility of holes are not considered here. Indeed, calculations of the refs. [Pasveer et al., 2005, Arkhipov et al., 2001b] yield the weak dependence of mobility on hole's density and field strength, if $p_h/N \le 10^{-4}$, $F_0 \le 5 \cdot 10^5$ V/sm, $T > 250$ K, $\sigma/kT \le 3$.

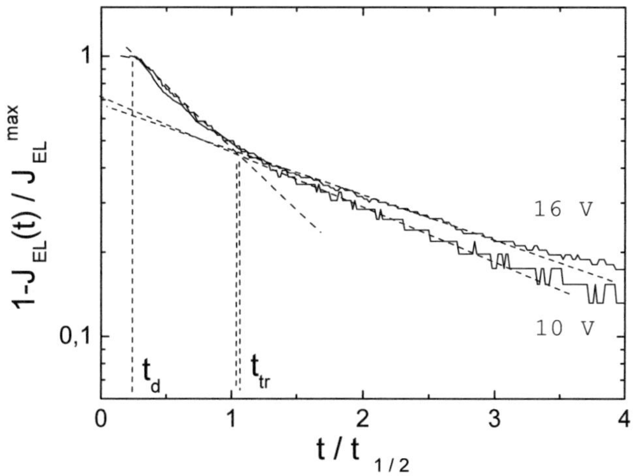

Fig. 12. Normalized electroluminescence transients, obtained for the structure ITO:PANI/co-PPV (100 nm)/Ca:Al. Dashed lines show how the delay and transit times are defined.

Normalized TrEL signals calculated for injection- limited (IL) and SCL regimes of hole transport are shown in the Fig. 11 as dashed and solid curves, respectively. Calculations are carried out for two values of applied voltage, 8 and 16 V, and film thickness 100 nm, providing that $\tau_e = 3t_1$, where t_1 is calculated from Eq. (33); $\phi(t) = \phi_0$ =0.3 (SCL regime); $\sigma = 0.075$ eV, $T = 295$ K, $N = 4.6 \cdot 10^{21}$ cm^{-3}, $2\gamma N^{-1/3} = 10$. Time is normalized by the half-rise time, $t_{1/2}$. The simplest case, $\phi = const$, is assumed for the IL regime. Built-in voltage $V_{bi} = 2$ V is taken in account, so the field lies in the range from $6 \cdot 10^5$ to $14 \cdot 10^5$ V/cm. Fig.11 shows an approximate universality in both regimes. As field increases, the dispersion parameter $W_R = (t_{1/2} - t_d)/t_{1/2}$ (Nikitenko & von Seggern, 2007), varies from 0.64 to 0.76 and from 0.52 to 0.58 for IL and SCL regimes, respectively. Obviously, TrEL raises steeper in SCL regime. The delay times, t_d, are defined as it is shown in Fig.11. Both these variations are much less than it is predicted by the formula, $W_R = \sqrt{\pi D/\mu F_0 L}$, which can be derived in analogy with the TOF by the use of Eq. (28), providing the time- independent FAD coefficient $D_{Feq} \propto F_0^2$. A reason of the universality is the non- stationary FAD coefficient, in analogy with TOF experiments. If the Einstein's relation, $D/\mu = kT/e$, is the case, then one obtains variation of W_R from 0.11 to 0.07 contrary to both the calculated and experimental (see below) results.

The results of the calculations are compared with experimental data. Single-layer OLEDs were fabricated on ITO glass substrates covered with polyaniline (PANI) as a hole injecting layer followed by a 100-nm- thick co-PPV layer as active material where co-PPV is poly[(p-phenylenevinylene)-alt-(2-methoxy-5(2-ethylhexyloxy)-p-phenylenevinylene)] from Sigma-Aldrich. A Ca cathode and Al protecting layer were thermal deposited in vacuum. TrEL measurements were performed using a Keithley source-measure unit and photomultiplier tube (Nikitenko et al., 2008).

The built-in voltage for this structure is $V_{bi} = 2\ V$ and holes are the fastest charge carriers (Scott et al., 1999). Fig. 12 shows the semi logarithmic plot of experimental TrEL intensities, $J_{EL}(t)$, subtracted from its long- time value, $J_{EL}(\infty) = J_{EL}^{\max}$, and normalized by the $J_{EL}^{\max}$. Obviously, the transit time can be determined by the method of the work (Pinner et al., 1999), see dashed lines, and this time is very close to the half- rise time of TrEL, $t_{1/2}$ (see also Fig. 13 of the cited work).

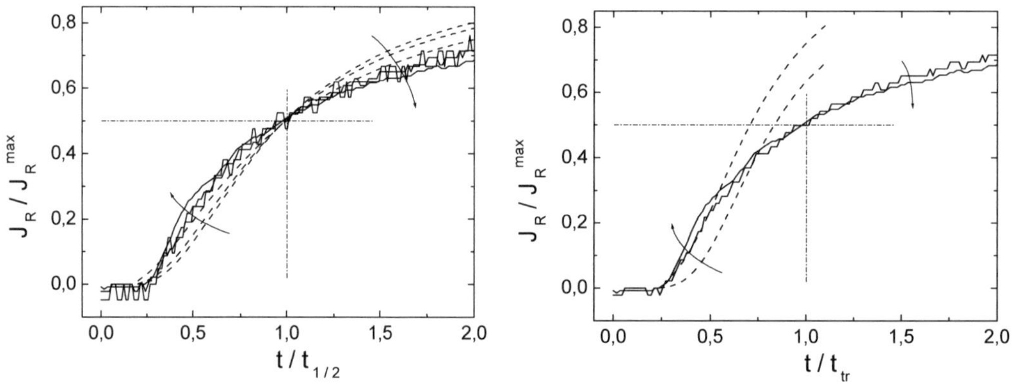

Fig 13. Normalized experimental TrEL signals (solid lines), compared with results of calculations (dashed lines). Recombination current calculated for IL regime (left panel, applied voltages are 8, 10, 16 volts) and for SCL regime (right panel, 10 and 16 volts). On the right panel, $t_{tr}=t_{1/2}$ and $t_{tr}=t_1$, see Eq. (33), for experimental and calculated curves, respectively. Increase of voltage is shown by arrows.

One can obtain parameters of $\phi(t)$, see Eq. (26), from the long- time exponential asymptotic, namely $\phi_0 \approx 0.3$ and $\tau_e(V=10) = 2.2\ t_{1/2}$, $\tau_e(V=16) = 2.7\ t_{1/2}$.

Fig. 13 shows the universality of normalized experimental TrEL intensities. The results are in good agreement with calculations in the IL regime (left panel), see the dashed lines, although initial rise of experimental curves is somewhat steeper. Both calculated and experimental data are normalized to the steady-state level, time is normalized to the theoretical transit time of holes and half-rise time, respectively. Both times coincides practically. Again, one can identify the latter with the transit time of holes, while the delay time of TrEL is much smaller than the transit time.
The difference of the workfunction of ITO and HOMO level of co-PPV yield the energy barrier 0.5 eV. The assumption about IL- regime of hole transport is questionable, however, especially at highest voltage. TrEL is calculated in SCL regime for the same set of parameters and compared with experiment in the right panel of Fig. 13. Obviously, the initial slope of the 16V- curve is reproduced by calculations better, than in the left panel of Fig. 13. One can conclude that the transition from IL to SCL regime of hole's transport occurs with the increase of applied voltage from 10 V to 16 V. Subsequent rise of calculated curves is unreasonably steep, however, suggesting that the accuracy of the approximate Eq. (32) is insufficient at $t \geq t_{tr}$. The steepness of the initial rise of TrEL in SCL regime increases together with the electric field. Rise of TrEL is moderated, on the other hand, by the increase of $\phi(t)$, which reflects an electron's kinetics, hence the calculated $t_{1/2}$ underestimates the transit time at low voltages not considerably.

6. Conclusion

It has been shown that in energetically and spatially random hopping systems, there is a time domain in which the transport is neither fully dispersive nor quasi- *equilibrium*. It is referred to as a quasi- dispersive regime. It is the time domain in which the charge carriers in the top portion of the density of states distribution that contribute most to the current are already equilibrated while the entire ensemble of photoexited carriers still relaxed towards the bottom states. Previous Monte- Carlo simulations delineated that field- assisted diffusion increases at long time domain although the carrier mobility has saturated already (Pautmeier et al., 1991; Borsenberger et al., 1993b). The present analytic theory is able to account for the quasi- dispersive features, i. e. scaling of normalized transient currents with anomalously large tails at different values of sample thickness and field strength as well as almost equilibrated transport borned out by the plateau in the $j(t)$ dependence. It also provides a quantitative explanation for the experimentally observed and simulated spread of the transit times, quantified by the dispersion parameter $W(L,\sigma/kT,F_0)$ as a function of sample thickness, energy disorder parameter and electric field strength (Borsenberger et al., 1993a,b). The theory applies to the case of moderate electric field and field dependence of mobility is not considered here.
Hirao et al., 1995; 1999 attempt to interpret experimental data on the field dependence of carrier mobility under weak field, based on the assumption that the transport is quasi-equilibrium at all times. Simple analytic expression for $j(t)$ inthese works is a consequence of Eq. (20), assuming that the charge density $p(x,t)$ is a Gaussian function characterized by time- independent mobility and diffusion coefficient of charge carriers. These values defined by fitting of experimental $j(t)$ dependencies. This procedure, in spite of its success to

explain the temperature dependence of the charge carrier mobility, cannot reproduce the spatial spread of TOF transients at variable sample thickness for large and small values of σ/kT, see the Fig. 6 of the work (Hirao et al., 1999). It implies $W \sim L^{-1/2}$ for the both cases, at variance with experiment on systems with moderately strong energetic disorder, i. e. $\sigma/kT \geq 3$.

Effects of anomalous field-assisted dispersion on initial TrEL kinetics cannot be ignored, basing on arguments following from both theoretical and experimental data. Transit time of fastest charge carriers (holes) can be identified rather with half- rise time of TrEL (in analogy with half- decay time of TOF signal (Bässler, 1993), than with the delay time. The latter is a measure of a time of flight of fastest fraction of holes which hopping paths include only the states with energies shallower than the mean energy of occupied states in quasi-equilibrium regime, $-\sigma^2/kT$. One can overestimate the mobility (in the case of our experimental device, by a factor 4) if the delay time is taken as a transit time. The same conclusion was made in the work Pinner et al., 1999. The method of this work is appropriate in our case as well (see Fig. 12). In general, the method of half- rise time seems to be more appropriate if the long-time TrEL kinetics is not pure exponential and the steady- state level can be observed clearly.

Most of recent studies of charge transport are focused on behaviour of carrier mobility; this chapter is focused on less studied problem of dispersion of charge carriers in space. The objective was to emphasize that a carrier's non-equilibrium manifestations are much wider than effects of dispersive transport. Results of this chapter provide options for analytic modeling and correct determination of material's parameters from data of time- of- flight and transient electroluminescence measurements.

7. References

Arkhipov, V.I. & Rudenko, A.I. (1982a). Drift and diffusion in materials with traps. I. quasi-equilibrium transport regime. *Phil. Mag. B*, 45, 177-187, ISSN 0141-8637

Arkhipov, V.I. & Rudenko, A.I. (1982b). II. Non-equilibrium transport regime. *Phil. Mag. B*, 45, 189-206, ISSN 0141-8637

Arkhipov, V.I. & Nikitenko, V.R. (1989). Dispersive transport in materials with a nonmonotonic energy distribution of localized states. *Sov. Phys. Semicond.* 23, 6, 612-615, ISSN 0038-5700

Arkhipov, V.I. &, Bässler H. (1993a). A model of weak-field quasi-equilibrium fopping transport in disordered materials. *Phil. Mag. Lett.*, 67,5, 343-349, ISSN 0950-0839

Arkhipov, V.I. &, Bässler H. (1993b). An adiabatic model of dispersive hopping transport. *Phil. Mag.B.*, 68, 425-434, ISSN 0141-8637

Arkhipov, V.I.; Wolf, U. &, Bässler H. (1999). Current injection from a metal to a disordered hopping system. II. Comparison between analytic theory and simulation. *Phys. Rev. B*, 59, 11, 7514-7520, ISSN 0163-1829

Arkhipov, V.I.; Emelianova, E.V. & Adriaenssens, G.J. (2001a). Effective transport energy versus the energy of most probable jumps in disordered hopping systems. *Phys. Rev. B*, 64, 125125, 1-6, ISSN 0163-1829

Arkhipov, V.I.; Heremans, P.; Emelianova, E.V. & Adriaenssens, G.J. (2001b). Space-charge-limited currents in materials with Gaussian energy distributions of localized states. *Appl. Phys. Lett.*, 79, 25 4154-4156, ISSN 0003-6951

Baranovskii, S.D.; Cordes, H.; Hensel, F. & Leising G. (2000). Charge-carrier transport in disordered organic solids. *Phys. Rev. B.* 62, 12, 7934- 7938, ISSN 0163-1829

Barth, S.; Müller, P.; Riel,H., et al. (2001). Electron mobility in tris(8-hydroxy-quinoline) aluminum thin films determined via transient electroluminescence from single- and multilayer organic light- emitting diodes. *J. Appl. Phys.*, 89, 7, 3711-3719, ISSN 0021-8979

Bässler H. (1993). Charge transport in disordered organic photoconductors. *Phys. Status Solidi B*, 175, 15-56, ISSN 0370-1972

Blom, P. W. M. & Vissenberg, M. C. J. M. (1998). Dispersive hole transport in poly(p-phenylene vinylene). *Phys. Rev. Lett.*, 80, 17, 3819-3822, ISSN 0031-9007

Blom, P. W. M. & Vissenberg, M. C. J. M. (2000). Charge transport in poly(p-phenylene vinylene). *Mater. Sci. and Eng.*, 27, 53-94, ISSN 0927-796X

Borsenberger, P. M.; R. Richert, R. & Bässler, H. (1993a). Dispersive and nondispersive charge transport in a molecularly doped polymer with superimposed energetic and positional disorder. *Phys. Rev. B*, 47, 8, 4289-4295, ISSN 0163-1829

Borsenberger, P. M.; Pautmeier, L.T. & Bässler, H. (1993b). Scaling behavior of nondispersive charge transport in disordered molecular solids. *Phys. Rev. B*, 48, 5, 3066-3073, ISSN 0163-1829

Borsenberger, P.M. & Bässler, H. (1994). Tall broadening of photocurrent transients in molecularly doped polymers. *J. Appl. Phys.*, 75, 2, 967-972, ISSN 0021-8979

Crone, B.K. et al. (1999). Device physics of single layer organic light-emitting diodes. *J. Appl. Phys.*, 86, 10 5767-5774, ISSN 0021-8979

Hirao,A.; Nishizawa, H. & Sugiuchi, M. (1995). Diffusion and drift od charge carriers in molecularly doped polymers. *Phys. Rev. Lett.*, 75, 9, 1787-1790, ISSN 0031-9007

Hirao,A.; Tsukamoto, T. & Nishizawa, H. (1999). Analysis of nondispersive time-of-flight transients. *Phys. Rev. B.*, 59, 20, 12991-12995, ISSN 0163-1829

Fishchuk, I.; Kadashchuk,A.K.; Bässler, H. & Weiss, D.S. (2002). Nondispersive charge-carrier transport in disordered organic materials containing traps. *Phys. Rev. B.*, 66, 205208 1-12, ISSN 0163-1829

Friend, R.H.; Gymer, R.W.; Holmes, A.B. et al. (1999). Electroluminescence in conjugated polymers. *Nature*, 397, 121-128, ISSN 0028-0836

Gartstein, Yu. N. & E. M. Conwell, E.M. (1996). Field-dependent thermal injection into a disordered molecular insulator. *Chem. Phys. Lett.* 255, 93-98, ISSN 0009-2614

Many, A. & Rakavy, G. (1962). Theory of transient space-charge-limited currents in solids in the presence of trapping. *Phys. Rev.*, 126, 6, 1980-1988, ISSN 0163-1829

Miller, A. & Abrahams, E. (1960). Impurity conduction at low concentrations. *Phys. Rev.*, 120, 3, 745-755, ISSN

Monroe, D. (1985). Hopping in exponential band tails. *Phys. Rev. Lett.* 54, 2, 146-148, ISSN 0031-9007

Nikitenko, V.R. (1992). Theoretical model of dispersive tunnel transport in disordered materials. *Sov. Phys. Semicond.* 26, 8, 807-811, ISSN 0038-5700

Nikitenko, V.R.; Tak, Y.H. & Bässler, H. (1998). Rise time of electroluminescence from bilayer light emitting diodes. *J. Appl. Phys.* 84, 4, 2334-2340, ISSN 0021-8979

Nikitenko, V.R.; von Seggern, H. & Bässler, H. (2007). Non-equilibrium transport of charge carriers in disordered organic materials. *J. Phys.: Condens. Matter*, 19, 136210 1-15, ISSN 0953-8984

Nikitenko, V.R. & von Seggern, H. (2007). Nonequilibrium transport of charge carriers and transient electroluminescence in organic light-emitting diodes. *J. Appl. Phys.*, 102, 103708 1-9, ISSN 0021-8979

Nikitenko, V.R.; Tameev R.A. & Vannikov A.V. (2008). Initial rise of transient electroluminescence in organic films. *Mol. Cryst. Liq. Cryst.*, 496, 107-117, ISSN 1542-1406

Novikov, S.V.; Dunlap, D.H.; Kenkre, V.M.; Parris, P.E. & Vannikov, A.V. (1998). Essential role of correlation in governing charge transport in disordered organic materials. *Phys. Rev. Lett.*, 81, 20, 4472-4475, ISSN 0031-9007

Pasveer, W.E. et al. (2005). Unified description of charge- carrier mobilities in disordered semiconducting polymers. *Phys. Rev. Lett.*, 94, 206601, 1-4, ISSN 0031-9007

Pautmeier, L.; Richert, R. & Bässler, H. (1991). Anomalous time-independent diffusion of charge carriers in a random potential under a bias field.*Phil. Mag. B*, 63, 3, 587-601, ISSN 0141-8637

Pinner, D.J.; Friend, R.H. & Tessler, N. (1999). Transient electroluminescence of polymer light emitting diodes using electrical pulses. *J. Appl. Phys.* 86, 9, 5116-5130, ISSN 0021-8979

Rubel, O.S.; Baranovskii, S.D.; Thomas., P. & Yamasaki, S. (2004). Concentration dependence of the hopping mobility in disordered organic solids. *Phys. Rev. B* 69, 014206 1-5, ISSN 0163-1829

Schmechel, R. (2002). Gaussian disorder model for high carrier densities: theoretical aspects and application to experiments. *Phys. Rev. B* 66, 235206,1-6, ISSN 0163-1829

Scott, J.C.; Malliaras, G.G.; Chen, W.D., et al. (1999). Hole limited recombination in polymer light- emitting diodes. *Appl. Phys. Lett.*, 74, 11, 1510-1512, ISSN 0003-6951

Shklovskii, B.I. & Efros, A.L. (1984). *Electronic Properties of Doped Semiconductors*, Springer, ISBN 0387129952, Heidelberg

5

Preparation of Polyimide Thin Films by Vapour Deposition and Solid State Reactions

Anton Georgiev[a*], Erinche Spassova[b], Jacob Assa[b] and Gencho Danev[b]
[a]University of Chemical Technology and Metallurgy, Department of Organic Chemistry*
8 St. "Kliment Ohridski" Blvd, 1756 Sofia, Bulgaria, e-mail: antonchem@abv.bg
[b]Bulgarian Academy of Science, Central Laboratory of Photoprocesses
bl. 109 "Acad. G. Bonchev" Boulevard, 1113 Sofia, Bulgaria

Abstract
In this chapter we describe the preparation of polyimide thin films by physical vapour deposition and comment on their potential application as a pure material or a thin layer matrix for producing nanocomposite layers. Their superb properties, such as a low dielectric constant, high thermal- and photo-stability, high chemical resistance and high optical transmittance predetermine their wide- spread applications as a casts and layers used as insulators, protective or capsulation layers, mechanical or diffusion barriers, in opto- and microelectronics. The bulk properties of the polyimide allowed the preparation of nanocomposite materials with organic chromophores as a "guest" (the embedded in the matrix nanosized particles). Moreover, some of the "guest" could bind to the polyimide chain. There are numbers of aromatic polyimides which are broadly used as thin layers in nanotechnology.
Vapour deposition of the precursors and solid state reactions of imidization are of a greater priority than the spin coating and dipping methods. These as-deposited films by the vacuum deposition process consist of a dianhydride and diamine mixture, which by solid state reactions is converted to polyimide by thermal treatments or by combined microwave and thermal treatments. The physical vapour deposition as a "dry" method provides high purity for producing thin polymer films of controlled thickness, ratio of precursors and composition control of the so prepared layers. In this chapter we suggest possibilities for the practical application of vapour deposition of precursors and the following solid state reactions.
By the used spectral method- Fourier Transform Infrared Spectroscopy for analysis of the investigated kinetics of imidization reactions and microstructure of the layers are studied. The relationship between vapour deposition conditions and the presence of regular chains leading to the appearance of infrared bands is discussed. Polymers are also capable of forming a range of conformations depending on the backbone structure. The conditions for preparation by physical vapour deposition and solid state reaction of polyimide or nanocomposite polyimide layers are discussed.

Key words: Polyimides, thin polymer layers, solid state reactions, vapour deposition, FTIR spectroscopy

1. Application of polyimides in nanotechnology as thin layer matrix for nanocomposites

Polyimides (PI) are a class of organic compounds containing imide bond in their molecule. Aromatic polyimides are well-known polymers and due to the attractiveness of their properties such as a low dielectric constant, high thermal stability, high chemical resistance, high optical transmittance as well as very good mechanical properties. They are used in opto- and microelectronics, as well as in nanotechnology as a matrix in the production of nanocomposite layers (Francisko Raymo, 2007; Strunskus,Y and Grunze,M, 1994; Osvaldo N. Oliveira et al, 2005; Mitchell Anthamatten et al., 2004; C.P. Wong, 1993). Nanocomposite materials represent combinations of substances - polymers, chromophores, metals, etc. in which one component is the matrix and the other one - the "guest", embedded in the matrix as nanosized particles. There is no chemical interaction occurring between the matrix and the "guest". The space volume between the individual molecules allows for the "guest" molecules to be embedded in the matrix pores and a thickening of the layer achieved during the following thermal process.

In Table 1 the initial precursors and the respective PI, which find wide-ranging applications in opto- and microelectronics as modulators, barrier layers, etc. are presented (E. Mazoniene et al., 2006; Steve Lien-Chung Hsua et al., 2003).

Dianhydride *(precursor 1)*	Diamine *(precursor 2)*	Polyimide
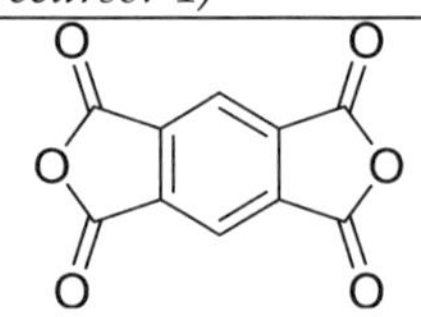 PMDA *pyromellitic dianhydride*	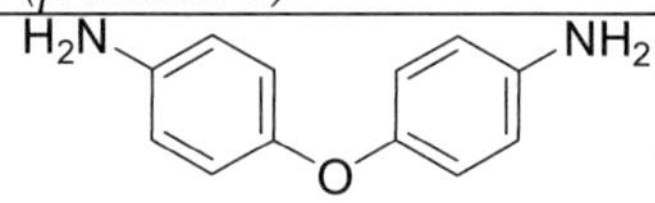 ODA *4,4'-oxydianiline* 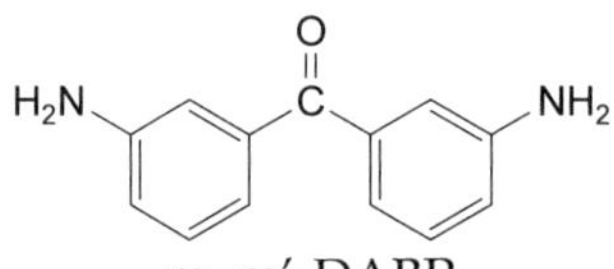 m, m'-DABP *m,m'-diaminobenzophenone* 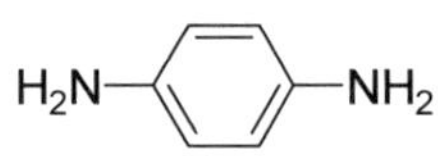 PDA *phenilendiamine*	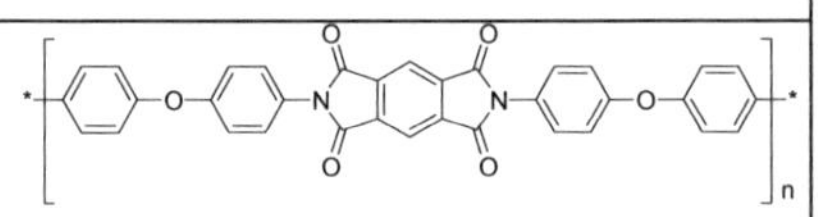 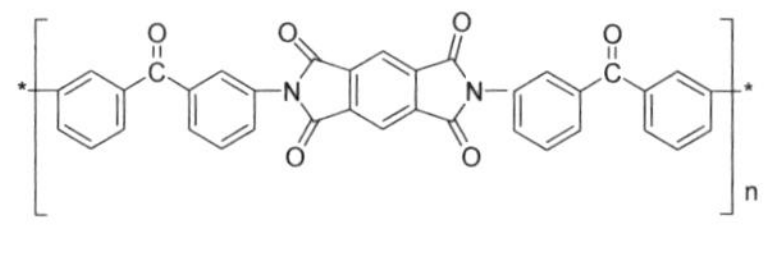 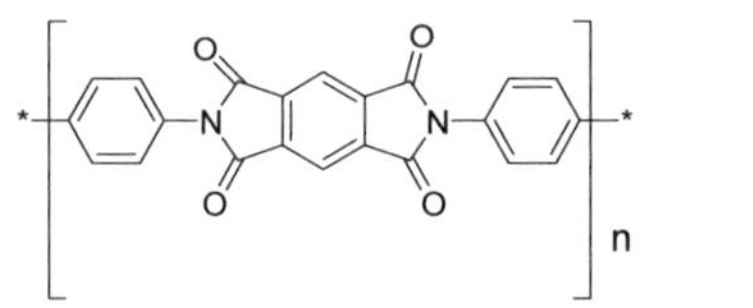

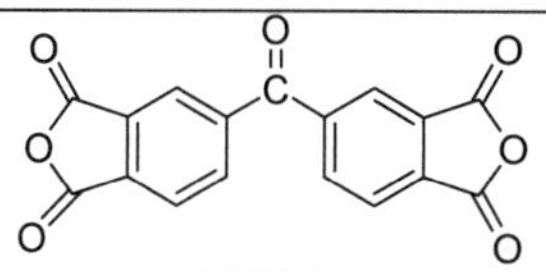 BTDA *3,3',4,4'-benzophenone tetracarboxylic dianhydride*	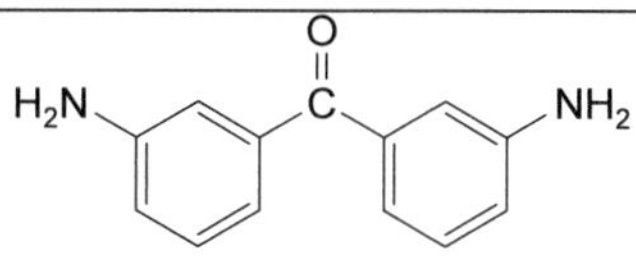 m, m'-DABP	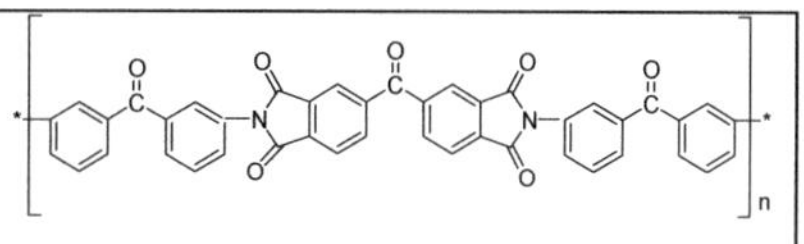
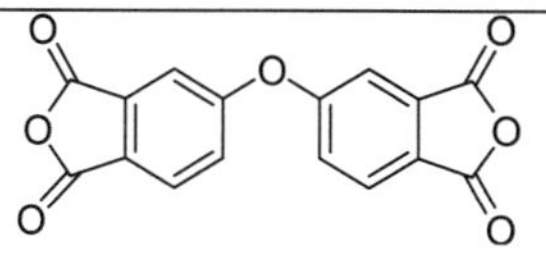 ODPA *4,4'-oxydiphthalic anhydride*	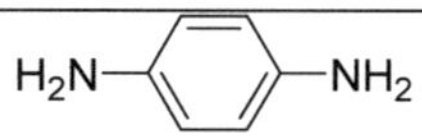 PDA	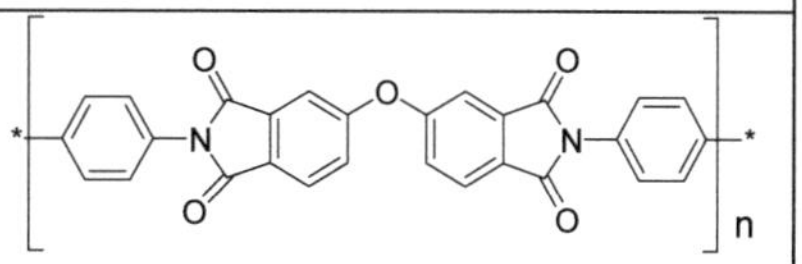
	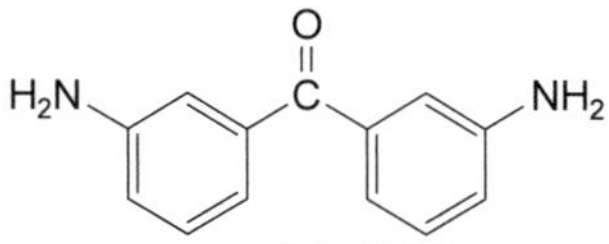 m, m'-DABP	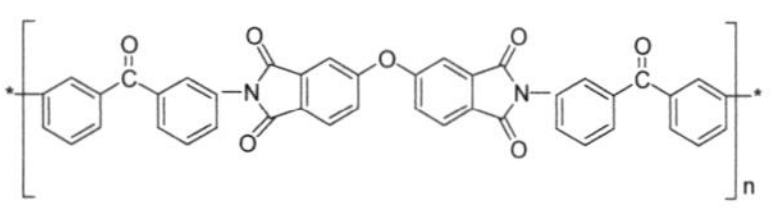
	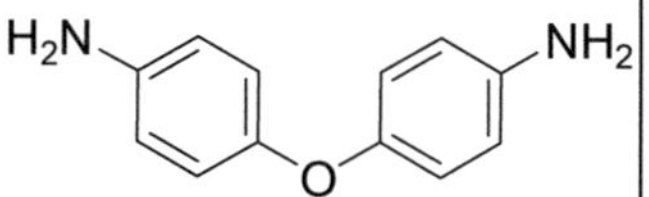 ODA	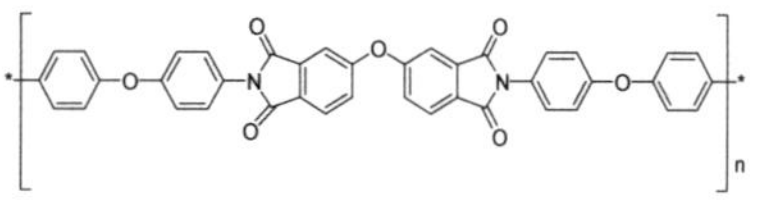
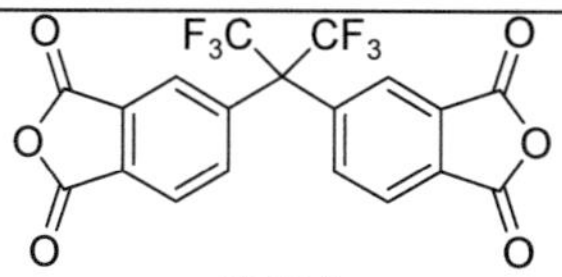 6FDA *2,2'-Bis-(3,4-Dicarboxyphenyl) hexafluoropropane dianhydride*	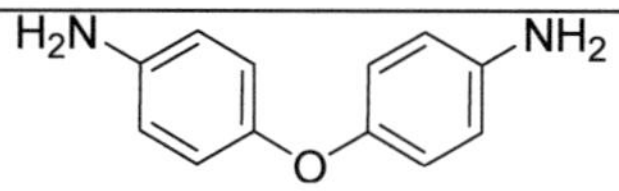 ODA	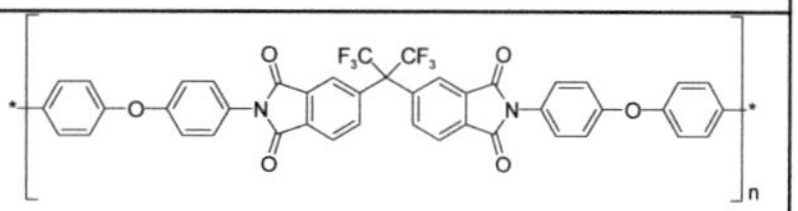
	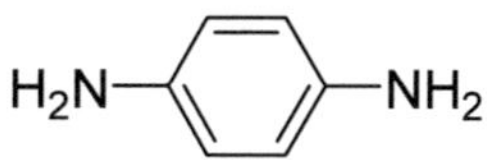 PDA	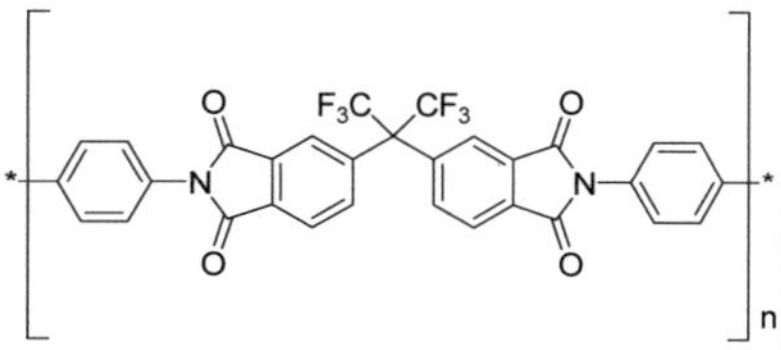

Table 1. Initial precursors and the respective polyimides finding wide-ranging applications in nanotechnology.

The high thermal and chemical stability of PI is interpreted by two factors:

(i) the high resonance energy of the benzene rings due to delocalization of the π-electrons and the great number of resonance structures;

(ii) strength of the imide bonds, resulting from the competitive *n*-π conjugation between the carbonyl group and the non pair electron couple from the nitrogen atom as well as from the conformation state of the 5- member imide ring. The lack of Baer's angular torsion is due to the fact that all C- and N- atoms are in a *sp*^{2} hybrid state with valency angle of 120° and planar conformation of the ring.

Thermal destruction of the PI obtained from the precursors PMDA (pyromellitic dianhydride) and ODA (4,4'-oxydianiline) is only observed at temperature above 420-450 °C the mechanism studied by R. Ginsburg and J.R. Susko and proven with mass spectrometry (Fig. 1) (R. Ginsburg and J.R. Susko, 1984).

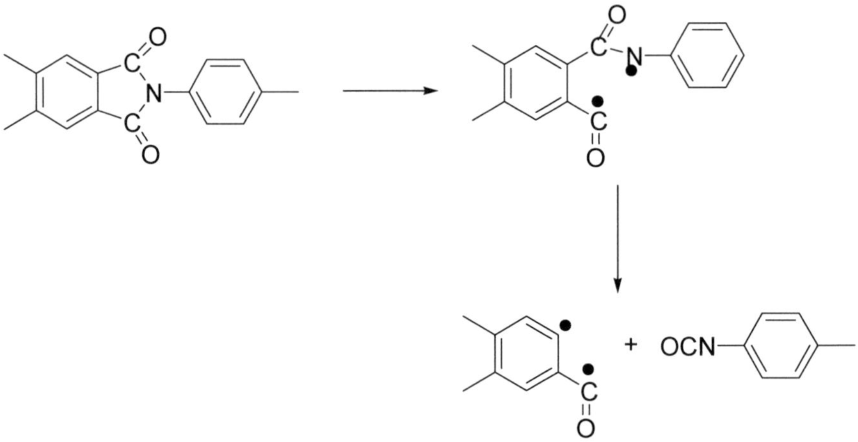

a) *homolytical cleavage of the C-N bond of the imide ring and formation of isocyanate and acylic radical.*

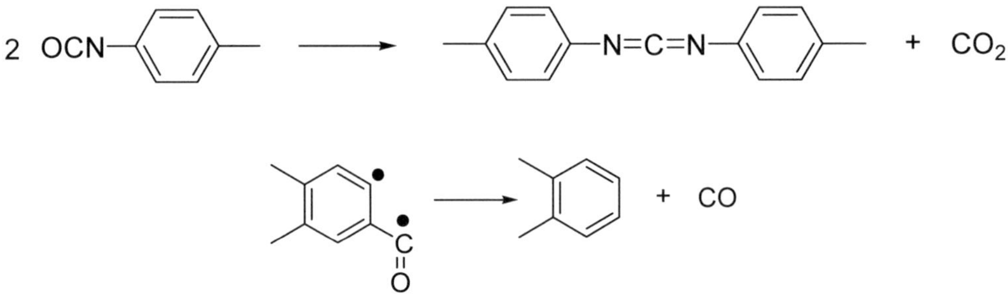

b) *recombination and release of CO_2 and CO.*

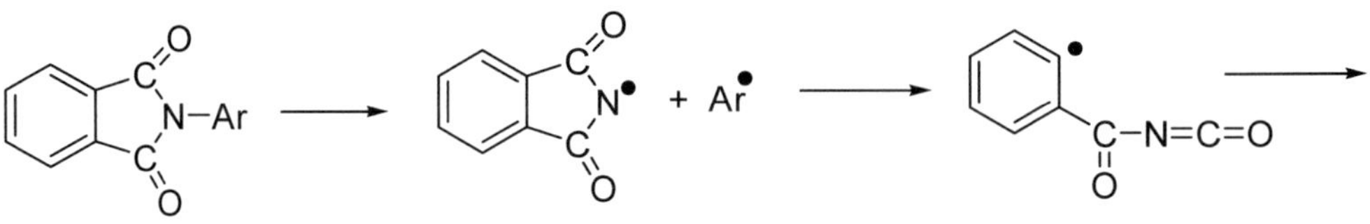

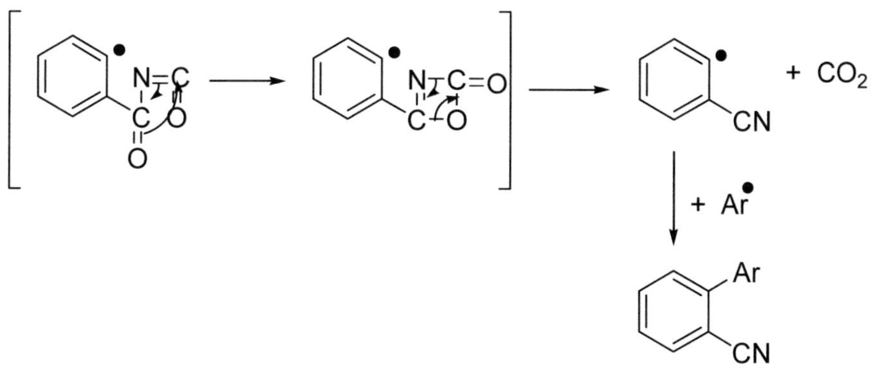

c) *decomposition via a rearrangement of the imide ring and CO_2 releasing.*

Fig. 1. Mechanism and principal stages of the thermal destruction of PI: **a)** homolytical cleavage of the C-N bond of the imide ring; **b)** release of CO_2 and CO; **c)** decomposition via a rearrangement of the imide ring.

Aromatic polyimides display attractive properties such as chemical resistance, thermal stability and stability to photo-ageing. They have the capacity to perform the matrix role in the formation of nanocomposite layers with an embedded chromophore as "guest" and are materials of good prospects for applying in contemporary and future nanotechnology.

2. Vapour deposition of thin polymer films

Obtaining of nanostructured polymer layers (from 2-4 nm to 4-5 μm thick) by deposition of their components from the gas phase renders opportunities for the production of novel materials in the field of nanotechnology. The thin layer composite materials obtained by using the vacuum technologies ensure one basic advantage - the absence of solutions and elimination of the necessity of complicated technical solutions for their removal (C.-C. Lee et al., 1993). The deposition in vacuum and the polycondensation between the precursors of the PI matrix a reaction taking place in a solid state represents an attractive method for the formation of thin polymer layers. Polyimides have the capacity of implementing nanocomposite matrix both due to the possibility to be deposited in vacuum and their chemical inactivity, high thermal stability and appropriate optical and dielectric properties (Strunskus,Y and Grunze,M, 1994; E. Spassova, 2003; Iijima M and Takahashi Y, 1986).
Most often conventional polyimides are produced from a solution of polyimide acid (PAA), obtained by polycondensation of dianhydride and diamine. The solution of PAA is deposited on a substrate and the solvent being removed by an ensuing thermal treatment and the PAA imidized to PI. This is the so called "wet" method for obtaining thin layers. The advantages of the wet methods are as follows:

(i) simplicity, fast, performance and the use of a comparatively cheap equipment;
(ii) thin films can be produced from substances hard to melt and sublimate as well as from such thermally unstable and easily decomposed which in vacuum deposition is impossible;
(iii) this is also valid for the compounds of a high molecular mass and low pressure of the saturated vapours in this way "wet methods" being the only alternative for the thin layer formation.

The shortcomings of the „wet" methods are as follows:

(i) the exigency of a solvent or a combination of solvents;
(ii) in the case of using a solvent the latter should be inert chemically to the substance and form a solution with it as well as to be easily removable;
(iii) in the elimination of the solvent the film is deformed and the surface morphology is uneven and with a number of defects;
(iv) „wet methods" cannot be used for obtaining of very thin films of even thickness as well as in the cases of substrates with a complex form;
(v) a shorter "shelf life". The solutions of the initial compounds are especially sensitive to an increased moisture;
(vi) in using strongly volatile and toxic solvents there is an augmentation of the risk to the environment and people's health.

The overcoming of the „wet methods" shortcomings is achieved by the introduction of the method of physical vacuum deposition. Vacuum deposition displays number of opportunities for broadening the spectrum of novel materials of suitable electrical, optical and mechanical properties. In it, the different precursors are evaporated in a high vacuum and are deposited at comparatively low temperature on appropriate substrate followed by a thermal treatment. In the case of PI the purpose of the thermal treatment is the run of polycondensation reactions in solid state till completion of the PI formation. As a consequence of these reactions leading to a release of water and imidization also a certain pack of the layer is achieved. Some more substantial advantages of the vacuum deposition are as follows:

(i) as compared with the „wet methods" the possibility for side reactions to take place is minimum (Strunskus,Y and Grunze,M, 1994; E. Spassova, 2003; Salem J et al., 1986);
(ii) an even surface morphology and thickness in the interval between nanometers and microns are much easier to achieve without the presence of any rough defects;
(iii) a possibility for a precise control of the multitude of the process parameters: degree of evacuation, mode of evaporation (thermal, electron beam, magnetron, etc.) and a computerized control of the source temperatures with feedback, guarantee for a manageability of the deposition rates and the ratios of the precursors in the flux, etc. These constitute substantial prerequisites for the production films of reproducible composition, structure and thickness, but also demands considerable resources and highly qualified specialists;
(iv) the risk to environment and people's health is reduced to minimum.

The method of physical vapour deposition for thin-layer production also displays certain disadvantages that go as follows:

(i) not all substances are susceptible to evaporation since some of the decompose prior to reaching an equilibrium pressure over their saturated vapours or in cases when it is too low. In such a case an evaporation of even rate is very hard to be achieved in practice;
(ii) the precursors have to close evaporation temperatures so that a vapour flow of an even temperature could be formed;
(iii) a costly technical equipment is necessary and as mentioned above a highly trained technical staff is exigent.

The quality and structure of the obtained films to great extent depend on method used for their production. The deposition of the precursors in high vacuum with following thermal processing and the reactions of polycondensation and imidization taking place is a „dry" method since solvents are not implemented. Besides, the method is compatible with the technologies demanding a high degree of purity, low moisture and allows for the formation of uniform layers on substrates of complex configuration which is hard to achieve if not impossible with the „wet methods" (Salem J et al., 1986; K.S. Sree Harsha, 2006). The vacuum deposited layer of dianhydride and diamine is treated in two steps at defined temperature (from 180 to 350°C) to obtaining PI.
In the usage of the „wet methods" the nanocomposite layers are produced by a complex combination of processes leading to obtaining of the initial solution containing matrix components and „guest" particles. Also, the latter would not yield quality nanostructured layers not only due to the shrinkage of the film after the evaporation of the solvent and the impossibility for its thorough removal but because of the uneven distribution of the „guest" in the matrix as well. In the case of the vacuum evaporation an automatic, computerized regulation of the "guest" deposition rate and the precursors of the polymer matrix are attained as well as the movement of the substrate against the evaporation flux. In that way conditions for obtaining quality thin layer nanostructured materials of even distribution and precise concentration control of the „guest" in the polymer matrix are ensured (Salem J et al., 1986; K.S. Sree Harsha, 2006).
The considered methods for deposition of thin polymer layers constitute a part of the technological process of formation of layers with good parameters- optical transmittance, reflexive capacity, conductivity, etc. The comparison of the two methods with their advantages and disadvantages bring about to the further development of the process of quality film formation. Our view is that the method of vacuum deposition provides for a greater degree of purity in the thin film production, opportunity for controlling and computerizing the processes of heating, imidazion and layer formation designed for obtaining standard PI or nanocomposite products of reproducible composition, thickness, structure and properties.

3. Solid state reactions in polyimide films formation

In the vacuum deposition of the precursors PMDA (pyromellitic dianhydride) and ODA (4,4'-oxydianiline) at temperature of 120-145°C reaction of polycondensation to PAA (polyamide acid) with opening of the anhydride ring of PMDA takes place (Ac-S_N2 - reaction). These processes are to great extent accelerated and controlled in the thermal treatment of the condensed solid phase which represents PAA with regard to their transformation to PI by means of reaction of polycyclodehydration in solid state to linear PI. The reaction is presented in scheme 1. In Fig. 2 the FTIR spectra of the thin films of PMDA, ODA and PAA in the range of 1900-650 cm^{-1} are presented.

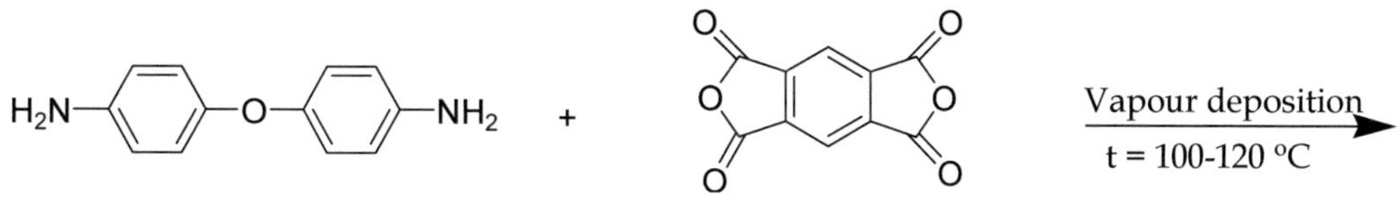

ODA (4,4'-oxydianiline) *PMDA (pyromellitic dianhydride)*

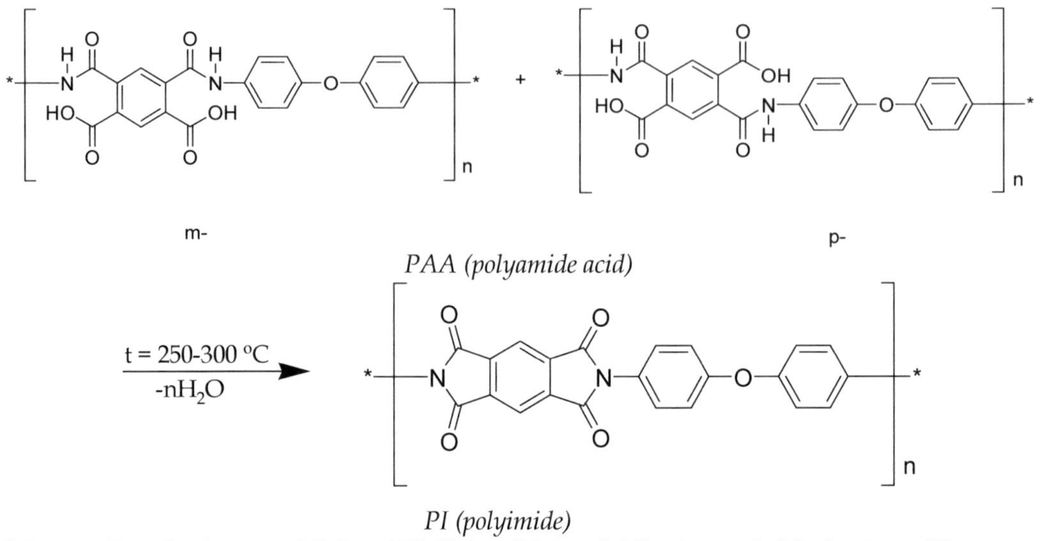

Scheme 1. Reaction between ODA and PMDA to PAA and following cyclodehydration to PI.

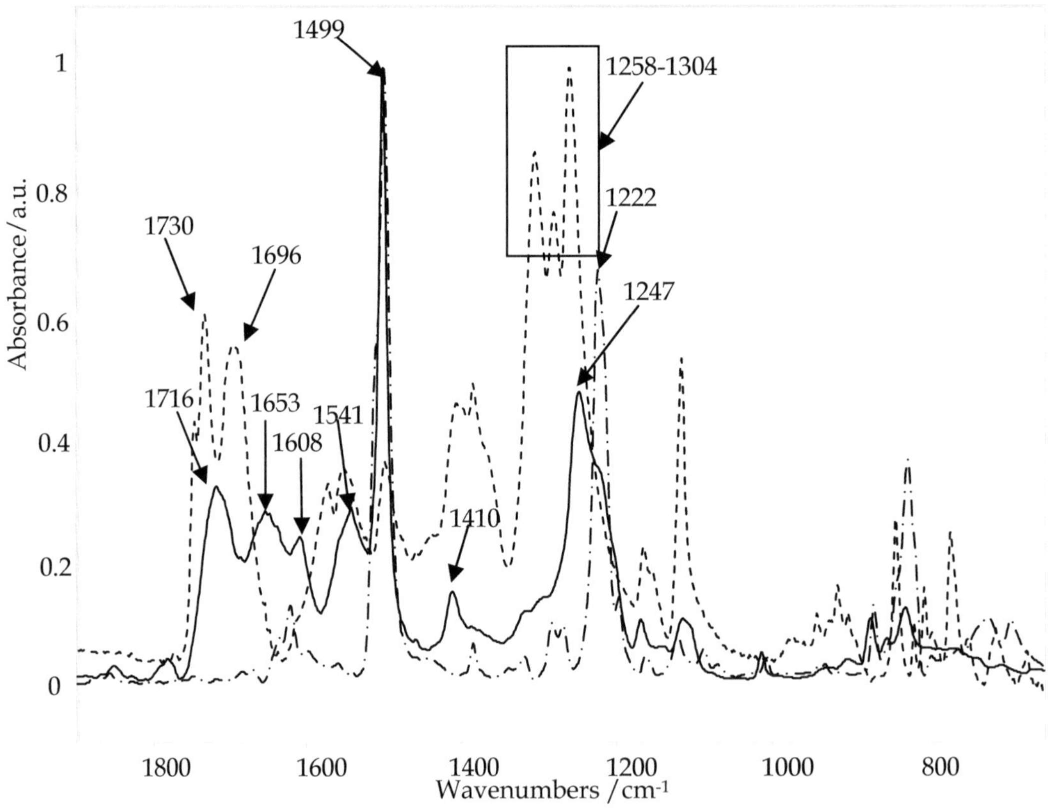

Fig. 2. FTIR spectra of the vacuum deposited films of: ------PMDA; —·—·— ODA; ——— simultaneously deposited both precursors in a mole ratio of PMDA : ODA= 1:1(PAA).

The band ν 1716 cm^{-1} for the >C=O group in the PAA spectrum is related to acid. The presence of hydroxyl group of the acid (C-OH) is corresponded at δ 1247 cm^{-1}. The amide bond is identified by the bands at ν 1653 cm^{-1} (>C=O amide I band) and δ 1608 cm^{-1} (N-H amide II band). The bands characterizing PMDA and ODA are presented in Table 2 (Gerd Kaupp, 2002; M.B. Saeed and Mao-Sheng Zhan, 2006).

	Absorbance bands/cm^{-1}
PMDA	ν_{as} 1746 >C=O (anhydride) ν_s 1730 >C=O (anhydride) ν 1696 >C=O (acid - hydrolysis of the anhydride) ν 1499 C-C (aromatic ring) δ 1258-1304 C-O (anhydride ring and C-OH acid)
ODA	δ 1610 N-H (NH_2) ν 1499 C-C (aromatic ring) δ 1222 C-N (C-NH_2)

Table 2. Assigning of the main bands for PMDA and ODA.

Compared to the classical methods for producing films from PAA, in which the acid is preliminarily obtained in a solution and after that deposited as a thin film in the vacuum deposition method this process is performed in only one step. According to the kinetic theory of the collisions the rate of the reaction mainly depends on the energy factor i.e. the number of the collisions between the particles of the reagents. These collisions are called effective and chemical interaction takes place only between the particles taking part in them. The rate of the reaction also depends on the so-called factor of orientation (a possible space volume factor) between the reagents, particles. It expresses the probability for their appropriate spatial orientation needed for the accomplishment of effective collisions between them. The reactions between vacuum deposited precursors of PMDA and ODA take place with less by products since the molecules have sufficient spatial accessibility for effective compound as compared to the reactions in a solution. In the course of the compound of PMDA and ODA in the process of vacuum deposition a gradual and even deposition of PAA takes place which ensures the production of high quality thin layers. The obtained PAA layer is subjected to a thermal treatment for imidization to PI with a reaction of polycyclodehydration taking place (the closure of the ring after the Ac-S_N2 - mechanism). It has been established that the imidization commences at temperature of over 170 °C and the reaction takes place under kinetic control. Save for the chemical reaction process of thickening the layer also occurs which leads to obtaining a quality thin layer matrix devoid of defects on the surface (M.B. Saeed and Mao-Sheng Zhan, 2007). The FTIR spectra of the layers of PAA and PI are presented in Fig.3.

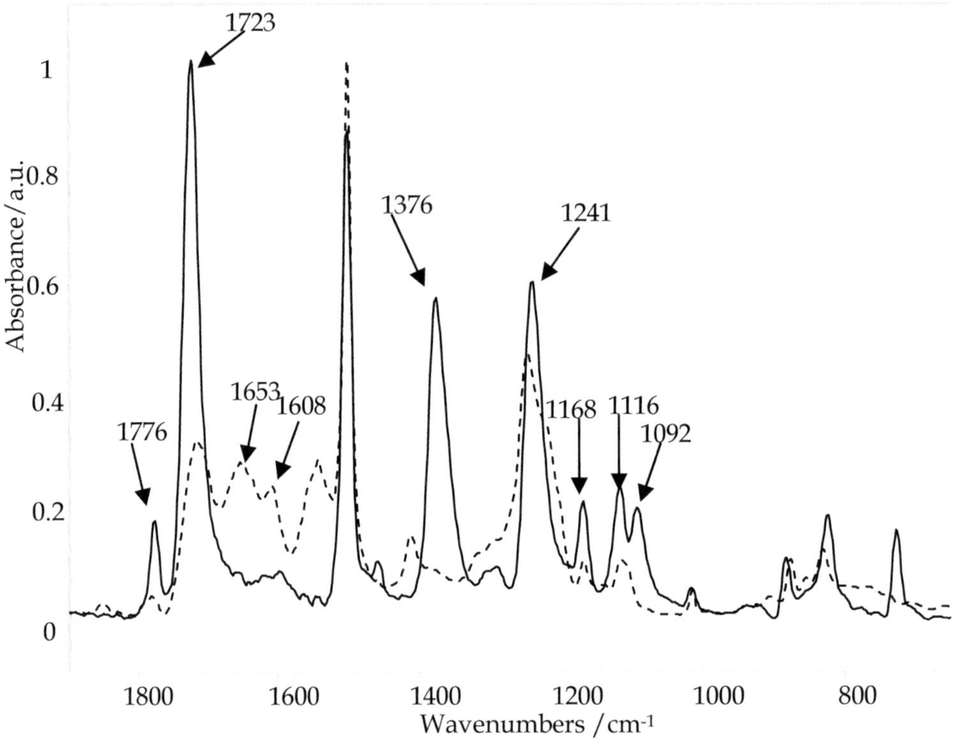

Fig. 3. FTIR spectra of vacuum deposited films with thickness 200 nm and PMDA:ODA(PAA)=1:1 - - - - - thermally untreated; ——— thermally treated 1 h at 300 °C.

The results show that following the thermal treatment the imidization has been completed to the production of PI since compared with the thermally untreated layers the bands for amide I and amide II in the PI spectrum are missing. The bands at v_s 1776 cm^{-1} and v_{as} 1723 cm^{-1} characterize the >C=O groups of the imide ring. The imidization is confirmed by the v_{C-N} 1376 cm^{-1} imide III band where a minimum is observed in the spectrum of the untreated layer at this frequency. The emergence of new bands in the area of the deformation vibrations for the C-O and C-N bonds respectively at 1241, 1168, 1116 and 1092 cm^{-1} is observed. The band at 1376 cm^{-1} is used for the qualitative and quantitative determination of the imidization degree. The imidization degree is determined by the absorbance FTIR spectra according to the Lambert- Beer's Law and represents the ratio of the number of imidized groups to the number of all imidizable groups (Equation 1) (Gerd Kaupp, 2002; M.B. Saeed and Mao-Sheng Zhan, 2006):

$$\textbf{Imidization Dergee} = \frac{\textbf{Number of imidzed groups - ni}}{\textbf{Total number of imidizable groups - na}}$$

Equation 1. Determining imidization degree.

Where, n_i is the number of imide groups and n_a is the number of amide groups.

The FTIR (Fourier-Transform InfraRed) spectroscopy is one of the methods for qualitative and quantitative analysis of organic polymers and nanocomposites. FTIR spectroscopy in the range 4400-650 cm^{-1} is used both for identification of functional groups and for the determining the microstructure of polymer and nanocomposite films. It is a non-destructive qualitative and quantitative method for analyzing thin organic nanocomposite layers. The reliability, quickness and accessibility of the method are advantages as compared with the other spectral methods for analysis (A. Georgiev et al., 2008; Barbara Stuart, 2004). The date from the FTIR spectral analysis of organic polymers and composites depend on the mode of their obtaining and the physical properties of the used materials. Mainly the experiment is related to the technique of taking the spectra and the whole range of spectroscopic techniques which serve for the mathematical processing of the primary results. The selection of the technique not only depends on the physical nature of the material but on the reproducibility of the measurements as well. The specific peculiarities in the polymer spectrum are also related to the method for the sample preparation. Thus, for example due to the uneven thickness of the film the direct quantitative measurements are quite difficult especially in the cases of the "wet" methods for the formation of the films under study (Liliane Bokobza, 2002; V.P. Tolstoy et al., 2003; John M. Chalmers and Neil J. Everall, 2002). The quantitative FTIR spectroscopy is based on the Lambert- Beer's Law (Equation 2).

$$A = \lg\left(\frac{Io}{I}\right) = a_{\upsilon}bc$$

where:

c - *concentration of the substance [mol/l];*

b - *thickness of the layer [cm];*

a_{υ} – *molar absorptivity [l/mol.cm];*

I_0 – *intensity of the incident light [J/s.m²];*

I - *intensity of the transmitted light [J/s.m²];*

Equation 2. The Lambert- Beer's Law

Quantitative measurements can also be taken of polymer films as well by using the method of the internal standard and normalization of the spectrum in the suitable range. In the polycyclodehydration of PAA the imide band at 1376 cm^{-1} grows intensively. The FTIR absorbance bands used for qualitative and quantitative analysis of PI and PAA are presented in Table 3 (Gerd Kaupp, 2002; M.B. Saeed and Mao-Sheng Zhan, 2006).

	Absorbance band/ cm^{-1}
Polyimide	ν_s 1776 >C=O ν_{as} 1723 >C=O ν 1376 C-N
Polyamide acid	ν 1716 >C=O (COOH) ν 1653 >C=O (CONH) δ 1608 C-NH
Standard	ν 1501 *p*-substituted benzene rings

Table 3. FTIR absorbance bands used for the qualitative and quantitative defining of PI and PAA

The peak at ν 1501 cm^{-1} [C-C (Ar)] characterizes *p*-substituted benzene rings. It is used as an internal standard and does not change at the imidization time. That is why the degree of imidization can be indirectly defined by the ratio of the absorptions (the corrected areas of the bands) at 1723 or 1376 cm^{-1} to 1501 cm^{-1} (Gerd Kaupp, 2002; M.B. Saeed and Mao-Sheng Zhan, 2006; Vasilis G. Gregoriou and Sheila E. Rodman, 2002).

Nanotechnologies due to the fact that they present the greatest modern technical challenge have been the subject of intensive studies and development. The solution of the problems accompanying the different trends in this technology of the future imposes the creation and implementation of novel specific sources of energy impact. As mentioned above, the additional heating has been commented on as aimed at acceleration and raising the degree of the imidization process as a pre-condition for the formation of PI. Logically enough a question is posed: as the thermal treatment or most generally speaking the energy impact on the precursors leads to PI formation would it be not appropriate for another type of energy treatment to be implemented for the same purpose? Another specific source of energy action is the microwave (MW) interaction (from 300 MHz to 30 GHz) with matter (Komarneni S and Katsuk H, 2002; Michael D and Mingos P, 2005). The task of applying a MW action in the thin layer systems on the basis of which the entire structure of microelectronics and micromechanics is built appears to be exceedingly attractive. The opportunities for activation and control of the chemical processes in the solid and liquid phases also implies the need of creating new conditions in the preparation of novel materials and the development of waste-free technologies. These opportunities encompass the macro-, micro- and nano- levels of action.

Such are the expectations based on these results:

(i) a considerable reduction of the time for the additional processes taking part in the technologies and their accomplishment in situ without the release of harmful and toxic substances;
(ii) development of the methods for obtaining nanocomposite thin layers;
(iii) a high selectivity in the production of an impact applying the MW irradiation in complex thin layer systems.

The microwave technology displays potential advantages in the polymer material production both in the bulk production and as thin films (Michael D and Mingos P, 2005; D. Lewis et al., 1995). MW synthesis is a method of good prospects for synthesizing of polymers in a thin film making it possible for the temperature and time for imidization to be significantly reduced. At the same time the application of this method in nanotechnologies for thin layer production for the needs of opto- and microelectronics render an opportunity for optimization until quality nanocomposite films are obtained.

Studies of ours (D. Dimov et al., 2007; D. Dimov et al., 2006) and others (Michael D and Mingos P, 2005; D. Lewis et al., 1995) on the imidization reactions in the solid state of thin PI films are also focused on establishing the opportunities for the application of the MW synthesis. The main target of these purposed studies is to find conditions for lowering the temperature and reducing the time for imidization which is of utmost importance in the production of nanostructured layers with an embedded chromophore as a "guest". Most chromophores at high temperatures and a prolonged thermal influence lose part of their optical properties or destructed (F. Kajzar and J. D. Swalen, 1996; V. Degiorgio and C. Flytzanis, 1993). In the case of MW synthesis the molecules receive additional energy and the rate of the reaction grows (Michael D and Mingos P, 2005). In the case of the combined treatment (MW and thermal) the imidization reaction takes place for 5 to15 min which is confirmed by the spectra in Fig. 4. The quality of the obtained films is identical with the one of PI obtained only after a thermal treatment for 1h at 300 °C. In the spectrum of the PI layer treated with MW for 5 min and thermally treated for 15 min at 300 °C the intensity and the area of band at 1382 cm^{-1} are smaller than that at 15 min MW and 15 min thermal treatment at 300 °C. It is clearly seen that same value of the temperature and time of the thermal treatment the time of MW treatment considerable influence on the imidization. Consequently, duration of the MW treatment and the temperature values are importance for the imidization degree and the quality of the obtained films.

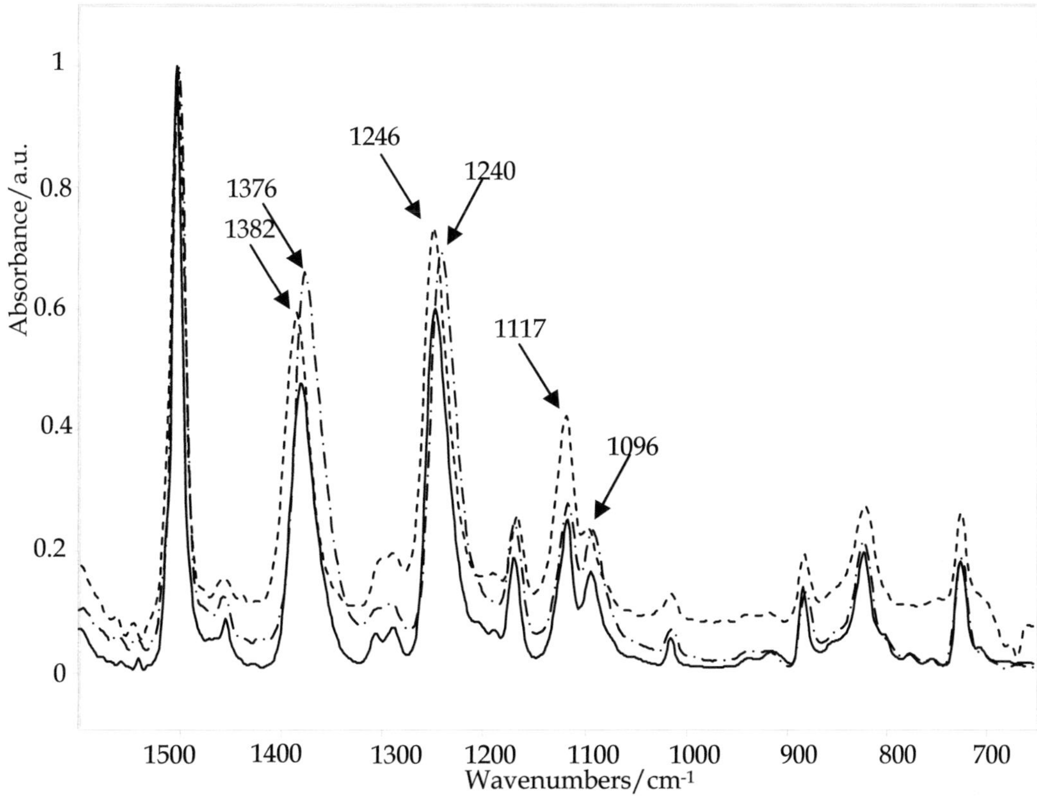

Fig. 4. FTIR spectra of vacuum deposited films with thickness 200 nm of: ——— PI, MW treated for 5 min and thermally treated for 15 min at 300 °C; - - - - - -PI, MW treated for15 min and thermally treated for 15 min at 300 °C; —·—·— PI, thermally treated for 1 h at 300 °C.

Polyimides	Band area at 1501 cm^{-1}/a.u.	Band area at 1382 cm^{-1}/a.u.	Imidization degree
PI treated for 5 min MW and thermally treated for 15 min at 300 °C	14.05	13.16	0.96
PI treated for 15 min MW and thermally treated for 15 min at 300 °C	16.30	17.24	1.05
PI thermally treated for 1 h at 300 °C	15.93	19.34	1.21

Table 4. Bands area at 1501 and 1382 cm^{-1} and imidization degree of the PI obtained after MW and thermal treatment.

The results from the calculations of the imidization degree are presented in Table 4. The PI treated for 15 min MW and thermally treated for 15 min at 300 °C displays a higher degree of imidization than that treated for 5 min MW and thermally treated for 15 min at 300 °C. These results are compared with the results about the PI obtained after thermal treatment for 1h at 300 °C. The experiments carried out allowed for the method for PI production to be optimized by the introduction of a MW treatment of the vacuum deposited layers from ODA and PMDA.

A part of our efforts are made with regard to the creation of azo-polymer of the "main chain" type as well in which the azo- chromophore is covalently bound to the polymer matrix. We have used the precursors DAAB (4,4'-diaminoazobenzene) and PMDA (pyromellitic dianhydride) as initial monomers (Petrova TS et al., 2003; A.Georgiev et al., 2008). The covalent bonding of the chromophore by a "side-chain" or a "main-chain" is of considerably greater perspective for the production of organic nanostructured layers due to their greater stability and uniform density as compared with the "host-guest" system. The azo-benzene derivatives have been the subject of extensive investigations for a long time (Osvaldo N. Oliveira et al, 2005; Cristina Cojocariu and Paul Rochon, 2004). In the aromatic azo-compounds the azo-group is formed by σ- and π- bonds between two nitrogen atoms. Such structure determined planar configuration of the σ- skeleton at the functional group and the angular localization of substitutes to it. The existence of the π-bond between the nitrogen atoms in the DAAB molecule explains the availability of the π- diastereoisomery. E- diastereoisomers are thermodynamically more stable than Z- diastereoisomers (Fig. 5). When the polymer nanocomposite is irradiated by linearly polarized light optical anisotropy is observed as a result from the photo-isomerization and photo-orientation of the azo-chromophore perpendicular to the direction of the polarized beam. Later this effect has been recorded also in other azo-polymer films that have found application in the development of devices for preservation of reversible optical information. Photoisomerization is observed when the chromophores pass from the low-energy trans-form to the cis-form after absorption of light (Fig. 5). The reverse process is accompanied by emitting of thermal energy but it could be also induced light. Trans-cis-trans isomerization cycle leads to shifted of the $\lambda_{max\ in}$ the absorption UV-VIS spectrum and the isomers posses different properties – dipole moments, refraction indexes and space volume (Osvaldo N. Oliveira et al, 2005;

Cristina Cojocariu and Paul Rochon, 2004). These properties of the aromatic azo-compounds give possibility for detail investigation with the aim of optimization of their properties and the creation of new nanostructure films of better physico-chemical properties.

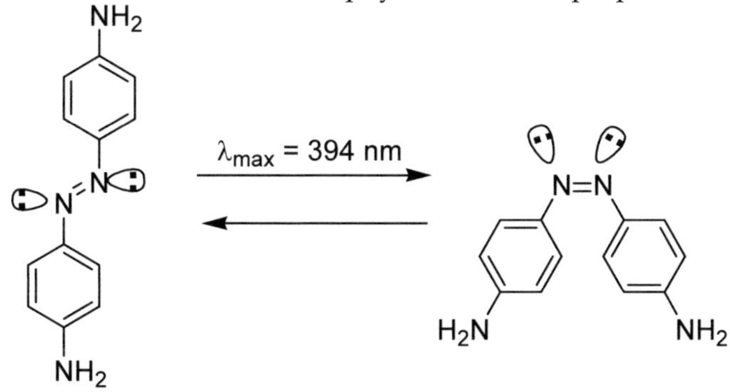

Trans (E)- Cis (Z)- isomers (diastereoisomers)

Fig. 5. Photo-isomerization of 4,4′-diaminoazobenzene (DAAB).

Scheme 2 illustrates the polycondensation between PMDA and DAAB and the following cyclodehydration to linear azo-polyimide (Azo-PI). The covalently bound to the polymer chain chromophores are preferable to the „guest-host" system, because the covalent bonding ensure a uniform and dense layer without defects in the polymorph structure and the physico- chemical properties of the layer are significantly improved.

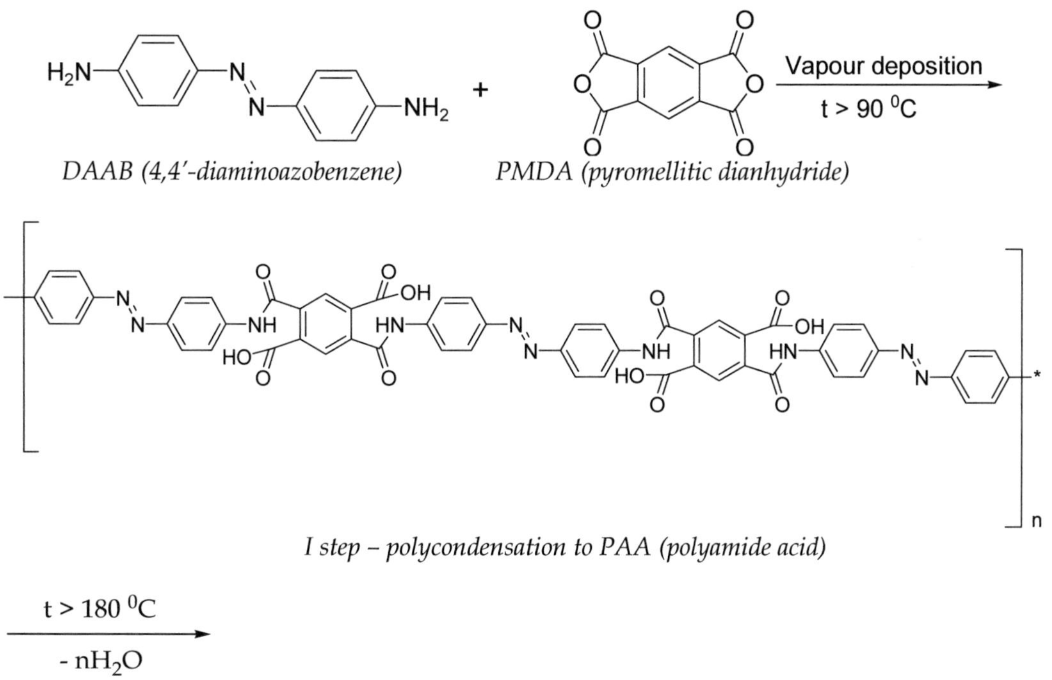

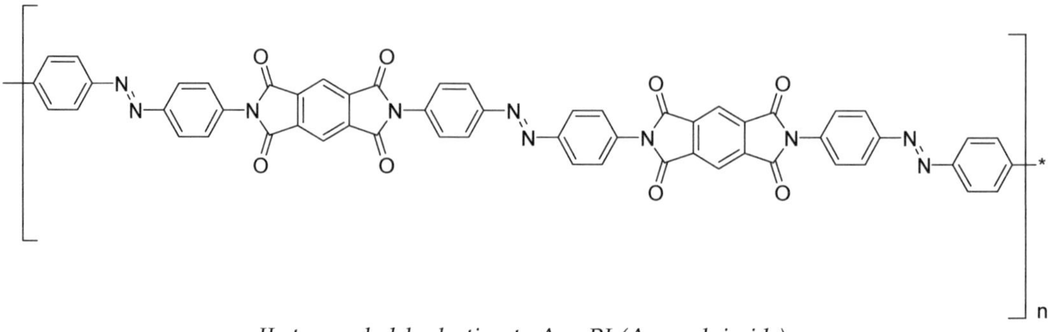

II step-cyclodehydration to Azo-PI (Azo-polyimide)

Scheme 2. Reaction between PMDA and DAAB to Azo-PI.

Having in mind the structure and the spectral properties of the Azo-PI there is perspective to be successfully employed as a polymer matrix in the formation of nanostructure layers with potential application as optical modulators, optical recording media and other optical devices (Petrova TS et al., 2003, Valtencir Zucolotto et al., 2004).

We suppose that PAA (polyamide acid) is formed since the simultaneous vacuum deposition of the two precursors DAAB and PMDA as thin deposited film. Our assumption is confirmed by the FTIR spectrum presented in Fig. 6. Reaction of polycyclodehydration to Azo-PI takes place following the MW treatment of the film for 5 min ensued by the thermal at 30 min at 300 °C.

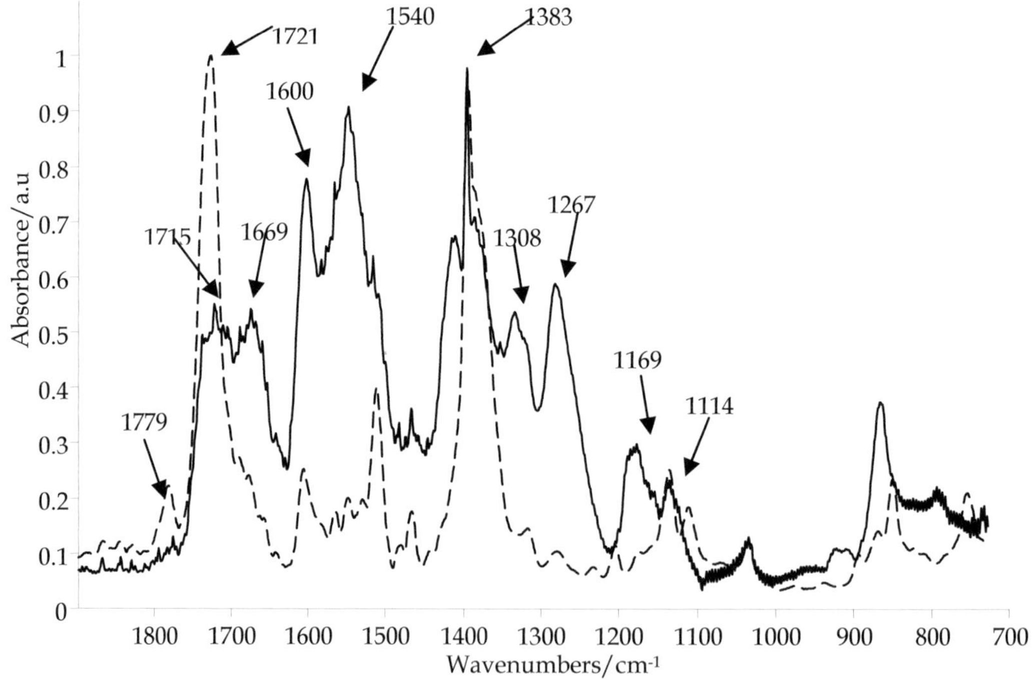

Fig. 6. FTIR spectra of vacuum deposited films with thickness 250 nm: —— DAAB:PMDA ratio 1:1(PAA); - - - - - - Azo-PI, obtained after 5 min MW treatment and 30 min thermal treatment at 300 °C.

In the spectrum of deposited films (Fig. 6) the typical bands characterizing PAA for amide I 1669 cm^{-1} ($\nu_{C=O}$) and amide II (δ_{N-H}) are identified at the broad and complex band at 1540 cm^{-1}, which overlaps with the second band of stretching vibration of C-C from the benzene ring. The broad band in the carbonyl area at 1715 cm^{-1} related to the >C=O group of acid. The bands at 1308 and 1267 cm^{-1} related to the deformation vibrations of C-OH (an acid). After the MW and thermal treatments process of imidization to Azo-PI takes place. The band at 1383 cm^{-1} is typical for imidization process. The typical bands for >C=O from the imide ring are identified at ν_s 1779 cm^{-1} and ν_{as} 1721 cm^{-1} (narrow and intense band). In the Azo-PI spectrum a minimum at 1669, 1540, 1308 и 1267 cm^{-1} is observed which confirms the imidization process. The stretching vibration for C-N at 1114 cm^{-1} are shifted towards the lower frequencies due to the lowering of the force constant for the stretching vibration in the imide ring as compared with the spectrum of the untreated layer (1169 cm^{-1}). The presence of azo-group is confirmed by the optical spectra of the corresponding films. Transmission at λ_{min} = 394 nm characterizing the azo-group is observed in the UV-VIS spectra of vacuum deposited films from DAAB and DAAB and PMDA (Fig. 7).

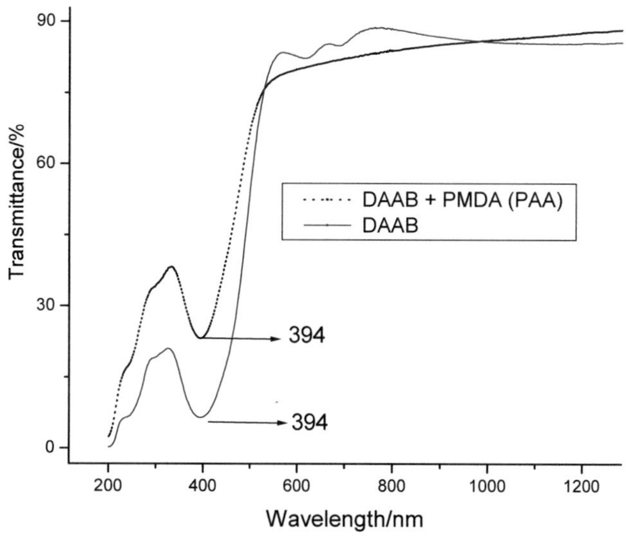

Fig. 7. UV-VIS spectra of vacuum deposited, untreated films from DAAB and DAAB: PMDA=1:1

The UV-VIS spectrum of Azo-PI (Fig. 8) is typical in which hyperchromic and hypsochromic effects are observed following the imidization and transmission minimum being weak at λ_{min} = 300 nm. The energy of the electron transitions grows significantly due to the decrease of π-conjugation after the imidization.

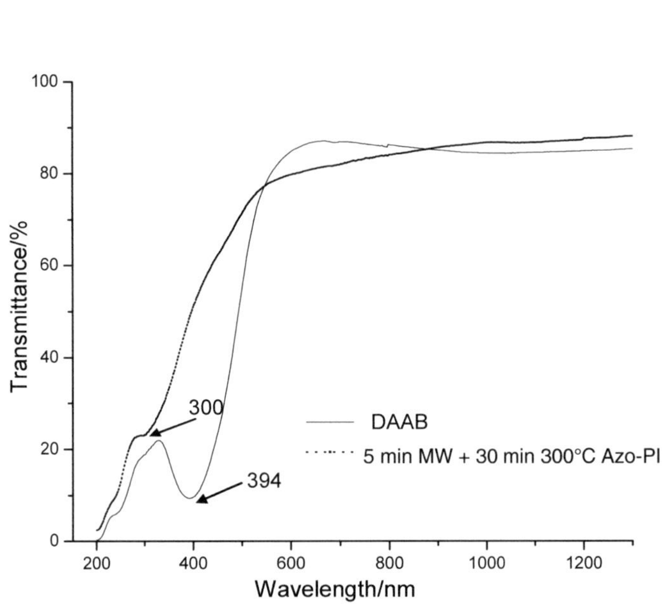

Fig. 8. UV-VIS spectra of vacuum deposited films from DAAB and Azo-PI.

The azo-polymer of the "main-chain" type, where the azo-chromophore is covalently bound to the polymer matrix represents material of good prospects in the formation of nanostructured layers for the purposes of micro- and optoelectronics, media for optical recording and other devices.

4. Conclusion

Systemized results from a study carried out with the aim for production of PI (polyimide) films by vacuum deposition and solid state reactions of diamine (ODA- 4,4'-oxydianiline) and dianhydride (PMDA- pyromellitic dianhydride), following transformation to PI by thermal treatment are presented in this chapter. It has been applied combined (thermal and MW) treatment with the purpose of obtaining layers of a smooth, defectless surface and reproducible composition and thickness. The established technical conditions allow for the production of PI layers of a proven composition and high imidization degree as prerequisites for acquiring a number of attractive properties of theirs- high transmission, high chemical resistivity and well pronounced dielectric properties at high thermal stability. FTIR spectroscopy was used in these investigations of the films. It has been shown that the determination of optimal parameters is importance for the production of PI layers with high quality.

A new approach in the preparation of "main-chain" polyimide films containing an azo-group is developed. After MW and thermal treatment the deposited layers from the precursors DAAB (4,4'-diaminoazobenzene) and PMDA (pyromellitic dianhydride) are transformed to Azo-PI by solid state reaction. The imidization reactions has been confirmed by FTIR spectroscopy. The approach provides the opportunity by the reaction of both

precursors to obtain Azo-PI and to overcome the necessity of evaporation of compounds by the classical fabrication of dye containing composites with PI matrix.
As a result from the research carried out and the solutions forwarded defectless films of a reproducible composition, smooth surface, high chemical stability, desired levels of dielectric properties with the preservation of high thermal resistance have been obtained. These properties of theirs as well as the possibility for them to be used as an appropriate thin layer matrix for embedding of metals, salts and dyes they could contribute to the creation of novel nanocomposite materials i.e. the development of material science also in the nanomaterial reality.

Acknowledgements
The financial support of the National Fund of the Ministry of Education and Science, Bulgaria- contracts: BУ-TH- 205/06 and DO 02-254/18.12.2008 is gratefully acknowledged.

5. References

A. Georgiev, I. Karamancheva, D. Dimov, E. Spassova, J. Assa, G. Danev. Polyimide coatings containing azo-chromophores as structural units *Journal of Physics: Conference Science* 113:012032, Fifteenth International Summer School on Vacuum, Electron and Ion Technologies 2008, IOP Publishing.

Anton Georgiev, Ilyana Karamancheva, Dejan Dimov, Ivailo Zhivkov, Erinche Spassova. FTIR study of the structures of vapor deposited PMDA–ODA film in presence of copper phthalocyanine. *Journal of Molecular Structure,* Vol. 888, 2008, 214-223.

Barbara Stuart. *Infrared Spectroscopy: Fundamentals and Application,* Wiley, 2004, Chapter 3, pp. 45-50; Chapter 6, pp. 113-118.

C.- C. Lee, C.- H. Lee and S. Chao. Making aspherical mirrors by thin-film deposition. *Applied Optics* Vol. 32, Issuse 28, 1993, 5575-5540.

C.P. Wong. *Polymers for Electronic and Photonic Applications* Academic Press, San Diego California, USA 1993 pp 37-243.

Cristina Cojocariu and Paul Rochon. Light-induced motions in azobenzene-containing polymers. *Pure Applied Chemistry,* Vol. 76, 2004, 1479-1497.

D. Dimov, A. Georgiev, E. Spassova, I. Karamancheva, Y. Shopov and G. Danev. *Journal of Optoelectronics and Advanced Materials,* Vol. 9, Issuse 2, 2007, 494-497.

D. Dimov, V. Denishev, E. Spassova, Y. Shopov, G. Danev. *Microwave Processes as a TOOL for Material Engineering*. Proceedings of Fifteenth International Scientific and Applied Science Conference pp. 89-93 ELECTRONICS ET, Varna 2006,.

D. Lewis, S. LaMaire, A. Viehbeck. "Microwave Processing of Polyimide Thin Films for Microelectronics" in "*Microwaves: Theory and Application in Materials Processing III*", pp. 59 - 64, ed. D.E. Clark, D.C. Folz, S.J. Oda and R. Silberglit, Ceramic Transactions 433, 1995.

E. Mazoniene, J. Bendoraitiene, L. Peciulyte, S. Diliunas and A. Zemaitaitis. (Co)polyimides from commonly used monomers, and their nanocomposites. *Progress in Solid State Chemistry* Vol. 34, 2006, 201-211.

E. Spassova. Vacuum deposited polyimide thin films. *Vacuum,* Vol. 70, Issue 2, 2003, 551-561.

F. Kajzar, J. D. Swalen. *Organic Thin Films for Waveguiding Nonlinear Optics.* CRC Press, 1996, pp. 112-135.

Francisko Raymo. Nanomaterials Synthesis and Application: Molecule-Based Devices. *Handbook of Nanotechnology"* Springer Berlin Heidelberg 2007, pp. 13-42.

Gerd Kaupp. Solid-state reactions, dynamics in molecular crystals. *Current Option in Solid State & Materials Science,* Vol. 6, Issuse 2, 2002, 131-138.

Iijima M, Takahashi Y. *US Patent 4624867,* 28:11, 1986.

John M. Chalmers and Neil J. Everall. Qualitative and Quantitative Analysis of Plastics, Polymers and Rubbers by Vibrational Spectroscopy. *"Handbook of Vibrational Spectroscopy".* John Wiley & Sons Ltd 2002, Volume 4, pp.2423-2628.

K.S. Sree Harsha. *Principles of Physical Vapor Deposition of Thin Films.* Elsevier, San Jose State University, CA, USA, 2006 first edition. pp. 11-27; pp. 367-369, pp. 961-963.

Komarneni S., Katsuk H. Nanophase materials by a novel microwave-hydrothermal process. *Pure Applied Chemistry,* Vol 74, Issuse 9 2002, pp. 1537-1543.

Liliane Bokobza. Application of Vibrational Spectroscopy for the Analysis of Rubber Composites. *Handbook of Vibrational Spectroscop,* John Wiley & Sons Ltd 2002, Volume 4, pp. 2629-2649.

M.B. Saeed, Mao-Sheng Zhan. Effects of monomer structure and imidization degree on mechanical properties and viscoelastic behavior of thermoplastic polyimide films. *European Polymer Journal,* Vol. 42, Issuse 8, 2006, 1844-1854.

M.B. Saeed, Mao-Sheng Zhan. Adhesive strength of partially imidized thermoplastic polyimide films in bonded joints. *International Journal of Adhesion and Adhesives,* Vol. 27, Issuse 1, 2007, 9-19.

Mitchell Anthamatten, Stephan A. Letts, Katherine Day, Robert C. Cook, Anthony P. Gies, Tracy P. Hamilton and William K. Nonidez. Solid-state amidization and imidization reactions in vapor-deposited poly(amic acid). *Journal of Polymer Science: Part A: Polymer Chemistry,* Vol. 42, Issuse 23, 2004, 5999-6010.

Michael D., Mingos P. Theoretical aspects of microwave dielectric heating. *Microwave Assisted Organic Synthesis,* edited by Tierney J., P., & Lidstroem P., Blackwell Publishing Ltd 2005 pp. 63-179.

Osvaldo N. Oliveira Jr., David S. dos Santos Jr., Debora T. Balogh,Valtencir Zucolotto and Cleber R. Mendonca. Optical storage and surface-relief gratings in azobenzene-containing nanostructured films. *Advances in Colloid and Interface Science,* Vol. 116, Issues 1-3, 2005, 179-192.

Petrova, TS, Mancheva I, Nacheva E, Tomova N, Dragostinova V, Todorov T, Nikolova L.

New azobenzene polymers for light-controled optical elements. *Journal of Materials Science-Materials in Electronics,* Vol. 14, Issuse 10-12, 2003, 823-824.

R. Ginsburg and J.R. Susko. High-temperature stability of a polyimide film. *Journal Research Development,* Vol. 28, Issuse 6, 1984, 735-740.

Salem J, Sequeda F, Duran J, Lee W, Yang R. Solventless Polyimide Films by Vapor Deposition. *Journal of Vacuum Science and Technology part A,* Vol. A4, 1986, 369-374.

Strunskus Y, Grunze M. *Polyimides – fundamentals and applications.* In: Crosh M, Mittal K, editors. New York: Marcel Dekker, 1994. pp. 187–205.

Steve Lien-Chung Hsua, Ulin Wanga, Jinn-Shing Kingb and Jyh-Long Jengb. Photosensitive poly(amic acid) organoclay nanocomposites. *Polymer,* Vol. 44, 2003, 5533-5540.

V.P. Tolstoy, I.V. Chernyshova, V.A. Skryshevsky, Handbook of Infrared Spectroscopy of Ultrathin films, Wiley Interscience, Wiley, 2003, Chapter 3 pp. 140-176.

V. Degiorgio, Christos Flytzanis. *Nonlinear Optical Materials: Principles and Applications* IOS Press, 1993, pp. 230-248.

Vasilis G. Gregoriou and Sheila E. Rodman. Vibrational Spectroscopy of Thin Organic Films. *Handbook of Vibrational Spectroscopy.* John Wiley & Sons Ltd 2002, Volume 4, pp.3010-3033.

Valtencir Zucolotto, Newton M. Barbosa Neto, Jose J. Rodrigues Jr., Carlos L. Constantino, Sergio C. Zilio, Cleber R. Mendonsa, Ricardo F. Aroca and Osvaldo N. Olivera Jr. Photoinduced phenomena in layer-by-layer films of poly(allylamine hydrochloride) and Brilliant Yellow azodye. *Journal of Nanoscience and Nanotechnology,* Vol. 4, Issuse 7, 2004, 855-60.

Thin films of aromatic polyazomethines

Weszka J.
Centre of Polymer and Carbon Materials of Polish Academy of Sciences
ul. M. Curie-Skłodowskiej 34, 41-819 Zabrze
Institute of Engineering and Biomedical Materials, Silesian University of Technology,
ul. Konarskiego 18, 44-100 Gliwice
Poland

1. Introduction

Aromatic polyazomethines, which are known also as the Schiff bases (Yang & Janekhe, 1991) belong to alternately conjugated polymers having nitrogen atoms incorporated in ordered way into the backbone. Their backbone conformation is derivative of aromatic di-amines and di-aldehydes structures, as they are prepared via condensation polymerization. Since many years this polymer family have been of interest for their good thermal stability and interesting opto-electronic properties (Yang & Janekhe, 1991; Yang & Janekhe, 1991; Kubono & Okui, 1994). Some poly(azomethines) have been prepared by solvent-based polymerization and their thin films have been deposited by spin-on technique (Yang & Janekhe, 1991; Yang & Janekhe, 1991; Kubono & Okui,1994; Thomas & Ingnes, 1998; Luzny et l., 1999; Suhl & Shim, 2000; Jung et al., 2002). However, applicability of this technique is limited only to soluble polymers and in order to prepare thin films of insoluble poly(azomethines), the thermal vacuum evaporation or chemical vapor deposition (CVD) methods based on polycondensation have been used (Kubono & Okui, 1994; Takahashi et al., 1991; Weaver & Bradley, 1996; Rohlfing & Bradley, 1998; McElvain et al., 1998). It is worth mentioning that thermally vacuum evaporated films, similarly like those prepared by CVD, are free from unintentional impurities that cannot be avoided in the case of solvent-based polymerization or spin-on coating techniques (Kubono & Okui,1994; Takahashi et al.,1991; Weaver & Bradley, 1996; Rohlfing & Bradley, 1998; McElvain et al., 1998). The simplest representative of poly(azomethines) is 1,4-phenylene-methylidynenitrilo-1,4-phenylenenitrilomethylidyne, which hereafter will be referenced as PPI (Yang & Janekhe, 1991). PPI has the backbone consisting of alternately repeating *para*-phenylene and azomethine dimers. Being an isoelectronic counterpart of poly(*para*-phenylene vinylene) (PPV), PPI has been expected to be suitable for photovoltaic and other optoelectronic applications (Kubono & Okui,1994; Takahashi et al., 1991; Weaver & Bradley, 1996; Rohlfing & Bradley, 1998; McElvain et al., 1998) (Kubono & Okui,1994; Thomas & Ingnes, 1998; Takahashi et al.,1991; Weaver & Bradley, 1996; Rohlfing & Bradley, 1998; McElvain et al., 1998). The aim of this work is to show how the electronic properties of PPI thin films formed by CVD based polycondensation can be taylored by technological conditions of their preparation or chain modifications. Our interest in studying the CVD prepared PPI thin films is also to better understand mechanisms of condensation polymerization process running at the interface of the vapor/solid phase, structure and morphology of as-deposited films as well as the electronic structure of polymers having nitrogen atoms in the backbone.

2. Polycondensation based thin film preparation

Polycondensation is polymerization process, in which polymer chain progression is running through chemical reactions between the functional groups of two bifunctional monomers with releasing low weight molecule (Flory, 1971). While preparing thin films of polyazomethines, the polycondensation process is based on chemical reaction between aldehyde and amine groups of aromatic di-aldehyde and di-amine, so that larger bi-functional species, having aldehyde and amine end groups and azomethine -CH=N-linkage between phenylene rings, is formed and one water molecule being released. Such species having formed can combine with one another or with relevant monomer by means of chemical reaction between aldehyde and amine groups with forming a larger bi-functional oligomer and releasing water molecule at each stage.

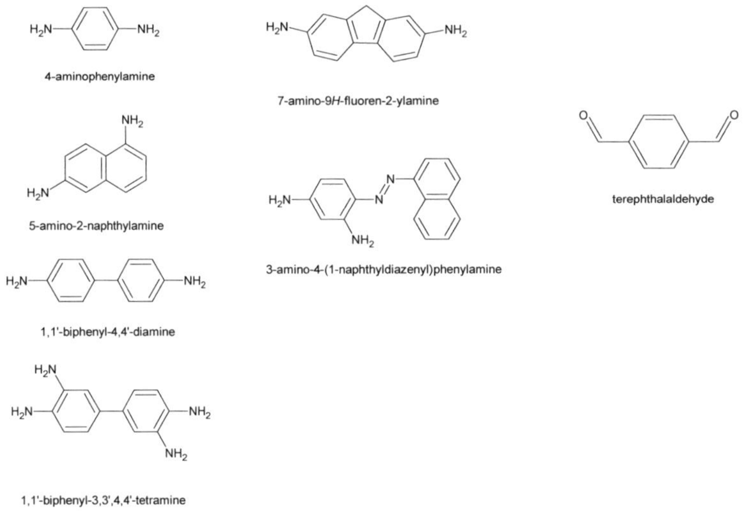

Fig. 1 Diamines and therephthal aldehyde used in polycondensation based process of polyazomethine thin film growth.

As a result many different polymer chains can progress upon the surface of the substrate and subsequently upon layers having already been formed. However, one should remember that removing this low weight molecule may be essential for polymer chain progression. While preparing poly(1,4-phenylenemethilidynenitrilo-1,4-phenylenenitrilomethilidyne) this polymerization process is illustrated in Fig. 2. According to Flory rule (Flory, 1971), which states that chemical reactivity of end groups does not depend on the length of polymer chain they are attached to, one may expect that polyazomethine thin film growth is connected closely with progressing the polyazomethine chains all over the substrate surface simultenously. Molecules of monomers arriving at the surface can be attracted at different active points of the surface starting off polymer chains progressions, which can proceed along many chains simultaneously. This kind of polymerization is thought to be very suitable to prepare thin films of polyazomethines with the use of such techniques as chemical vapor deposition (CVD) or thermal vacuum evaporation (TVE) from two sources. The essential and common feature of the two techniques is that thin films grown on various substrates are free from any unintentional doping and the two methods are especially very convenient for preparing thin films of polyazomethines insoluble in organic solvents.

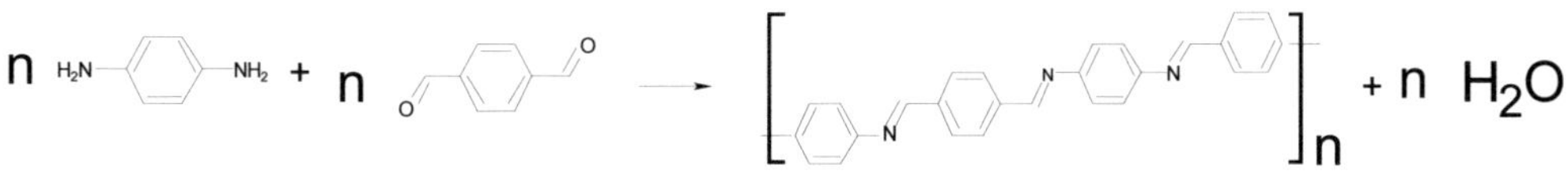

Fig. 2. Polycondensation reaction resulting in PPI polymer chain progression

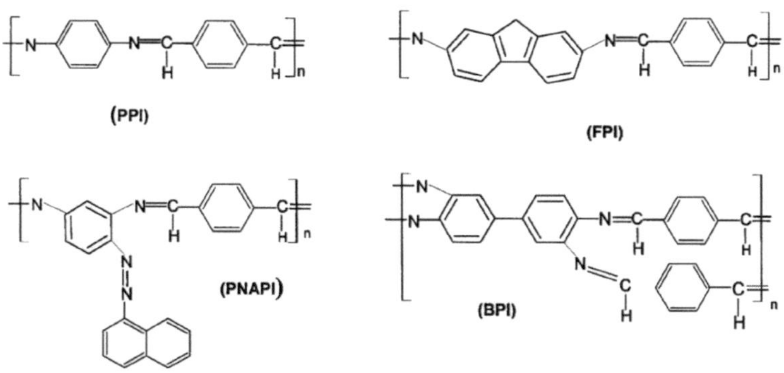

Fig. 3. Polyazomethines prepared by CVD and thermal vacuum evaporation.

2a. Chemical vapor deposition of polyazomethine thin film preparation

Chemical vapor deposition is a technique of thin film preparing, where all the reagents involved in chemical reactions driving a film growth are being transported towards the substrate wafer area with the use of carrier gas agent (Weszka et al., 2008). Here, because the process of thin film formation is based on condensation polymerization running between molecules of bifunctional aromatic diamines and dialdehydes, neutral gaseous argon is used as a carrier agent to transport reagent molecules to the reaction chamber where the substrate is mounted. The process of CVD deposition of polyazomethine thin films has been made in the technological equipment, which was built up in horizontal and vertical geometries (Weszka et al., 2008; Hajduk et al., 2009). The equipment for thin film deposition in horizontal geometry is schematically illustrated in Fig. 4, though general ideas outlined in this figure are applicable for vertical geometry of the CVD equipment illustrated in Fig. 5, too. How it can be seen in Fig. 4, argon is being dosed into the CVD system with the pressure regulator mounted on an argon reservoir while its flow rate are measured with a rotameter, similarly as it has been described for preparing PPI thin films (Weszka et al., 2008). The Ar stream, with a fixed flow rate controlled by a pressure regulator, is forked into two or three equal streams, one flowing over a boat containing PPDA and the others over a boat with TPA, while the third one, if used, it has been argon itself. The third channel has been used while preparing PPI thin films doped with ferrous three chloride ($FeCl_3$) in the vertical CVD setup. It has been found out while preparing thin films of the simplest (for chemical structure) aromatic polyazomethine PPI that such technological conditions like temperatures of monomers therephtal aldehyde, T_{TPA}, and p-phenylenediamine, T_{PPDA}, their difference, ΔT, transport agent flow rate and the distance from the inlet into the reaction chamber to the substrate, d, are very important parameters determining deposition conditions. They have been found to influence the structure, surface morphology and optical properties of deposited films of polyazomethines (Weszka et al., 2008). Because thin films of polyazomethines have been prepared via condensation polymerization, one may expect that the film growth process is controlled to large extent through this polymerization process by the rate water molecules are removed into flowing out argon stream. While comparing how each of technological parameters used while preparing PPI thin films discussed influence the film thickness and the time it was deposited it seems difficult to decide which one of them has the most important impact on the PPI thin film growth. To consider these technological issues one can have a look at Tab. 1 where are collected data presented in paper (Weszka et al., 2008). It can be seen that in the processes used the TPA source temperature, T_{TPA}, has been kept fixed at about 326 K, with one exclusion when it was 313 K. The source temperature of the other monomer was about 9 to 19K higher, with one exclusion when it was as low as 315 K, so nearly coinciding with that of TPA. If one compares data for films *a* and *b*, one can see that they were prepared under the same flow rate of about 80 l/min and the same T_{PPDA} and T_{TPA}, with forking argon stream into two equal ones, the film has grown faster at the substrate placed farther from the inlet. One can see that under the same conditions three-stream prepared PPI films grew slower than two-stream ones, which could be attributed to much lower reagent concentration in the stream entering the reaction chamber diluted by the third stream used.

Film	T_{PPDA} (K)	T_{TPA} (K)	Flow rate l/min	Time (s)	Distance (cm)	Number of streams	Thickness (nm)
A	335	326	80	120	5	2	200
B	334	326	80	180	4	2	200
C	340	326	150	60	5	2	200
D	334	326	80	420	4	3	170
E	343	327	80	90	5	3	180
F	315	313	60	1500	5	3	50

Table 1. CVD technological parameters of PPI thin film preparation (according to Weszka et al., 2008)

This effect is seen to be more pronounced in case of thin films prepared when the substrate was placed 4 cm from the inlet. But most pronounced influence is seen in the case of *e* film prepared with slightly lower flow rate (60 l/min) and relatively low and nearly coinciding temperatures of monomers (about 313 K). Then, depending on the regimes used, two or three relevant streams have merged into one Ar stream that transported mixed molecules of both monomers towards the inlet into the cylindrical reaction chamber of a much larger diameter, as it is schematically outlined in Fig. 4. The inlet is centered on the axis of the cylinder chamber and the substrate holder is fixed perpendicularly to the cylinder axis at a distance **d** = 4 or 5 cm from the inlet. Argon and reaction by-products or monomers that have not reacted have been removed from the reaction chamber with an exhaust rotary pump operating continuously during the deposition process.

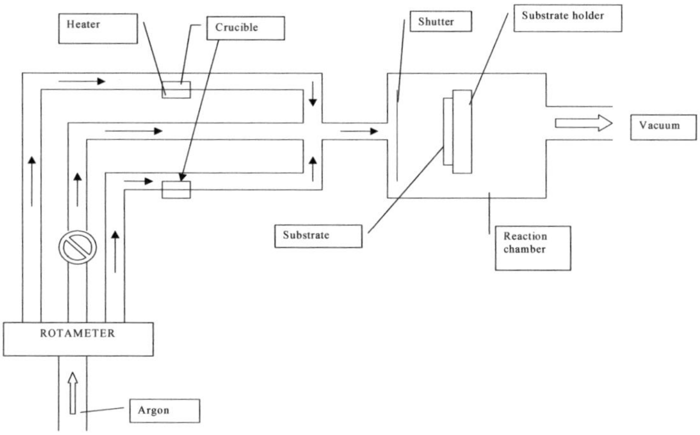

Fig. 4. Equipment for polycondensation made by chemical vapor deposition

In Fig. 5 one can see schematically drawn technological vertical setup for CVD deposition process. Essentially, this setup is similar to the precedent with the only difference that three streams were merged into one in the stream collimator, which after having passed the pupil is flowing upward towards the substrate mounted onto the rotary plate. This deposition setup allowed for five substrate to be mounted simultaneously, so that turning on the plate one can place subsequent substrate over the collimator pupile.

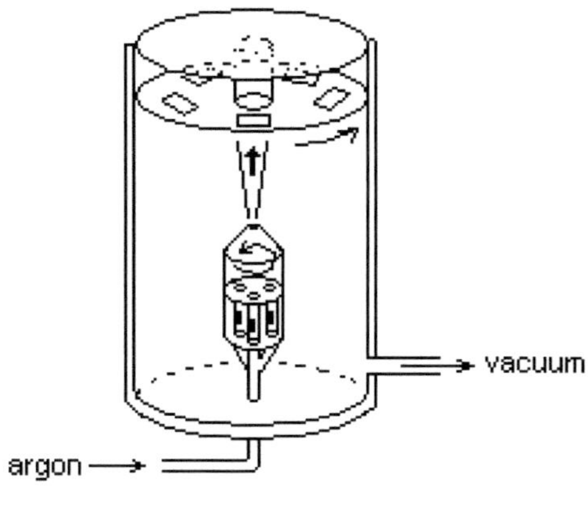

Fig. 5. Equipment for polycondensation by CVD in vertical geometry (Hajduk et.al, 2009)

Influence of technological parametrs of the deposition process on PPI thin films prepared in vertical geometry has been discussed in (Hajduk et.al, 2009) and these conditions are collected in Tab. 2. Four series of PPI thin films were prepared each for five different flow rates with two minute deposition time. The series were distinguished by temperatures of therephtal aldehyde (T_{TPA}) and p-phenylene diamine (T_{PPDA}) the source materials applied (Hajduk et.al, 2009). There are also given pressures in the reaction chamber during deposition process with the relevant flow rates. These experiments have shown that there are some parallel behaviors of monomers dependently on the source temperatures as well as on their difference.

No	T_{TPA}/T_{DA}	Flow rate	Time t	Pressure
1(a,b,c,d,e)	50/50	0.05 (a)	2	3,8 *10^-2^
2(a,b,c,d,e)	50/60	0.15(b)		1,5*10^-1^
3(a,b,c,d,e)	50/70	0.4 (c)		2,5*10^-1^
4(a,b,c,d,e)	50/80	1.0(d)		3,6*10^-1^
		3.0(e)		7,6*10^-1^

Table 2. Technical parameters of individual experiments (according to Hajduk et.al, 2009)

It was found that thin films, which were prepared under higher flow rates, were the thicker the higher flow rates were.

2b. Thermal vacuum evaporation

The thermal vacuum evaporation technique (Kubono & Okui,1994; Takahashi et al.,1991; Weaver & Bradley, 1996; Rohlfing & Bradley, 1998; McElvain et al., 1998) has been utilized to prepare thin films of more complex azomethines produced by polycondensation of TPA and fluorene diamine, benzidinediamine and fatt Brown 15R (Jarzabek et al.,2008). The technological conditions of the film deposition processes are collected in Tab. 3. It is seen there that they were prepared under vacuum of 2 x 10^{-5} Tr, while temperatures of diamines fluorine, benzidine and fatt Brown were of about 145, 160 and 165C, while TPA temperatures being 51, 59 and 48 C, respectively (Jarzabek et al.,2008). Thus, one can see that the temperature used to evaporate these complex diamines have appeared nearly twice as large as that o PPDA in CVD process (Weszka et al., 2008). Thin films of FPI, BPI and PNAPI were thermally evaporated under vacuum of 2×10^{-5} Tr onto quartz substrates. X-ray diffraction examination of as-prepared films have revealed amorphous character of polyazomethine thin films and their surface examination with atomic force microscope (AFM) indicate that the surfaces are homogenous, plane with rather small granularity, rather charecteristic for polymer thin films, which is manifested by rather low value of root-mean square roughness (rms) $\sigma_{rms} \approx 3$ nm. Technological parameters of the examined polymer thin films, i.e. evaporation time, source temperatures and thicknesses of the as-prepared films are collected in Tab. 3. The as-prepared films are transparent and pale orange or pale yellow and their transmission is as high as 80 - 96%, while their reflctance is below 20% in the area of low absorption.

Layer	t (s) (± 1 s)	Amine T (°C) (± 2°C)	Aldehyde T (°C) (± 2°C)	d (nm) (± 20 nm)
FPI	300	145	51	100
PNAPI	300	160	59	90
BPI	300	165	48	220

Table 3. Technological parameters of polyazomethine thin films preparation (according to Jarzabek et al.,2008).

3. Kinetics of polyazomethine thin film growth

The process of PPI thin films preparation by CVD method via polycondensation of PPDA and TPA can be considered to consist of three essential stages. The first one is taking up molecules of the two monomers by argon streams flowing over boats filled with source PPDA and TPA monomers, respectively. Then, the two streams transporting monomers, and sometimes also the third of pure argon, merge together into one stream, which is associated with mixing up molecules of the complementary monomers and their transport towards the reaction chamber where the substrate is mounted. Finally, the third stage is consists of impinging the molecules onto the substrate, their adsorption at the substrate (Weszka et al., 2008). All the CVD process of the PPI film forming is thought to be due to polycondensation reactions running within the adsorption layer at the vapor/solid interface and resulting in the polymer chain progression (Kubono & Okui, 1994; Weszka et al., 2008). While monomers mixture being carried on by Ar stream towards the substrate, the polycondensation process should be postponned, which is equivalent to the positive Gibbs free energy ($\Delta G > 0$). In fact, being equal to about 50 °C, the temperature of transported species has appeared sufficient for the reaction to be practically postponed as walls of the pipe transporting the reagent mixture remained clean. As switching between the reaction postponning (monomers transport) and reaction progression (film deposition) conditions is connected with the tememperature decreasing from 50°C down to 25°C, one can expect rather low absolute values of ΔG at the two temperatures. Further, as the condensation polymerization runs easily the height of the reaction barrier equivalent to enthalpy of reaction (ΔH) is thought to be low. Thus, it is expected that the equilibrium constant of the polycondensation process is rather low (comparable to 1) and PPI film formation connected with the reaction progressing is maintained by removing released H_2O molecules from the reaction volume. Then, rather high rate of reaction should be attributed to high enough concentrations of PPDA and TPA molecules in the stream (Veinshtein et al.,1979; Pross, 1995), so that monomers molecules are rather near apart from one another making conditions favorable for sticking impinging molecules with complementary amine and aldehyde functional groups upon the substrate. The molecules that have not reacted and released H_2O molecules leave the substrate surface back to the vapor phase to be subsequently exhausted from the reaction chamber. It is thought that kinetics of polycondensation reaction driving PPI growth is some manifestation of the Flory rule (Flory, 1971) stating that reactivity of the end groups is independent of the chain length. Additionally, temperature T_{PPDA} being higher by 10 °C or more than T_{TPA} assures nearly 1:1

ratio of the number of PPDA and TPA monomers molecules during the polycondensation process. This, together with high density of incident monomers adsorbed at the substrate surface, is expected to make reaction between aldehyde and amine end groups to run simultaneously at large number of centers. Then, a great number of oligomers being close to each other can be formed, so that interactions between fragments of PPI chain may result in stacking planarized segments of chains in the solid state. However, when the temperature difference is smaller than 10 °C or even $T_{PPDA} < T_{TPA}$, conditions become non-stoichiometric and excess monomers, escaping back to the vapor phase, perturb the polycondensation process. These results sometimes in nucleation running seemingly in the gaseous phase and such grown grains fall down onto the surface and are incorporated into a film growing on the surface. In the three-stream regime, the third Ar stream makes the reagent concentration diminish, so that unreacted excess molecules easily come back to the vapor phase without perplexing the polycondensation process. However, while the source temperature of monomer species is quite low, stacking of planarized PPI chains is expected not to be the case. It is thought that kinetics of PPI thin film growth on the substrate in the vertical setup is very similar to that in horizontal geometry. Moreover, it is expected that merging the streams in the collimator is running at the temperature sufficient to preserve film formation on the collimator wall, so that the deposition process is running on the substrate kept at room temperature just over the pupile.

4. Structure and morphology of polyazomethine thin films

Real structure of PPI thin films was examined by X-ray diffraction and X-ray diffraction patterns taken on thin films prepared under different conditions (samples *a-c* and *d-f* have been formed in the two stream- and three stream regimes, respectively) shown in Fig. 5.

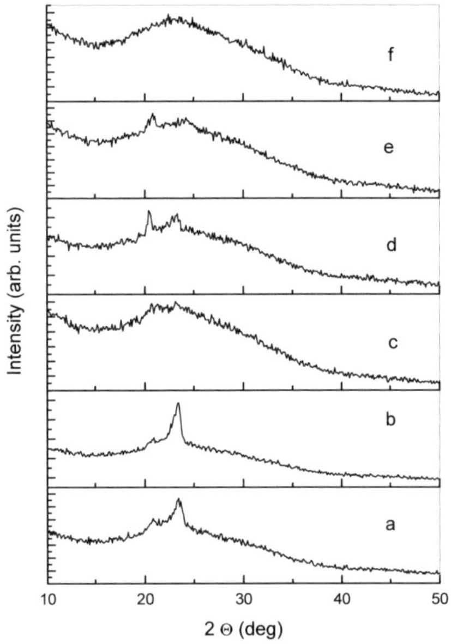

Fig. 6. X-ray diffraction patterns of PPI thin films CVD prepared in the horizontal geometry (according to Weszka et al., 2008).

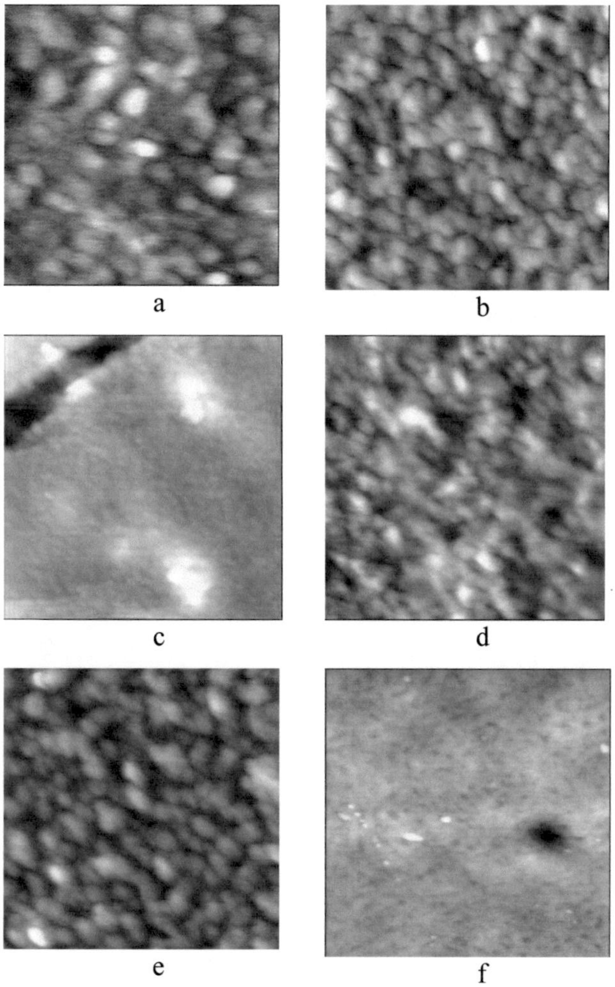

Fig. 7. Topographic AFM images of PPI thin films each in a scan area of 2.5□m x 2.5 □m. The gray scales are 48, 76, 64, 64, 70 and 155 nm for (a), (b), (c), (d), (e) and (f), respectively (according to (Weszka et al., 2008)

(Weszka et al., 2008) reveal predominant amorphous structure with some manifestation of ordered areas, most pronounced in films *a* and *b*. Strength of the features related to ordered areas is seen to be higher while d = 4 cm, independently on whether the two-stream- or three-stream regime is used. It is worth mentioning that the films prepared in the two-stream regime reveals essentially one dominant feature peaking at about 24°. In contrast, the films prepared in the three-stream regime revealed two features at about 21 and 24°, the former being more visible. The above mentioned variations of the recorded X-ray diffraction pattern may be related to the technological conditions given in Table 1. The peak locations correspond rather well to those at 20.56° and 23.86° observed for bulk PPI prepared from solution without, however, the third peak at 29.2° (Luzny et al., 1999), which is absent in the presented X-ray patterns. These angles are equivalent to the interplane distances between consecutive atomic planes being equal to 0.434, 0.372 and 0.305 nm, respectively. 3D topographic images of surfaces of PPI samples *a-f* are shown in Fig. 7 and the data deduced from these images are presented in Table 2. It is clearly seen that essentially all the presented images reveal granular morphology with grains sizing 0.1 - 0.4 µm. However, the images taken on films *c* and *f* reveal larger grains randomly distributed on the film surface, larger even than 1 m, with rms roughness attaining 32.5 nm. Some variations of the data given in Table 1 may be attributed to varying technological conditions. However, the large grains seen for films *c* and *f* are suggested to be amorphous rather than ordered ones in accordance with amorphous character of their X-ray diffraction patterns. Thus, there is a correspondence between the diffraction patterns of examined films shown in Fig. 6 and their AFM images from Fig. 7. The images shown in Fig. 8 were taken on PPI thin films prepared in the vertical setup under various stream flow rates in the vertical geometry shown in Fig. 4. These topographic AFM images reveal granular surface morphology, but in case of films deposited for $\Delta T=0^0C$ and 20^0C one can see surface covered uniformly with grains of nearly the same size, while for $\Delta T=10^0C$ one can see larger grains distributed uniformly on the surface, while for $\Delta T=30^0C$ there rarely and randomly distributed large grains upon uniform film surface.

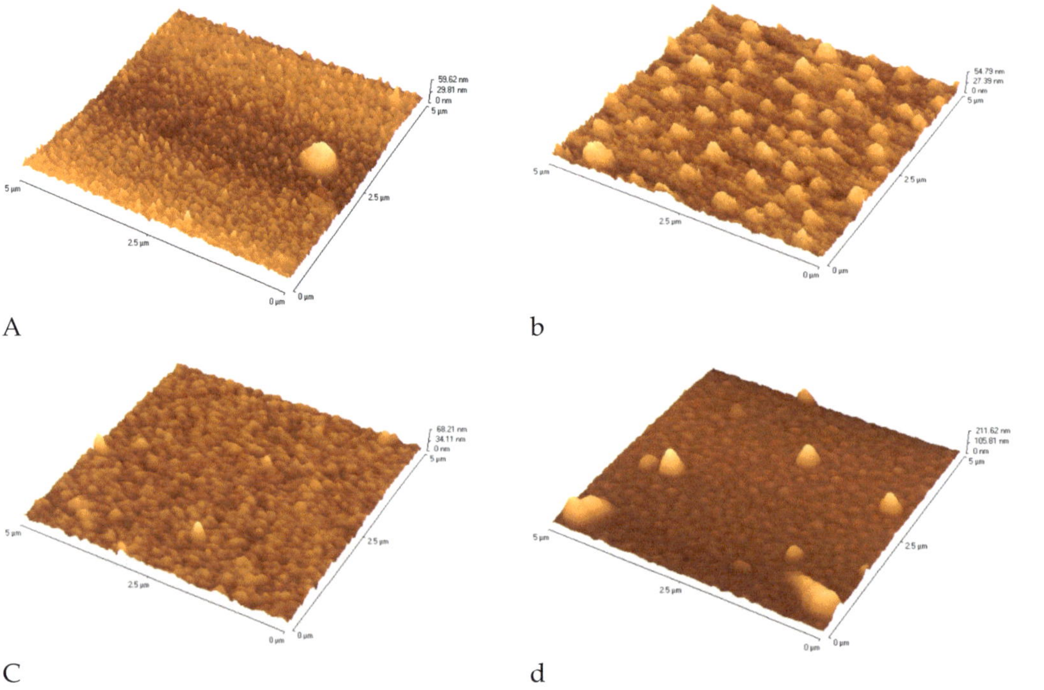

Fig. 8. AFM topographic images a, b, c and d taken on PPI thin films CVD deposited in vertical geometry with 3 Tr*l/s stream for ΔT equal to 0, 10, 20 and 30C, respectively (according to Hajduk et al., 2009).

5. Influence of technological conditions on optical spectra of polyazomethine thin films

PPI thin films

It is interesting to have a look at and to compare optical spectra taken on PPI thin films prepared via polycondensation in both horizontal and vertical geometry (Weszka et al., 2008; Hajduk et al., 2009) under various conditions. It has been mentioned in (Weszka et al., 2008)

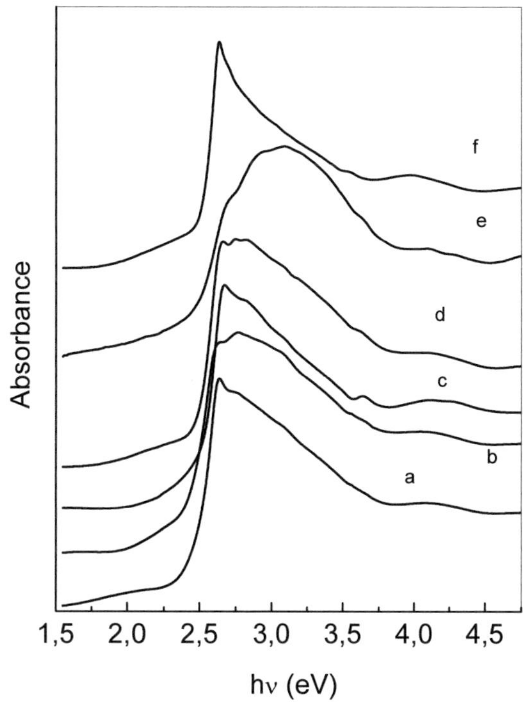

Fig. 9. Absorption spectra taken on PPI thin films prepared in horizontal geometry. Designation of the spectra coincide with film designation in Tab. 1 and Fig.6. (according to the data given in Weszka et al., 2008)

that source temperatures and the carrier agent flow rates have rather important impact on the shape of the spectra. In the case of the PPI thin films prepared in the horizontal geometry one could see that high flow rates resulted in films whose spectra reveal sharp feature superimposed onto low energy wing of the lowest energy vibronic band, which can be attributed to exciton of rather low binding energy, Wannier-Mott exciton. In case of PPI thin films prepared in the vertical geometry one can see that their thickness is growing with an increase in the flow rate of the transport agent and thin films prepared under the lowest flow rates are revealing shapes more round with less visible vibronic progressions. The latter one appears to be more pronounced in case of thin films deposited under larger flow rates. It is worth mentioning that temperature of the monomers and their difference have rather important impact on the shape of the recorded spectra UV-Vis. When temperature difference was as large as 30^{0}C, an excitonic peak has appeared on low energy wing of the low energy band in the spectrum of PPI thin film prepared with the highest flow rate.

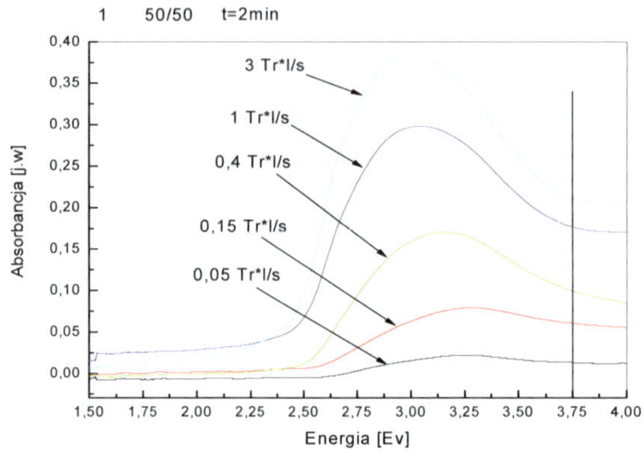

Fig. 10. Absorption spectra of PPI thin films prepared in the vertical geometry for T_{TPA}/T_{PPDA} 50⁰C/50⁰C with various flow rates (according to Hajduk et al.,2009)

Additionally, one can see that the strongest low energy band attributed to interband transitions linking delocalized states is observed at higher energy for films prepared at low flow rates in contrast to films prepared under higher temperature, which are distinctly moved towards lower energies. UV-Vis absorption spectra taken PPI thin films making part of the series prepared under conditions given in Tab.2. When comparing influence of carrier gas flow rates on optical spectra of PPI thin films, it seems that this influence has been more pronounced in the case of thin films deposited in horizontal geometry. It is thought that these effects should be linked with the rate of film growth rather than with flow rates only.

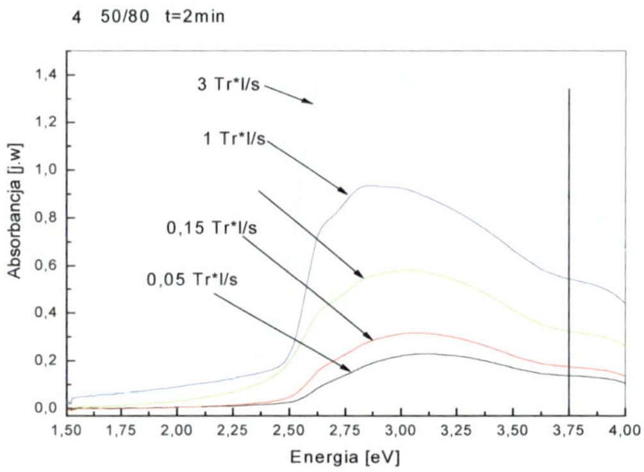

Fig. 11. Absorption spectra spectra of PPI thin films prepared in the vertical geometry for T_{TPA}/T_{PPDA} 50⁰C/80⁰C with various flow rates(according to Hajduk et al.,2009)

Polyazomethines prepared from different diamines

Absorption spectra of polyazomethines shown in Fig. 3, i.e. poly(1,4-phenylene-methylidynenitrilo-2,7,9H-fluorenenitrilomethylidyne) (FPI), poly(1,4- phenylene-methylidynenitrilo-1,1'-biphenylene-3,3',,4'-phenylenemethylidyneimine) (BPI), and poly(1,4-phenylenemethylidynenitrilo-1,3-phenylene-(4-(1-naphtylenediazenneol-yl)-nitrilomethylidyne (PNAPI), are illustrated in Fig. 12 (Jarzabek et al., 2008).

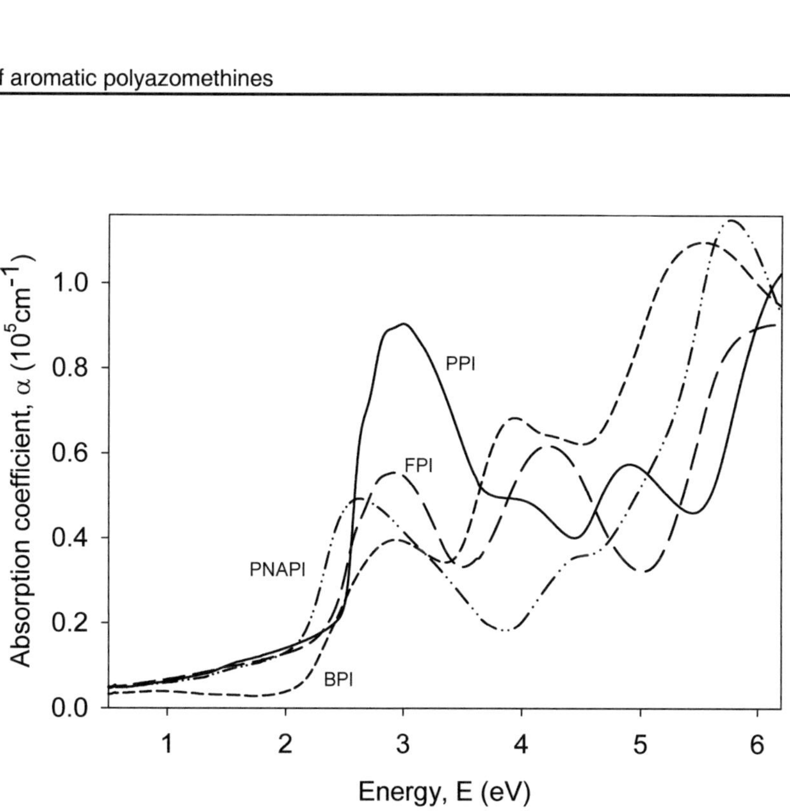

Fig. 12. Absorption in FPI, PNAPI and BPI reffered to PPI thin films (according to Jarzabek et al., 2008)

How it is seen in this figure all the spectra essentially reveal similar structure like of PPI but absorption strength distribution appear to be different. The spectra taken on BPI and FPI thin films reveal features at energies close to those where bands are observed in PPI spectrum. The spectrum taken on polyazomethine PNAPI seems to be similar but its absorption distribution is thought to much different from that in the case of the two precedent films. These similarities and variations are seen to correspond rather well with conformations of their polymer chains. FPI and BPI have backbones built up of bi-phenylenes and phenylene ring in contrast to PPI whose backbone is built up of phenylene rings only. However, in the case of FPI biphenylene part is expected to be planar due to presence of bridging aliphatic carbon 9, while in benzidine such bridge is absent and additionally, BPI has additional two azomethine links due to benzidine diamine being one of monomers. PNAPI has quite large conjugated system because there are naphthalene rings and azomethine units are in meta position rather than para asd it is the case for all the three polyazomethiones. Taking into account large conjugated systems one can see that only in the case of PPI planarity is the largest, which is supported by high intensity of this band relatively to nearest bands in its spectrum. To some extent PNAPI show off something similar to PPI character, in contrast to the other two, whose band due to transitions between delocalized states has smaller absorption strength than the feature that can be attributed to Frenkel localize excitons connected to the gap linking delocalized and localized states and vice verse.

6. Doping PPI with iodine and $FeCl_3$

PPI thin films were doped with iodine and $FeCl_3$ in rather different ways. The iodine doping was realized by placing PPI thin films deposited within iodine atmosphere during various periods, several days included. After removing them from the iodine atmosphere they were left for some time to remove eventual I_2 molecules covering the film surface (Hajduk et al., 2007). Then, they were annealed at 250C. However, it was found that the absorption of doped film did not change very much whether the spectrum was taken on a film uncovered or annealed (Fig. 10). $FeCl_3$ doping was being carried out during deposition process in the vertical CVD equipment, when the third stream flew around a crucible with dried $FeCl_3$.

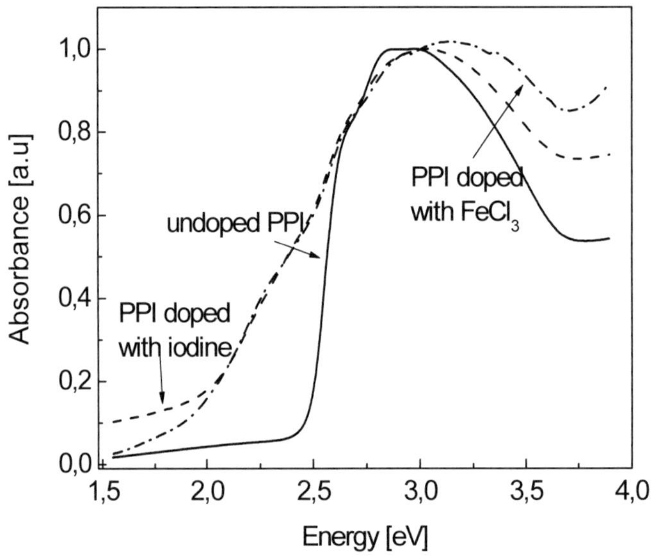

Fig. 13. Absorption spectra taken on thin films doped with iodine and $FeCl_3$ (according to the data in Hajduk et al..2007a, and Hajduk et al., 2007b)

How it can be seen in this figure the two spectra reveal nearly the same behaviour, which is expressed by appearance some increase in the absorbance strength below 2.4 eV, below absorption edge of PPI. Additionally one can see some decrease in intensity of the strongest low energy band due to interband transitions connecting delocalised states and some maximum shift towards higher energies, from some 2.7 eV to 3.0 eV. Then one cans see some increase in absorbance level at energies above 3.5 eV. Such behavior of the absorption spectrum is similar to optical spectra of PPV doped thin films (Wohlgenant, 2005). This can be attributed to polaron formation due to removing electron from PPI chain HOMO states, which is associated changing conjugation scheme along the chain from benzenoide into quinoide one. As a consequence two polaron states are formed within the HOMO - LUMO gap, the lower energy one resulting from PPI HOMO is occupied by one electron, while other one coupled with it coming from PPI LUMO being empty. Taking into account energy value indicating on some 0.6 eV shift with respect pristine PPI HOMO-LUMO gap, which could be attributed to transition from polaron level to empty LUMO (LUMO+1) state or from new HOMO (HOMO-1) to empty polaron level. However, it would be in contradiction with reported selection rules for transitions in doped polymers. Then, the shift of absorption band intensity on some 0.4 eV may suggest that the number of available HOMO and LUMO states is diminished due to doping and chain transformation into benzenoide modification whose size is dependent on electrostatic interactions between hole and electron residing on the extreme atom of the quionoid form. Additionally, interband transitions linking HOMO LUMO states are pushed to slightly higher energies. This is also some confirmation of rather distinct resemblance of π electron systems of PPI and PPV, moreover it indicates that doping mechanism in this polymer is connected with withdrawing electrons from the π electron system of PPI. Withdrawing π electrons is associated with electron transfer on iodine so that I_3^- counterions are formed. On the other hand, $FeCl_3$ doping mechanism is thought to be connected with Fe^{+3} reduction to Fe^{+2}, according to the chemical reaction

$$y(PPI) + 2\,FeCl_3 \longrightarrow y(PPI)^+ + FeCl_2 + FeCl_4^-.$$

That means Cl^- play the role of counterion in such doping.

7. Variation of electronic structure with polyazomethine structure

In the precedent sections it has been shown how electronic structure of aromatic polyazomethines is related with monomers used in polycondensation process of thin films deposition. There were used molecules having quite large conjugated systems, so that their π electron systemshave apeared to be quite developed. This is expected to give an account of their optical spectra whose structure resembles that of PPI, but the strength of absorption due to electronic states connecting their HOMO and LUMO levels is much lower compared to PPI, which might be attributed to rather bad conjugation. However, the precedent section proved that electronic structure of polyazomethine thin films can be modified by means of doping, chemical doping, while the spectra shown in Fig. 11 (Hajduk et al., 2007) were taken on thin film that could be seen as derivative of PPI, which were prepared by polycondensation of larger dialdehyde having oxygen atom to link neighbor phenylene rings.

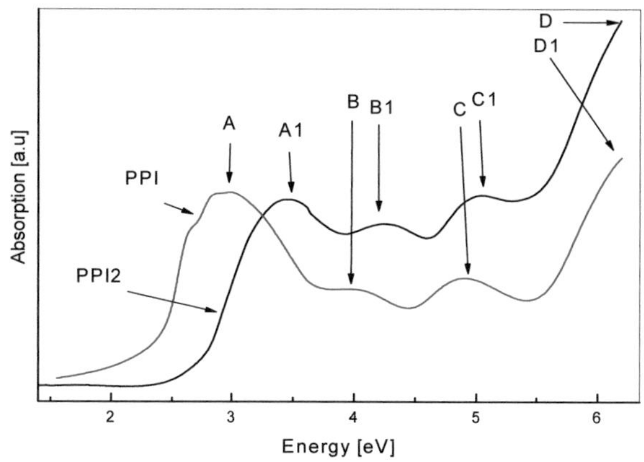

Fig. 14. Comparison of optical spectra taken on polyazomethine thin film having oxygen atoms in the backbone and that of pristine PPI

Thus, one can see that the backbone of PPI2 can be seen as modified by incorporating oxygen atoms. The spectra shown in Fig. 11.

8. Conclusions

Polyazomethine thin films can be prepared by chemical vapor deposition and thermal vacuum evaporation. It has been shown that structure, morphology and optical spectra of polyazomethine thin films are dependent on technological conditions they were deposited. Their optical spectra reveal features resembling optical spectra of polyparaphenylene vinylene. It is expected that low transport rate of monomers to the substrate independently of the setup used are favorable for small growth rates and optical spectra reveal features being resultant of distribution length of conjugated fragment in various chains within the whole thin film volume. High flow rates of transport agent have appeared favorable for excitonic feature of Wannier-Mott type to overlap low energy side of the absorption band due to transitions linking delocalized states. This could be attributed to geminate pairs or to stronger overlap of molecular orbitals of more planarized chains to form larger ordered areas where conjugation is not limited to conjugated fragments of individual chains but to some areas of thin film comprising several chains. On the other hand, one can see that polyazomethine spectra can be modified by using various more or less complex diamines aromatic with therephthal aldehyde. These spectra resemble those of PPV but one may expect that rather low intensity of low energy band due to interband transitions connecting delovcalized states due to lower degree of planarity, which to some extent recompense by relatively large π-electron system within such polymer monomer. Then, doping PPI thin films have exhibited features characteristic of PPV, which support similar polaron mechanisms driving doping PPI and PPV. This has appeared to go within this picture whether PPI thin films were iodine doped by diffusion in the iodine vapor or with the use of $FeCl_3$ in the process of thin film deposition. Additionally, it has been proved that incorporation of oxygen atoms into PPI chain, which result in dividing its conjugated fragments into quite short ones bringing about an increase in absorption gap but other features resemble those of PPI. The scope of this chapter has not allowed for electronic structure and atom dynamics of polyazomethine thin films to be discussed in details.

9. References

Flory, P. J. (1971), *Principles of Polymer Chemistry*; Cornell University Press: Ithaca, ch. 3.

Takahashi, Y.; Iijima, M.; Oishi, Y.; Kakinma M.; Imai, Y. (1991)"Preparation of ultrathin films of aromatic polyamides and aromatic poly(amide-imides) by vapor deposition polymerization", *Macromolecules*, **24**, 3543-3546.

Hajduk B., Weszka J., Jarzabek B., Jurusik J., Domański M., (2007a) "Physical properties of polyazomethine thin films doped with iodine" *Journal of Achievements in Materials and Manmufacturing Engineering*, **24**, 2, 67-70,

Hajduk B., Weszka J., Cozan V., Kaczmarczyk B., Jarząbek B., Domański M., (2007b), "Optical properties of polyazomethine with oxygen atom in the backbone", *Archives of Materials Science and Engineering* **32** , 2 , 85-88,

Hajduk B., Weszka J., Jurusik J., Jarzabek B., (2007c) „Thin PPI films moped with iron (III) chloride", *Scientific Paper Poznan University of Technology. Engineering and Production Management* No 4, 64-69, ISSN: 1733-1919

Hajduk B., Weszka J., Jurusik J., (2009) *Journal of Achievements in Materials and Manmufacturing Engineering*, **36**, 1, 41-48,

Jarzabek B., Weszka J., Domański M., Jurusik J., Cisowski J., (2006) "Optical properties of amorphous polyazomethine thin films" *Journal of Non-Crystalline Solids,* **352**, 1660-1662

Jarzabek B., Weszka J., Domański M., Jurusik J., Cisowski J., (2008) "Optical studies of aromatic polyazomethine thin films", *Journal of Non-Crystalline Solids,* **354**, 856-862

Jung, Sung Hyun; Lee, Tae-Woo; Kim, Yang Chul; Suh, Dong Hack; Chao, Hyun Nam, (2002), "Synthesis and characterization of fluorine-basesd poly(azomethines)", *Optical Materials* **21**, 169-173.

Kubono, A.; Okui, N. (1994) „Polymer thin films prepared by vapour deposition", *Prog. Polym. Sci.* **19**, 389.

Luzny, W.; Stochmal-Pomorzanska, E.; Pron A. (1999), „Structural properties of selected poly(azomethines)" *Polymer* **40,** 6611-6614.

McElvain, J.; Tatsuura, S.; Wudl, F.; Heeger, A. J. (1998)"Linear and nonlinear optical spectra of polyazomethines fabricated by chemical vapor deposition" *Synthetic Metals* **95**, 101-105.

Pross, A. (1995) *Theoretical and Physical Principles of Organic Chemistry*; John Wiley&Sons: New York,

Rohlfing, F.; Bradley, D. D. C. "Non linear Starc effect in polyazomethine and poly (p-phenylene-vinylene): The interconnections of chemical and electronic structure" (1998), *Chem. Physics* **227**, 133

Suhl, Sand Chul; Shim, Sang Chul (2000), "Synthesis and properties of novel poly(azomethine) the polymer with high photoconductivity and second-order nonlinearity" *Synthetic Metals* **114**, 91-95.

Thomas, O.; Ingnes, O.; (1998), "Synthesis and properties of a soluble conjugated poly(azomethine) with high molecular weight", *Macromolecules* **31**, 2676-2678.

Veinshtein, B. K.; Fridkin, V. M.; Indenbaum, V. L., (1979) *Sovremennaya Cristallografiya*", vol. 3, Izd. Nauka, Moscow (in Russian)

Wang, X.; Ogino, K.; Tanaka, K.; Usui, H. (2003), "Novel polyamine as electroluminescent material prepared by vapor deposition polymerization", *Thin Solid Films,* **438-439**, 75-79.

Weaver, M. S.; Bradley, D. D. C. "Organic elektroluminescence devices fabricated with chemical vapour deposited polyazomethine films" (1996) *Synthetic Metals,* **8**, 61.

Weszka J., Domański M., Jarzabek B., Jurusik J., Cisowski J., Burian A., (2008)" Influence of technological conditions on electronic transitions in chemical vapor deposited poly(azomethine) thin films" *Thin Solid Films* **516** , 3098 -3104

Wohlgenant M., (2005), "Polarons in π-conjugated semiconductors: absorption spectroscopy and spin-dependent recombination" in *Physics of organic semiconductors,* Editor: Brütting, Wiley-VCH Verlag GmbH&Co. KGaA, Weinheim, p. 235

Yang, C. Y.; Janekhe, S. A. (1991), „Conjugated aromatic poly(azomethines). 1. Characterisation of structure, electronic spectra, and processing of thin films from soluble complexes.", *Cheistry of Materials,* **3**, 878-887.

Yang, C. Y.; Janekhe, S. A. (1995) "Conjugated aromatic polyimines 2. Synthesis, structure and properties of new aromatic polyazomthines", *Macromolecules,* **28**, 1180-1196.

7

Investigations on pristine and swift heavy ion irradiated plasma polymerized aniline thin films

S. Saravanan[a], M. R. Anantharaman[b], S. Venkatachalam[c] and D. K. Avasthi[d]
[a]*Cell Technology, TATA BP Solar India Ltd, Bangalore - 560 100, India*
[b]*Dept. of Physics, Cochin University of Science and Technology, Cochin - 682 022, India*
[c]*Polymer and Special Chemical Division, VSSC - ISRO, Trivandrum - 695 022, India*
[d]*Inter University Accelerator Centre, New Delhi - 110 067, India*

1. Introduction

Organic polymers having extended π electron conjugation assume significance because of their special electrical properties, morphology and crystallinity compared with other polymers. The synthesis of conducting polymers, which are environmentally, stable, processable, and with good mechanical properties lead to the possibility of new applications. (B.Sanjai et al., 1997; W. R. Salaneck et al., 1991) Also conducting polymers were investigated extensively for understanding the underlying physics of the conduction process and for possible applications. Some of the potential devices based on these polymers are organic light emitting diodes, sensors, low power rechargeable batteries, super capacitors, photo voltaic cells and low dielectric materials in ICs (F. Garten et al., 1996; H. L. Wang et al., 1996; H. Sangodkar et al., 1996; Cartia Arbizzani et al., 1991; C. Joseph Mathai et al., 2002; S. Saravanan et al., 2004). Polyaniline occupies an important position among the conducting polymers (Alan G. MacDiarmid, 1997; H. L. Wang et al., 1996) as it is inexpensive, environmentally stable and exhibit high conductivity. Plasma polymerization is one of the novel techniques for preparing polymer thin films (C. Joseph Mathai et al., 2002; S. Saravanan et al., 2004). Thin films prepared by employing this technique are pin - hole free, uniform in thickness cross - linked and thermally stable. The method of plasma polymerization employs ac/rf/dc and pulsed techniques. Among these, rf plasma polymerization needs special mention since it yields conjugate structures of the polymers, which are supposed to be essential for making conducting thin films. Structural, optical and electrical properties can be modified when it is doped with appropriate dopants (Xing Rong Zeng et al., 1997; C. Joseph Mathai et al., 2002). Apart from adding dopants, irradiation of polymer thin films with swift heavy ions also modifies their various properties (L. Calcagno et al., 1991; Lynn B. Bridwell et al., 1991; J. Davehas et al., 1991; S. Saravanan et al., 2005; S. Saravanan et al., 2007). These swift heavy ions on bombardment results in the breaking of covalent bonds, promotion of cross linkages, formation of carbon clusters, liberation of volatile species and in some cases formation of new chemical bonds (A. Srivastava et al., 2002; Zhiyong Zhu et al., 1992). They also can induce changes in the electrical conductivity and optical band gap (H. S. Virk et al., 2001; M. Gaafar, 2001).

In this chapter fabrication and characterisation of RF plasma polymerised aniline thin films are described. The changes in structural, optical, morphological and electrical properties of pristine polyaniline are compared with the irradiated polyaniline.

2. Background and Experiments

Preparation of polyaniline thin films by employing AC plasma polymerisation and RF plasma polymerisation techniques are reported elsewhere (C. Joseph Mathai et al., 2002; S. Saravanan et al., 2004; U S Sajeev et al. 2006). Formation of polymeric materials under the influence of partially ionised gas is termed as plasma polymerisation. Since it involves the use of an electric glow discharge in vacuum it is also known as glow discharge polymerisation. Plasma polymerisation can take a variety of forms depending on the monomers and molecular fragments deposited on the surface. With appropriate monomer and operating conditions thin polymer films consisting of hydrocarbons, long polymeric chain consisting of linked carbon atoms and highly cross-linked carbon or hydrogen atoms will be deposited (C. Joseph Mathai et al., 2002; S. Saravanan et al., 2004; U S Sajeev et al. 2006).

Polyaniline thin films are prepared using RF plasma polymerisation technique. The experimental set up for the preparation of RF plasma polymerised aniline is shown in Figure 1. (S. Saravanan et al., 2004) It consists of a long glass tube of length 50cm and of diameter around 8cm with provisions for passing monomer vapour, dopants and for evacuation. Chemically and ultrasonically cleaned glass substrates was placed inside the glass tube exactly under the space separated by the aluminium foil electrodes which are capacitively coupled and wrapped around the glass tube separated by a distance of 5cm. The chamber was evacuated (10^{-2} Torr) and the monomer was admitted into the chamber. Glow discharge was obtained in between the electrodes by applying a high frequency (7-13MHz) and a current in the range of 60-80mA. The processing parameters are optimised and films were prepared under optimum conditions (S. Saravanan et al., 2004).

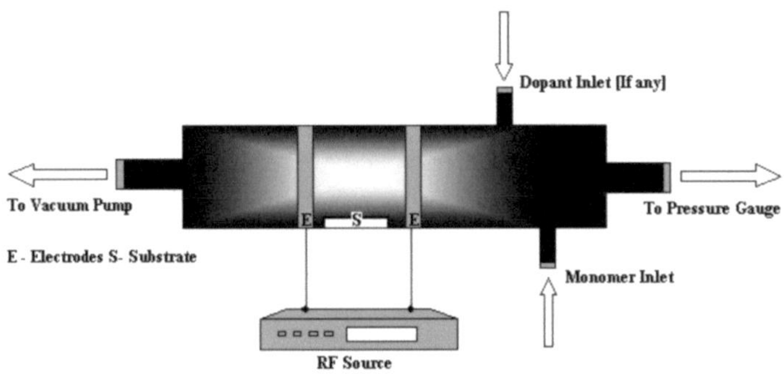

Fig. 1. RF plasma polymerisation set up (S. Saravanan et al., 2004)

These coated thin films were shifted to a metal coating unit for coating the counter electrode. The aluminium electrode was coated by evaporating high purity aluminium wire under high vacuum. (8×10^{-5} Torr) These films were in the sandwich form of cross sectional area

2.5 x 10^{-5} m^2. The thickness of the film was measured by a homemade setup employing Tolansky's interferometric method (A. Goswami, 1996).

RF plasma polymerised aniline thin films of area 1 x 1 cm^2 were exposed to 92 MeV Si ions in the material science irradiation facility of the Nuclear Science Centre, New Delhi. The ion beam current was ~ 3pna (particle nano ampere). The irradiation was carried out at room temperature and under high vacuum. The irradiation fluence was in the range of 10^{11}-10^{13} ions/cm^2, which depends on the time of irradiation and the incident ion current. (S. Saravanan et al., 2005)

The BRUKER EQUINOX 55 FTIR Spectrometer was used for taking IR spectra of monomer aniline, pristine polyaniline and irradiated polyaniline samples in the range of 4000 - 400cm^{-1}. UV Vis absorption and PL studies were carried out on these pristine and irradiated samples by using Hitachi U3300 spectrophotometer and He Cd laser with CCD array respectively. Capacitance and dielectric loss were measured by employing HP 4192A impedance analyser and a homemade dielectric cell in the frequency range of 100 Hz to 1 MHz at room temperature for pristine and irradiated polyaniline thin films. Dielectric constant was calculated from the known values of capacitance, thickness and the area of the sample. These measurements were carried out under dynamic vacuum.

3. Structural Studies

The FTIR spectra of aniline, polyaniline and irradiated polyaniline are as shown in Figure 1.A and B (S. Saravanan et al., 2005). The band assignments of the FTIR spectrum of polyaniline are given in Table.1. The peaks at 1656cm^{-1} and 1423cm^{-1} correspond to the retention of aromatic ring of polyaniline. The peak at 3207cm^{-1} shows the vibration of NH group (B. S. Furniss et al., 1998). The peaks at 2883cm^{-1} and 2834cm^{-1} are indicative of the CH stretch in polyaniline. The peak at 1059cm^{-1} is due to CH in plane deformation and CN stretching is observed at 971cm^{-1}. Substituted benzene ring is also detected from the peaks 783cm^{-1} and 676 cm^{-1}.

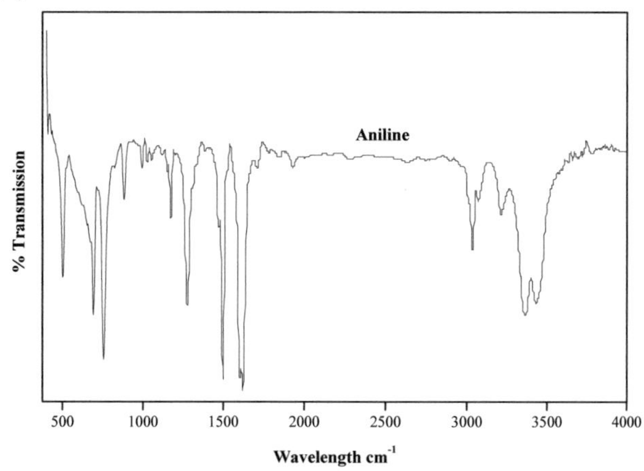

Fig. 2. A. FTIR Spectrum of Monomer aniline

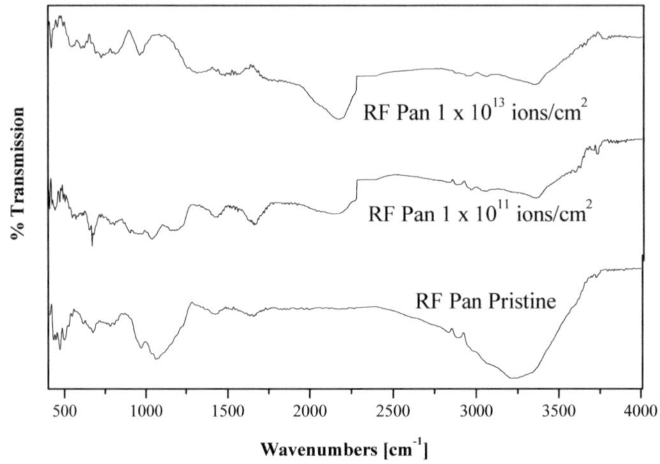

Fig. 2. B. FTIR Spectra of Polyaniline Pristine and irradiated polyaniline at different fluences

Assignment	RF Polyaniline		
	Pristine	Irradiated	
		1 x 10^{11} ions/cm^2	1 x 10^{13} ions/cm^2
N-H Vibration	3207	3353	3340
C-H Stretch	2883, 2834	3058, 2875	3058, 2873
$C \equiv C$	-	2132	2183
Ring Stretch	1656, 1423	1664, 1436	1565, 1548, 1444
CH in plane deformation	1059	1037	
C-N Stretch	971	973	975, 1326
CH Out of plane deformation	783	804	825

Table. 1. FTIR assignments of Polyaniline Pristine and irradiated polyaniline at different fluences

From the FTIR spectrum of polyaniline it is clear that the plasma-polymerised aniline is highly disordered. Based on the above discussion and from available literature it is inferred that the hydrogen abstraction is a possibility during plasma polymerisation. From FTIR analysis a plausible structure for plasma polymerised aniline is proposed and is shown in Figure 3 (S. Saravanan et al., 2008).

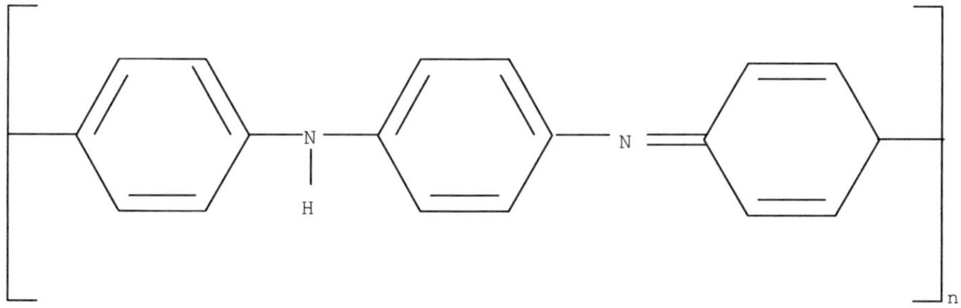

Fig. 3. Tentative structure of RF plasma polymerised aniline

From the FTIR spectrum, peaks assigned to the irradiated polyaniline are tabulated in Table 1. FTIR spectrum of polyaniline irradiated with swift heavy ions also shows broad peaks. From the spectrum, it can be seen that there is intense shift in NH stretching and CH stretching bands with respect to the polyaniline pristine. This indicates that the swift heavy ion irradiation disturbs the polymer chain. Also there is no intense shift in peaks corresponding to the aromatic ring, which is indirect evidence to the fact that the swift heavy ion irradiation doesn't affect the aromatic ring. Apart from these fundamental spectral bands, the bands at 2132 cm^{-1} and 2183 cm^{-1} are observed (Robert M. Silverstein, et al., 1980) for films bombarded with fluence of 1 X 10^{11} ions/cm^2 and 1 X 10^{13} ions/cm^2 respectively. This peak corresponds to the $C \equiv C$ group. The intensity of the peak increases with increase of fluence. Based on this a tentative structure for the irradiated polyaniline thin film is proposed and is shown in Figure 4 (S. Saravanan et al., 2008).

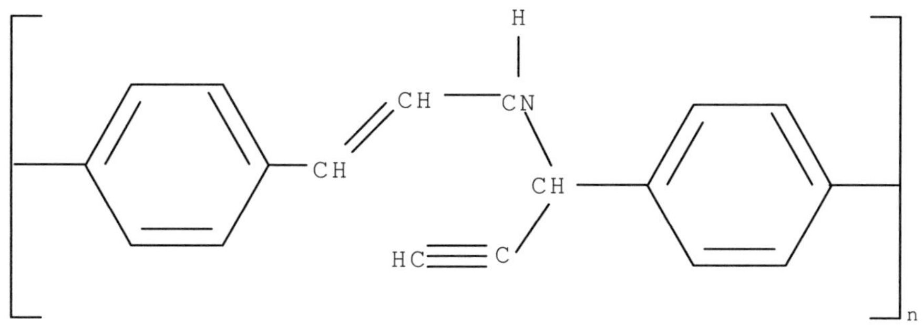

Fig. 4. Tentative structure of SHI irradiated RF plasma polymerised aniline

4. Optical Studies

4.1. UV Vis NIR Studies

The photon absorption in many amorphous materials is found to obey the Tauc relation (J. Tauc, 1970), which is of the form

$$\alpha h\nu = B\,(h\nu - E_{Opt})^n \tag{1}$$

Here α is the absorption coefficient, hν the photon energy, B a constant and the index n is connected with the distribution of the density of states. The index n = 1/2 corresponds to the direct allowed transition energy gap and n = 2 represents the indirect allowed transition energy gap.

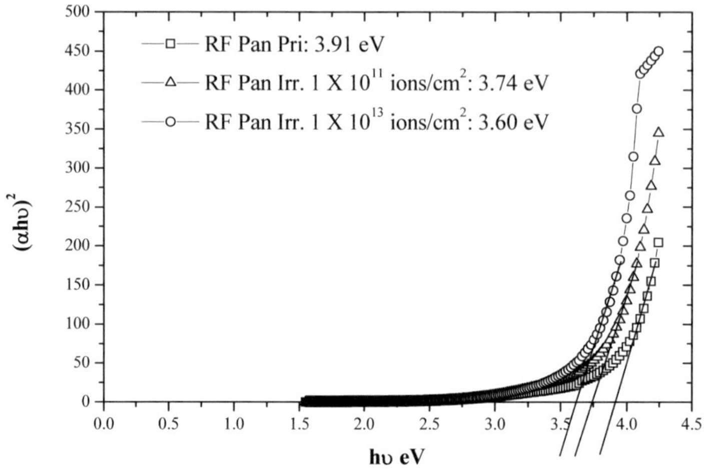

Fig. 5. UV VIS spectra of polyaniline pristine and irradiated polyaniline at different fluences

UV Vis absorption studies are carried out on pristine and irradiated polyaniline thin films using Hitachi U3300 spectrophotometer. The absorbance is plotted against the photon energy for polyaniline pristine and the irradiated polyaniline thin films and is depicted in Figure 5. The intercept of this plot on the photon energy axis gives the bandgap of the samples. From Figure 5. it is clear that there is a decrease in the optical bandgap of irradiated thin films. The bandgap decreases with increase of ion fluence. It can be seen that while pristine polyaniline exhibits a band gap of 3.91 eV. Polyaniline irradiated with fluence of 1 x 10^{13} ions/cm^2 exhibits a band gap of 3.60 eV. The bombardment of swift heavy ions on these thin film samples results in rearrangement and bond shifting which leads to ring opening in which $C \equiv C$ terminals are formed. In this process the resulting product have extended conjugated structures thus reducing the band gap.

4.2. Photoluminescence Studies

Photoluminescence spectra were recorded for plasma polymerised aniline samples before and after irradiation and is shown in Figure 6. The pristine sample exhibits the peaks at 575nm and 680nm. After irradiation the nature of PL spectrum remains similar but the peak intensity varied with the ion fluence. The peak intensities are compared for different ion fluences. From Figure 6. it is clearly seen that the intensity of the incident beam decreases and the shoulder peak intensity increases with increase of ion fluences.

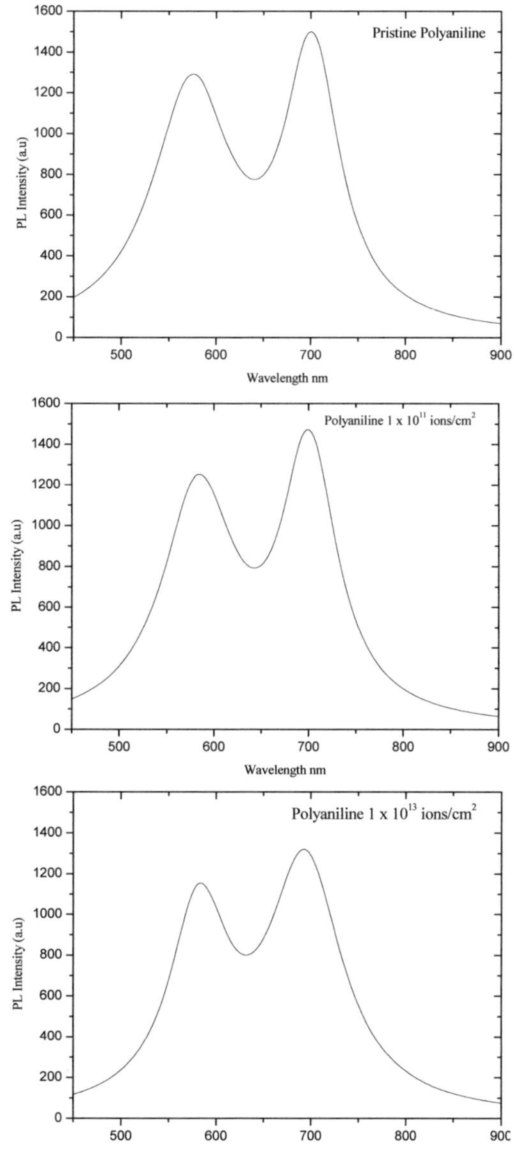

Fig. 6. Photoluminescence spectrum of polyaniline pristine and irradiated thin films

In general the effect of irradiation can be viewed in two ways.

1. The restructuring of the surface chemical species because of the energy deposited through electronic loss during the process of irradiation
2. Formation of radiation induced defects leading to non-radiative recombination centres.

FTIR spectra indicate that the pristine polyaniline retains the benzene ring and the ring is not opened up. But after irradiation, the ring is opened up because of high-energy ions, which induce structural changes. The increase of intensity of a shoulder peak may be attributed to the rearrangement of bonds as well as the increase of conjugation in the polymer. The incident peak intensity decreases with ion fluence. This could be due to opening of benzene rings. This is in conformity with FTIR results.

5. Electrical Studies

Capacitance and dielectric loss were measured by employing HP 4192A impedance analyser and a dielectric cell in the frequency range of 100 Hz to 1 MHz in the temperature range 300-373K. Dielectric permittivity was calculated from the known values of capacitance, thickness and the area of the sample. Further, from the measured values of dielectric constant, dielectric loss and frequency, ac conductivity was calculated by using an empirical relation $\sigma_{ac}=2\pi\varepsilon_0\varepsilon_r\tan\delta$. All these measurements were made under dynamic vacuum.

5.1. Capacitance and Dielectric Loss as a Function of Frequency and Temperature

The capacitance of the plasma polymerised aniline as a function of frequency at five different temperatures is shown in Figure 7.

From Figure 7. it is found that the capacitance is frequency dependent. A circuit model proposed by Goswami and Goswami (A. Goswami, et al., 1973) explains this type of behaviour. According to the model the capacitor system is assumed to comprise a frequency independent capacitive element C′ in parallel with a discrete temperature resistive element R, both in series with a constant low value resistance r. Basis on this model, the measured series capacitance C_s is given by

$$C_s = C^1 + \frac{1}{\omega^2 R^2 C^1} \tag{2}$$

Dielectric loss is given by

$$\tan\delta = \frac{(1 + r/R)}{\omega R C^1} + \omega r C^1 \tag{3}$$

Where ω is angular frequency. The temperature dependence of the model is represented by a thermally activated process and is given by

$$R = R_0 \exp\left(-{E_a}/{KT}\right) \tag{4}$$

Where R_0 is a constant and E_a is activation energy. Equation (2) predicts that C_s decreases with increasing ω_s and at higher frequencies C_s remains constant for all temperatures.

Equation (2) also envisages that, because of the decreasing value of R, C_s will increase with increase of temperature for any frequency. This effect is shown in Figure 7. and 8.

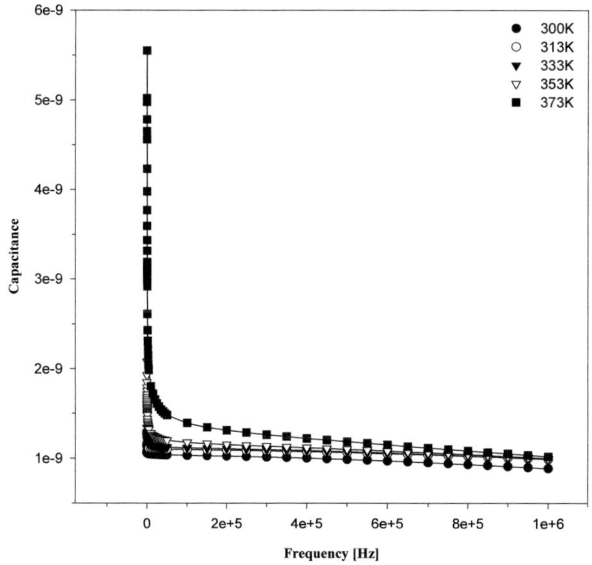

Fig. 7. Capacitance of plasma polymerised aniline thin film as a function of frequency at different temperatures

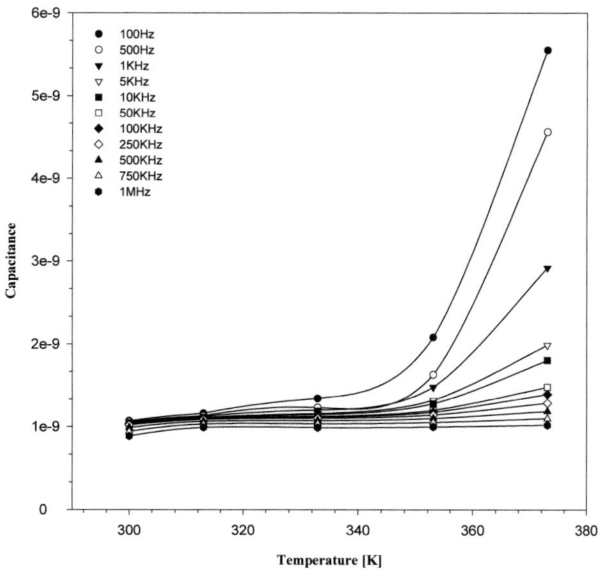

Fig. 8. Dependence of capacitance of polyaniline thin film as a function of temperature at different frequencies

The variation of loss with frequency for different temperatures is shown in Figure 9.

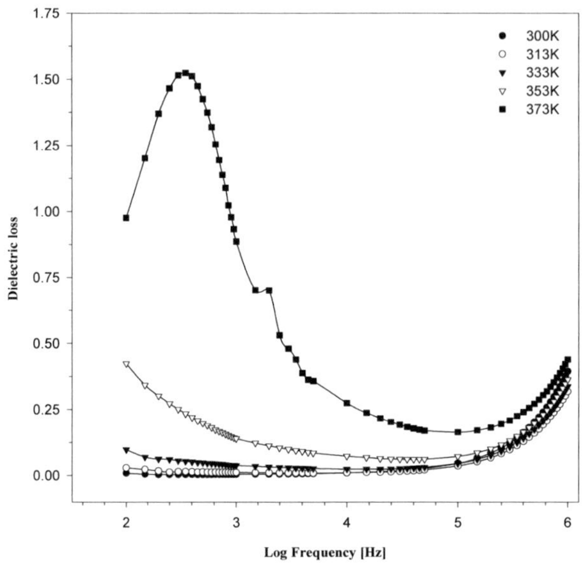

Fig. 9. Dielectric loss of polyaniline thin film as a function of frequency at different temperatures

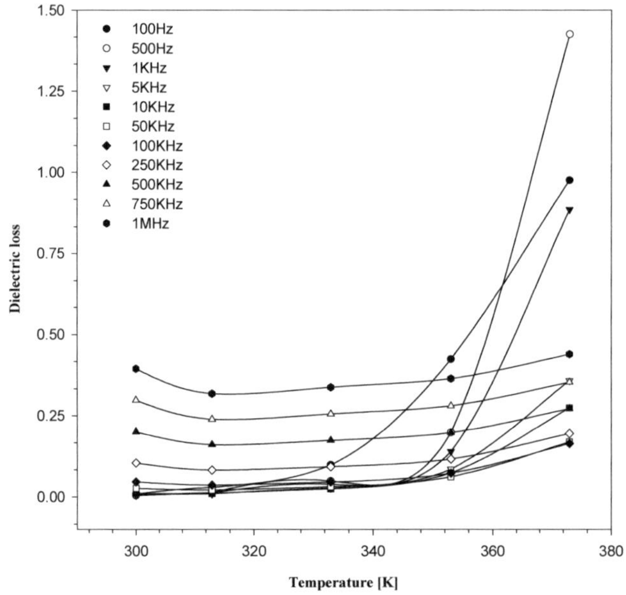

Fig. 10. Dielectric loss of polyaniline thin film as a function of temperature at different frequencies

As per equation (3) tanδ decreases with increase of frequency till the loss minimum is reached and after that tanδ increases with increase of frequency. From Figure 9. it is seen that at 373K there occurs a peak at 350Hz. It is reported that (H. Birey, 1978) similar kind of such peaks are expected at other temperatures and there occurs a peak shift. This could not be observed in our case since they might be beyond our ac measurement range. This increase of dielectric loss with decreasing frequency is usually associated with ion drift, dipole polarisation or interfacial polarisation (H. Birey, 1978). Variation of tanδ with temperature is shown in Figure 10. and is consistent with equation (3). In equation (3) the ω^{-1} term becomes dominant because of the decreasing value of R with temperature.

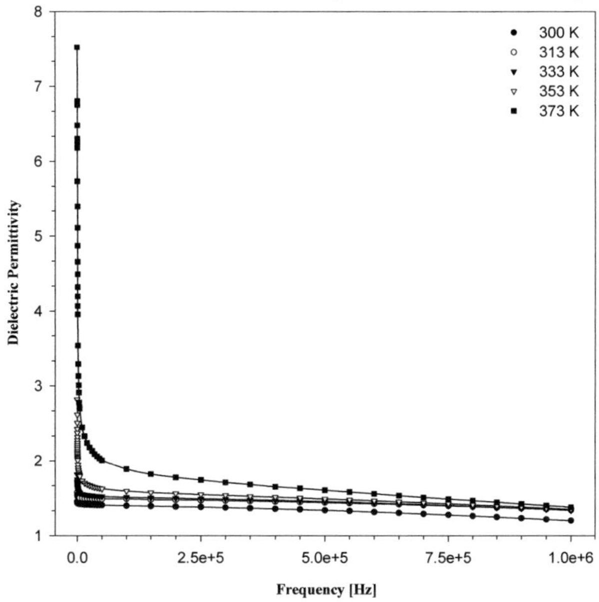

Fig. 11. Dielectric permittivity of polyaniline thin film as a function of frequency at different temperatures

5.2. Dielectric permittivity as a Function of Frequency and Temperature

The dielectric studies of plasma polymerised polyaniline thin film samples are calculated using the relation

$$C = \frac{\varepsilon_0 \varepsilon_r A}{d} \tag{5}$$

where C is the capacitance of the sample, A is the surface area of the sample, ε_0 is the permittivity of air and ε_r is the dielectric permittivity of the sample.

The dielectric measurement was carried out in the frequency range of 100Hz to 1MHz. The variations of dielectric permittivity with frequency for different temperatures were plotted and are shown in Figure 11. The dielectric permittivity lies in the range 7.52 and 1.38 for the entire frequency range for which the experiment was carried out (300K - 373K). The dielectric permittivity value lies between 1.45 and 1.20 at room temperature for the entire frequency range, which is considerably low. The characteristic dependence of the dielectric permittivity can be explained with the help of interfacial polarisation mechanism. Usually, interfacial polarisation is the type of polarisation found in the sandwich configuration. The space charge accumulations at the structural interfaces of an inhomogeneous dielectric material cause the interfacial polarisation and this was explained by Maxwell and Wagner in terms of a two-layer dielectric model.

In microelectronic circuits, RC time delay can be reduced by using this type of low dielectric permittivity materials as intermetallic dielectrics (G. Maier, 2001). The time delay depends on two factors one is due to resistance of the interconnections and the other is the capacitance of the dielectric media. RC delay can be calculated by the formulae (8),

$$RC = 2\rho k \varepsilon_0 \left[\frac{4L^2}{P^2} + \frac{L^2}{T^2} \right] \qquad (6)$$

where ρ is the resistivity, L the length of the interconnection, T the metal thickness, k the dielectric constant, ε_0 the permittivity of air and P=W (metal Width) + S (Space between metals). The dielectric permittivity of the RF plasma polymerised aniline thin film is 1.20. According to equation (6) it is found that the dielectric permittivity of polyaniline k=1.20 will reduce RC delay by about 70%.

5.3. AC Conductivity as a Function of Frequency and Temperature

The variation of ac conductivity σ_{ac} as a function of frequency for different temperatures of RF plasma polymerised aniline thin film is shown in Figure 12. From Figure 12. it is seen that the conductivity increases with increase in temperature and frequency. The conductivity increases rapidly at higher frequencies.

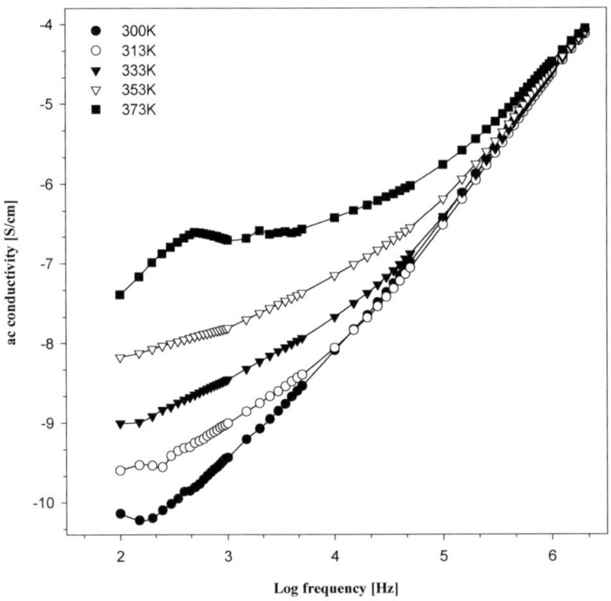

Fig. 12. Ac conductivity of polyaniline thin film as a function of frequency at different temperatures

This can be interpreted by involving the following empirical relation (N. F. Mott et al., 1971)

$$\sigma(\omega) \propto \omega^{n} \tag{7}$$

where ω is the angular frequency and n is the index used to understand the type of conduction mechanism in amorphous materials.

The values of n determined from Figure 12. lie between 0.5 and 1.1 for lower frequencies. The value of n in this frequency range is in accordance with the theory of hopping conduction in amorphous materials (M. Pollock, 1961).

The value of n gives the type of the dominant conduction mechanism in amorphous materials. This power law is an approximation of the Austin and Mott model, which describes the ac conductivity. Phonon assisted hopping of charge carriers through tunnelling from a localised site to another one is the basic physics behind the power law relation predicted by the Austin and Mott (A. N. Papathanassiou, 2002). Also Mott and Austin explained the dependence of ac conductivity at lower temperatures by the relation (A. N. Papathanassiou, 2002)

$$\sigma_{ac} = A\left(\frac{e^{2}}{\alpha^{5}}\right)\{N(E_{F})\}^{2} kT\omega\{\ln\left(\nu_{ph}/\omega\right)\}^{4} \tag{8}$$

This relation can be modified as follows by differentiating with respect to ω

$$\frac{d\ln\sigma_{ac}(\omega)}{d\ln\omega} = 1 - \frac{4}{\ln\left(\nu_{ph}/\omega\right)} \tag{9}$$

The value of n also determined by the phonon frequency and it depends on the ac frequency (A. N. Papathanassiou, 2002).

Figure 13. is the plot between $d\log\sigma_{ac}/d\log\omega$ and log frequency for three typical values of the parameter ν_{ph}: 10^{12},10^{13},10^{14} Hz. From the Figure 13. it is seen that, depending on the phonon frequency the value of $d\log\sigma_{ac}/d\log\omega$ lies between 0.71 - 0.88 for the entire frequency range. The predicted values of n lie within the values of n obtained from our experiments, provided that a single hopping ac conductivity mechanism operates in the plasma polymerised aniline thin films. Based on this it may be concluded that the conductivity is due to hopping. Also it is necessary to compare the experimental and the predicted values for finding out whether it is single or multiple hopping conductivity (A. N. Papathanassiou, 2002).

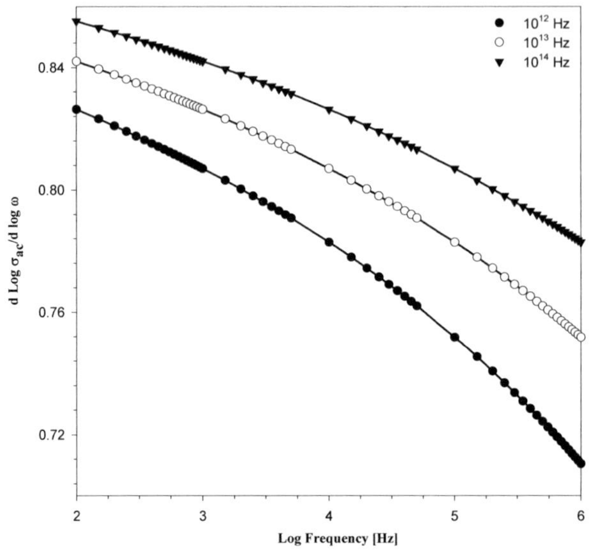

Fig. 13. dLog σac/dlog ω as a function of the frequency for 3 typical phonon frequencies

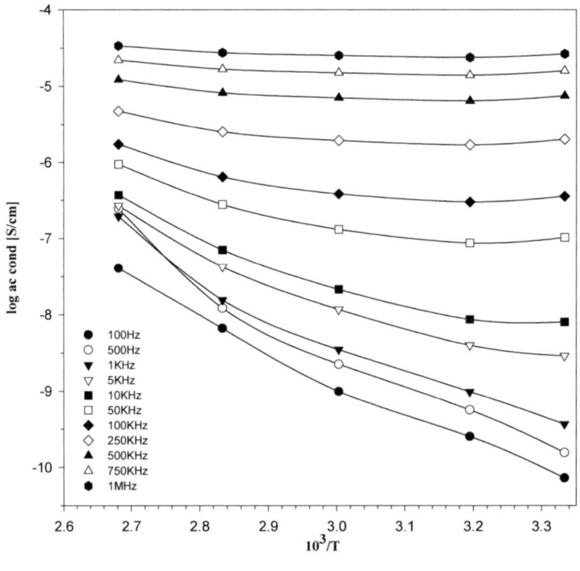

Fig. 14. Ac conductivity of polyaniline thin film as a function of temperature at different frequencies

The variation of ac conductivity with temperature as a function of different frequencies is shown in Figure 14. Activation energies were calculated and are found to be in the range 0.356 eV - 0.1435eV, which is considerably low. From Figure 14. it seen that the ac conductivity of the RF plasma polymerised polyaniline thin films is frequency dependent and it has very low activation energy. The low activation energies of these films indicate that hopping conduction mechanism occurs in RF plasma polymerised aniline thin films.

5.4. Effect of Swift Heavy Ions on Dielectric Permittivity

The dielectric measurements were performed on the irradiated polyaniline thin films and pristine polyaniline films and are shown in Figure 15. The dielectric permittivity for pristine films decreases with increase of frequency. The value of dielectric permittivity for the irradiated thin films is less than one. This violates the classical mechanical rule. It is also observed that the dielectric permittivity decreases with increase of ion fluence. This decrease in the dielectric permittivity can be explained on the basis of the decrease in the number of sites, which are available for polarization and flow of space charge carriers, which could be correlated to the production of defects created in the thin films due to the irradiation.

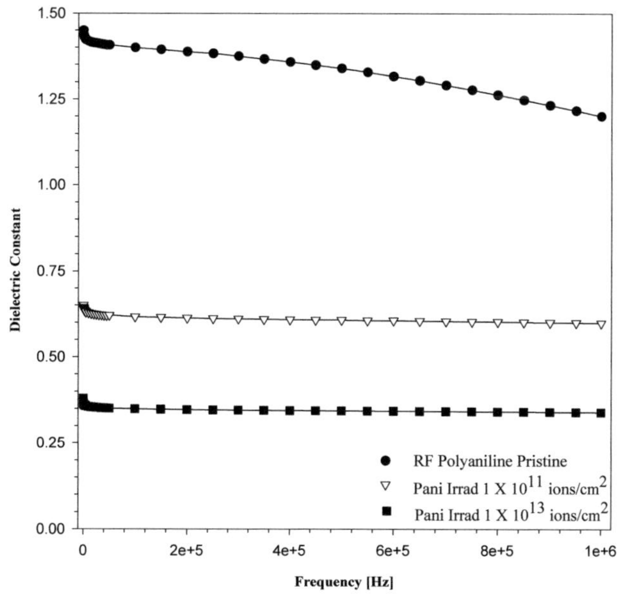

Fig. 15. Variation of Dielectric permittivity with ion Fluences

It may be noted here that irradiated samples exhibit ultra low dielectric permittivity (k<1). There is no immediate explanation for this observation and further investigations are necessary to ascertain the origin of ultra low permittivity in swift heavy ion irradiated thin films. In literature reports on ultra low are scanty. (R. I Mohamed, 2000)

6. Conclusion

RF plasma polymerised aniline thin films were prepared and irradiated with swift heavy ions. The effect of SHI on the structural, optical and electrical properties was studied. FTIR spectral results were compared with the standard data and based on the analysis a tentative structure for the pristine polyaniline and the irradiated polyaniline were proposed. The change in optical band gaps for irradiated samples was evaluated and found that the optical band gap reduces with increase of ion fluence. Electrical studies were carried out on the irradiated thin films and the results were compared with the pristine thin films.

7. Acknowledgments

One of the authors, SS thanks the Leadership Team and Prakash Suratkar, TATA BP Solar India Ltd for the support to complete this task. SS and MRA thank the Department of Space for financial assistance received in the form of a project and the Inter University Accelerator Centre for carrying out irradiation. Authors acknowledge the Institute of Physics and Elsevier for agreeing to reproduce the data.

8. References

Alan G. MacDiarmid (1997) Polyaniline and polypyrrole: Where are we headed, *Synth. Met.*, **84**, 27 - 34

Birey, H. (1978) Dielectric properties of aluminum oxide films, *J. Appl. Phys.*, **49 (5),** 2898 - 2905

Calcagno, L. & Foti, G. (1991) Ion irradiation of polymers, *Nucl. Instr. and Meth. in Phys. Res. B*, **59/60,** 1153 - 1158

Cartia Arbizzani; Marina Mastragostino & Bruno Scrosati (1997) *Hand book of Organic Conductive molecules and polymers,* Vol 4, John Wiley & Sons, Inc., ISBN: 978-0-471-96275-5, San Francisco, Edited by H S Nawa

Davehas, J. & Thevenard, P. (1991) Electronic structure characterization of ion beam modified polyimide by optical absorption and reflection, *Nucl. Instr. and Meth. in Phys. Res. B*, **59/60**, 1249 - 1252

Furniss, B. S.; Hannaford, A. J.; Smith, P. W. G. & Tatchell, A. R. (1998) *Vogel's text book of practical organic chemistry*, 5th Edition, Publishers, ISBN, Place

Gaafar, M. (2001) Ac-electrical conductivity of poly(propylene) before and after X-ray irradiation, *Nucl. Instr. and Meth. in Phys. Res. B*, **174**, 507 - 511

Garten, F.; Vrijmoeth, J.; Schlatmann, A. R.; Gill, R. E.; Klapwijk, T. M. & Hadziioannou, G. (1996) Light-emitting diodes based on polythiophene: influence of the metal workfunction on rectification properties, *Synth. Met.*, **76**, 85 - 89

Goswami, A. & Goswami, A. P. (1973) Dielectric and optical properties of ZnS films, *Thin Solid Films*, **16**, 175 - 185

Goswami, A. (1996) *Thin film fundamentals*, New age international (p) Ltd, Publishers, EAN: 9788122408584, New Delhi

Joseph Mathai, C.; Saravanan, S.; Anantharaman, M. R.; Venkatachalam, S. & Jayalekshmi, S. (2002) Characterization of low dielectric constant polyaniline thin film synthesized by ac plasma polymerization technique, *J. Phys. D: Appl. Phys.* **35**, 240 - 245

Joseph Mathai, C.; Saravanan, S.; Anantharaman, M. R.; Venkatachalam, S. & Jayalekshmi, S. (2002) Effect of iodine doping on the bandgap of plasma polymerized aniline thin films, *J. Phys. D: Appl. Phys.* **35**, 2206 - 2210

Lynn B. Bridwell; Giedd, R. E.; Wang, Y. Q.; Mohite, S. S.; Tamera Jahnke; Brown, I. M.; Bedell, C. J. & Sofield, C. J. (1991) Ion implantation of polymers for electrical conductivity enhancement, *Nucl. Instr. and Meth. in Phys. Res. B*, **56/57,** 656 - 659

Maier, G. (2001) Low dielectric constant polymers for microelectronics, *Prog. Polym. Sci.*, 26, 3 - 65

Mohamed, R. I. (2000) AC conductivity and dielectric constant of poly(vinyl alcohol) doped with $MnSO_4$, *J. Phys. and Chem. Solids*, **61**, 1357 - 1361

Mott, N. F. & Davis, E. A. (1971) *Electronic Processes in Non-Crystalline Materials*, Clarendon Press, ISBN: 0198512597, Oxford

Papathanassiou, A. N. (2002) The power law dependence of the a.c.conductivity on frequency in amorphous solids, *J. Phys. D: Appl. Phys.* **35,** L88 - L89

Pollock, M. & Geballe, T. H. (1961) Low-Frequency Conductivity Due to Hopping Processes in Silicon, *Phys. Rev.* **122,** 1742 - 1753

Robert M. Silverstein; Clayton Bassler, G. & Terence C. Morrill. (1980) *Spectrometric Identification of organic compounds*, Fourth Edition, 105, J. Wiley, New York

Sajeev, U. S.; Joseph Mathai, C.; Saravanan, S.; Rajeev R Ashokan; Venkatachalam, S. & Anantharaman, M. R. (2006) On the optical and electrical properties of rf and a.c. plasma polymerized aniline thin films, *Bull. Mater. Sci.*, Vol. 29 (2) (159–163)

Salaneck, W. R.; Clark, D. T.; Samuelsen, E. J. & Adam Hilger (1991) *Science and Application of Conducting polymers*, Taylor & Francis, ISBN: 0750300493, New York

Sangodkar, H.; Sukeerthi, S.; Srinivasa, R. S.; Lal, A. & Contractor, A. Q. (1996) Anal. Chem. **68**, 779

Sanjai, B.; Anasuya Raghunathan; Natarajan, T. S.; Rangarajan, G.; Solomon Thomas; Prabhakaran, P. V. & Venkatachalam, S. (1997) Charge transport and magnetic properties in polyaniline doped with methane sulphonic acidand polyaniline-polyurethane blend, *Phy. Rev. B*, Vol **55**, No **16**, 10734 - 10744

Tauc, J. (1970) *Optical Properties of Solids* Edited by A Ables, North Holland, Amsterdam

Saravanan, S.; Joseph Mathai, C.; Anantharaman, M. R. & Venkatachalam, S. (2004) Low k thin films based on RF plasma polymerised aniline, *New Journal of Physics* 64 (6)

Saravanan, S.; Anantharaman, M. R.; Venkatachalam, S. & Avasthi, D. K. (2008) Studies on the Optical Band Gap and Cluster Size of the Polyaniline Thin Films Irradiated with Swift Heavy Si Ions,*Vacuum* Vol 82 Iss 1 (56-60)

Saravanan, S.; Joseph Mathai, C.; Anantharaman, M. R.; Venkatachalam, S.; Avasthi, D. K. & Singh. F. (2005) Photoluminescence studies on Pristine and Swift Heavy Ion Irradiated Plasma polymerised polymer thin films, *Synth. Met.* 155 (311-315)

Srivastava, A.; Singh, T. V.; Mule, S., Rajan, C. R. & Ponrathnam, S. (2002) Study of chemical, optical and thermal modifications induced by 100 MeV silicon ions in a polycarbonate film, *Nucl. Instr. and Meth. in Phys. Res. B*, **192**, 402 - 406

Virk, H. S.; Chandi, P. S. & Srivastava, A. K. (2001) Physical and chemical changes induced by 70 MeV carbon ions in polyvinylidene difluoride (PVDF) polymer, *Nucl. Instr. and Meth. in Phys. Res. B*, **183**, 329 - 336

Wang, H. L.; Huang, F.; Mac Diarmid, A. G.; Wang, Y. Z.; Gebler, D. D. & Epstein, A. J. (1996) Application of aluminum, copper and gold electrodes in a.c. polymer light-emitting devices, *Synth. Met.*, **80**, 97 - 104

Wang, H. L.; Mac Diarmid, A. G.; Wang, Y. Z.; Gebler, D. D. & Epstein, A. J. (1996) Application of polyaniline (emeraldine base, EB) in polymer light-emitting devices, *Synth. Met.*, **78**, 33 - 37

Xing Rong Zeng & Tze Man Ko. (1997) Structure - conductivity relationships of iodine-doped polyaniline, *J. Polym. Sci. Part B, Polym. Phys.*, **35**, 1993 - 2001

Zhiyong Zhu; Youmei Sun; Changlong Liu & Yunfan Jin. (2002) Chemical modifications of polymer films induced by high energy heavy ions, *Nucl. Instr. and Meth. in Phys. Res. B*, **193**, 271 - 277

8

Tridimensional surface relief modulation of polymeric films

I. Apostol[1], N. Hurduc[2] and V. Damian[1]
[1]*National Institute for Laser, Plasma, and Radiation Physics, Atomistilor str. 409, Bucharest-Magurele, Romania*
[2]*Technical University of Iasi, Department of Natural and Synthetic Polymers, Bd. Mangeron 71, 700050-Iasi, Romania*

1. Introduction

The continuous need of miniaturization and highly integrated functionality related also with the need to use ecological processing methods resulted in the last decade in new processing methods. As an example, electronic components require conducting traces smaller than 100 μm, micro optical components and structures are integrated in the volume of the material in order to develop MEMS and MOEMS components, increasing the efficiency of the photo-luminescent diodes and solar cells or in the field of organic micro-lasers and biomedical materials. In this idea, diffraction gratings are important components due to their broad range of applications in integrated optics. The present chapter presents some results about single step surface relief modulation of the polymeric films in order to create integrated structures in materials with known properties. The studied polymeric materials are photoresist and azopolymers. The problem of tridimensional surface relief modulation is presented from two points of view: (i) technical possibilities to obtain surface relief gratings in a single step process on polymeric materials; (ii) physic-chemical processes responsible for single step surface modulation under the action of light and time stability. A general accepted technique for micro and nanoscale processing technologies in microelectronics is lithography, which is the exposure of a photoresist to a light pattern followed by a developing stage. This is a two step method, which is time consuming, non-localized and in the second stage (developing stage) is using non-ecological solvents.
A single step patterning method is laser ablation, with the advantage of the high processing speed and the possibility of small or high areas patterning. However, this method is locally changing the material properties, and it is important in some applications to preserve the composition and structure of the host material. A single step processing technique able to create surface relief modulation preserving material properties was already demonstrated. For the first time pure laser induced periodic structures (without any ablation or any larger structure) of submicron size (period and amplitude 0.2 μm) were obtained on polymer surfaces poly(ethylene terephthalate), poly(butylene terephthalate) and polystyrene, by irradiation with one thousand pulses of the polarized beam of the excimer laser (193 and 248

nm) (Matthias et al. 1992). By using the fourth harmonic of Nd:YAG laser (266 nm) periodic structures were created on the surface of poly(ethylene terephthalate), polyimide type Kapton (Hiraoka & Sendova, 1994). Under the 514 nm laser beam from an argon laser the surface of an azoaromatic polymer was optically altered in order to induce a highly efficient diffraction grating (Rochon et al. 1995). A single step processing method for surface modulation has been developed in order to obtain submicron structures for special applications and was reported (Dyer et al., 1996), (Castex et al., 2002). There were obtained surface relief gratings on PMMA and on the polymeric materials based on carbazole chromophores in order to develop a plastic blue laser emitting around 400 nm (Castex et al. 2006),. Single step recording of sinusoidal surface grating in hybrid sol-gel glasses was reported (Pelissier et al., 1999) and also holographic patterning of acrylamide-based photopolymer surface (Naydenova et al. 2005). Generally, the materials were polymers commercially available or special applications were developed and not a unitary explanation was produced for this effect. Recently, at the SPIE Conference: Technologies for the New Millennium in Dresden, may 2009, we have observed a growing interest about polymeric thin film structuration for applications in microelectronics, but a low understanding of the phenomena involved in the structuration process. In the idea to contribute to a better understanding of the capabilities of the single step surface modulation under the action of UV laser radiation we present our results obtained on different materials like commercial polymers (respectively photoresists), but also on laboratory synthesized azopolymers. The study of surface induced structuration under the action of UV laser radiation on polymers with very well controlled structures gives us the possibility to propose also mechanism for the laser induced effects.

2. Polymeric materials used for surface relief modulation

An important part of our studies is devoted to surface structuration of azo-polymeric materials. These materials are based on polysiloxanic chains, modified with azobenzenic groups and nucleobases (Hurduc et al. 2007) (Fig. 1). The presence of the nucleobases in the side-chain, are justified by the potential applications in biology. Two types of applications are possible: immobilization and laser nano-manipulation of biomolecules on the film surfaces and directional cells growth on the nanostructured surfaces. Details concerning the polymers synthesis and characterization were previously reported (Hurduc et al., 2007).

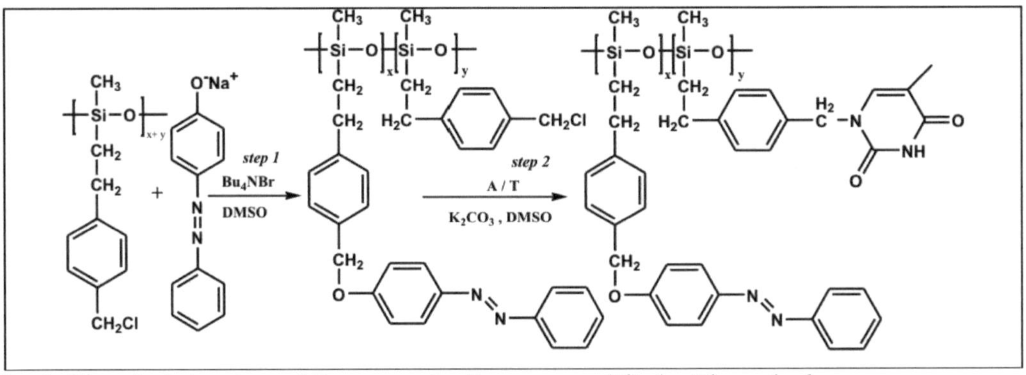

Fig. 1. Chemical structure of the azo-polysiloxanes modified with nucleobases

During the photo-chromic studies of the azo-polysiloxanes modified with nucleobases, a mechanism concerning the possibility to generate a fluid phase under UV/VIS irradiation was proposed. This mechanism is based on the concept of *conformational instability* state. This special state is a consequence of the continuous azobenzene groups *trans-cis-trans* photo-isomerization processes, which may be accompanied, as a function of chemical structure, by strong dipole-moment fluctuations along the chain (i.e. in the case of azobenzene, the dipole-moment is 0.1 D for *trans* configuration and 3.5 D for the *cis* one (Shishido, et al. 1997). If the azobenzene group is connected in the polymeric main- or side-chain, these modifications will impose conformational changes on the entire polymeric chain level, accompanied by strong dipole-moment fluctuations along the chain. This continuous photo-isomerization motion of the azo-groups induces a conformational instability on the entire polymeric chain that hinders the phase stabilization in a solid state.
The photoisomerisation kinetic curves are studied for some azopolymers (Enea et al. 2008 a), (Enea et al. 2008 b) (Fig. 2, 3). The high response speed of the azobenzenic groups at UV irradiation in the solid state comparing with the solution may be explained by the flexibility of the main chain with a polysiloxanic structure and by the amorphous film structure that assures a high free volume. The cis–trans azobenzenic groups'relaxation phenomena can take place thermal-activated only (in dark) but in this case the processes are much slower (in a time scale of days). Therefore, a big difference in the film surface response can be expected if the operational conditions are modified (presence or absence of visible light during UV irradiation).

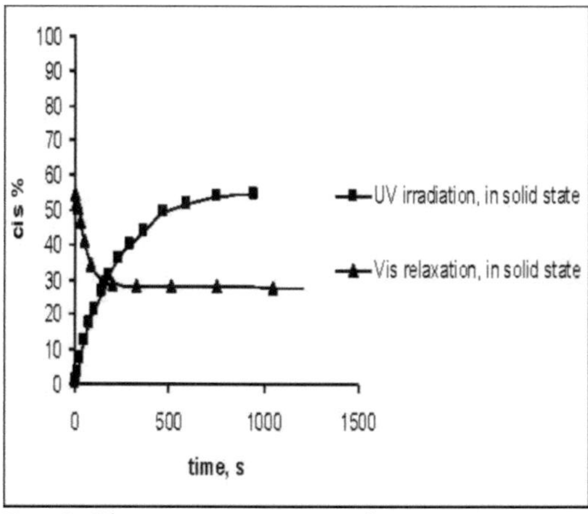

Fig. 2. Photo-isomerization kinetic curves obtained in the solid state (film): the trans–cis isomerization process under UV irradiation and in the presence of the natural visible light; the cis–trans relaxation process in dark (azo-polysiloxane modified with thymine units)

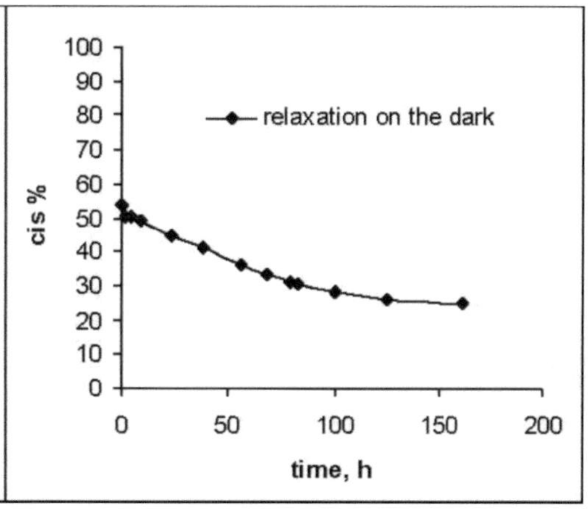

Fig. 3. Cis–trans relaxation process at the dark, in solid state (azo-polysiloxane modified with thymine units)

The absorption spectrum for a great number of azopolysiloxanes evidenced a maximum for the absorption spectrum in a region around of a wavelength of 350 nm, but also a tendency for higher absorption maxima at wavelengths lower than 200 nm (Fig. 4).

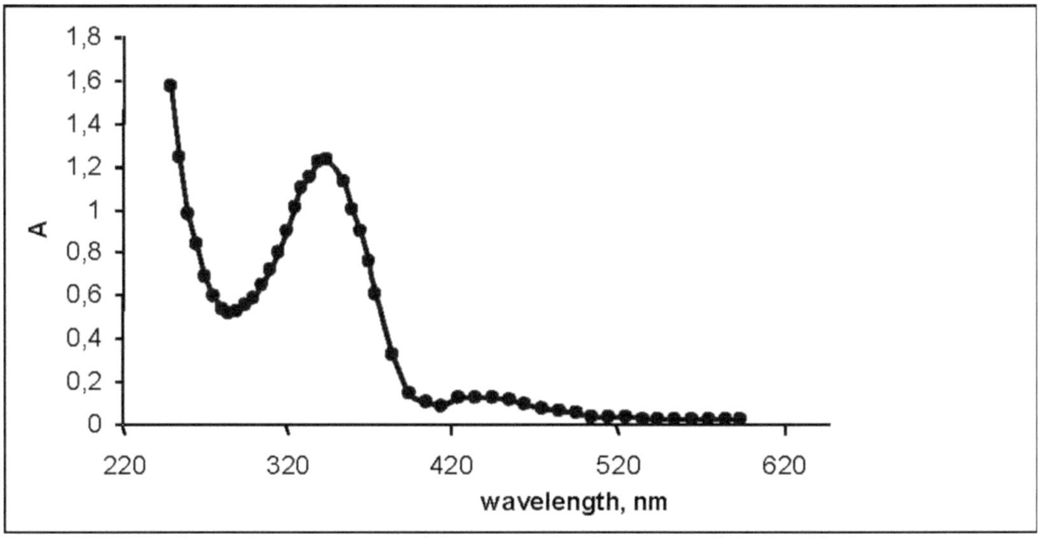

Fig. 4. UV-VIS absorption spectrum of the azo-polysiloxane modified with thymine

From the category of commercially available polymers we have used photoresists. They are well known and widely used in microelectronics. For their surface structuring microelectronics industry developed standard now technologies of their processing based on their characteristic to change their properties upon exposure to light. But all these processing methods are two step methods, time consuming and involving solutions considered dangerous from the ecological point of view.

3. Irradiation conditions for submicrometric single step surface relief modulation

Generally the leading idea was to obtain diffractive optical elements with controlled pitch, usually with submicron values in materials with special applications. In order to obtain surface relief gratings interferometic or holographic methods were proposed. The single step surface relief modulation method consists in the exposure of the polymer film surface to a light field with a controlled distribution of the light intensity in the irradiation pattern. To obtain a surface structuration without phase changes and discontinuities in the materials characteristics the maximum light fluence in the irradiation spot has to be lower than the ablation threshold of the material. A possibility is to create an interference field, generally in UV laser light. The majority of polymers have a maximum of light absorption at about 190 nm and also at about 350 nm, and as a consequence surface processing at these wavelengths is very efficient. Also the low wavelength in the UV region is recommended to create structures in the hundred of nanometers range. The 193 nm radiation is a wavelength emitted by ArF excimer laser. The problem is that the excimer laser presents a low spatial and temporal coherence to obtain interferometric pattern at this wavelength. To compensate the low coherence the interference pattern is produced between the diffracted beams of a phase mask (Dyer et al. 1996, Castex et al. 2002). The interference field pitch can be of the same order of magnitude or half as the pitch of the phase mask depending if interference take place between 0, ± 1 diffraction orders, respectively between +1 and -1 diffraction orders (Fig. 5). Unfortunately that type of interferometer introduces some inconvenient: difficulties to manipulation, necessity of a protection fused silica sheet, not a direct control of the interfering beams, secondary multiple beams. Those are overcome with an interferometer setup (Fig. 6) using as a beam splitter a diffraction grating. Laser radiation is diffracted and the beams are separated by an angle θ; at the interference, according to the interferometer scheme, in case two beams at θ angle, we obtain an interferometric image with the pitch equal with the pitch of the diffraction grating, or two beams separated by 2 θ angle, give us half the diffraction grating pitch.

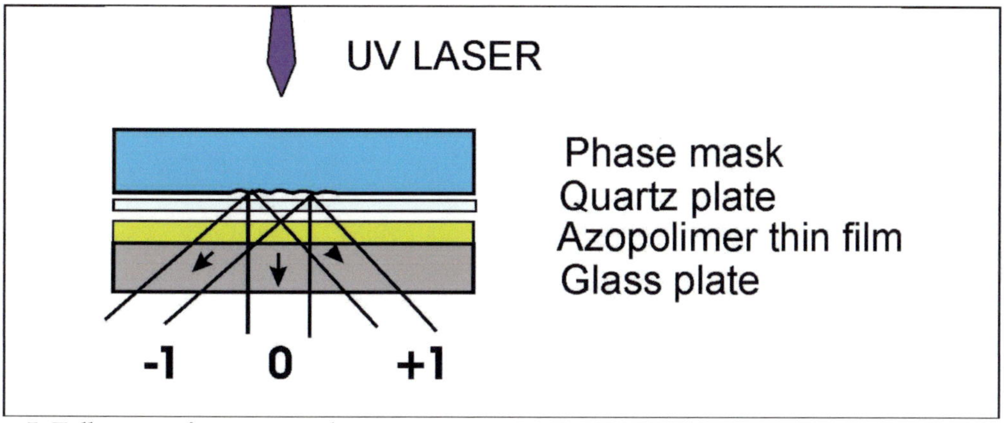

Fig. 5. Talbot interferometric scheme

In case of a grating with 1 μm pitch to obtain the interference pattern we use only the ±1 and 0 diffraction orders. The beams are used in pair (+1, 0) or (-1, +1) that can be selected with a beam selector (S).

To superpose the diffracted beams a pair of kinematic mirrors (M1, M2) that can be tuned according with the desired interference pitch grating and miss-alignments compensation are used. To compensate the phase difference between the beams, in case of (+1, 0) diffraction orders interference, we have used a fused silica plate introduced in the zero order beam pathway. The light source, of our setup, was an Nd:YAG laser working on his third harmonic at 355 nm and with a pulse length at FWHM of 5 ns, with a repetition frequency of 10Hz and a beam divergence of 0.6 mrad. The main advantage of the set-up is the possibility to know and to control the energy of the interfering beams and consequently the fluence of the laser field incident on the sample. For this, the laser energy was permanently monitored by an energy meter (PM).

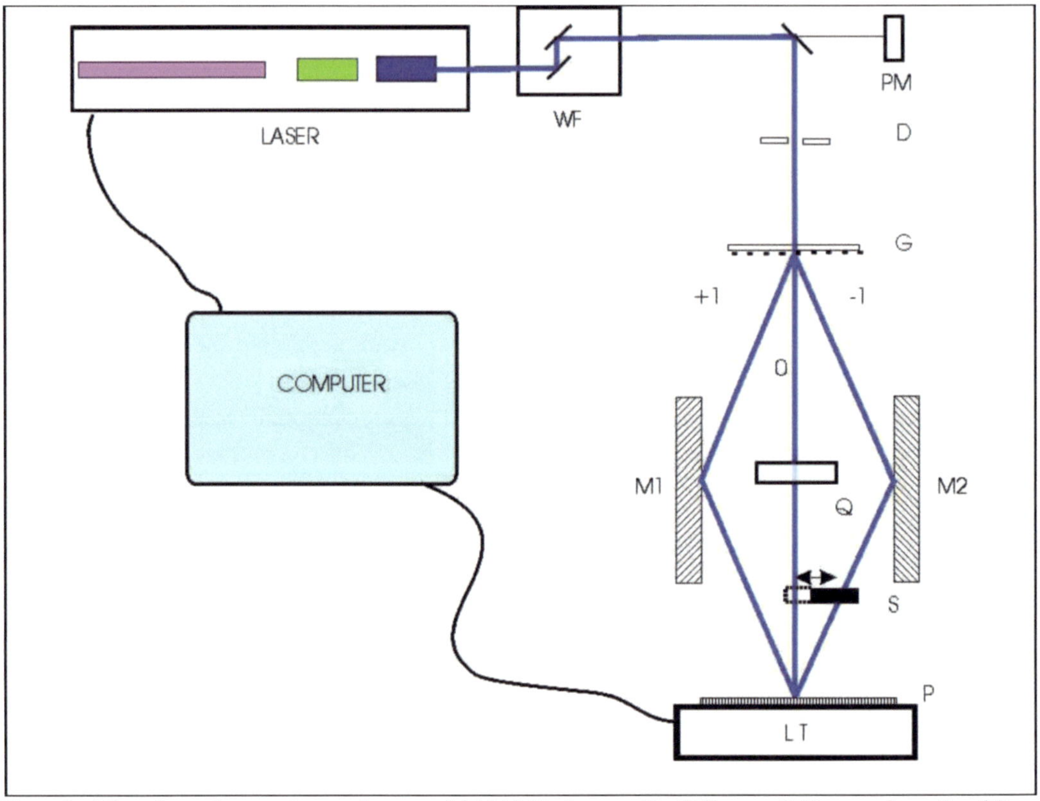

Fig. 6. The interferometer scheme: Nd:YAG laser @ 355nm, WF-wavelength filter, PM-Energymeter, D - aperture, G-diffraction grating, M1,2 -kinematics mirrors, Q - fused silica phase compensator, S - beam selector, P - sample, LT - translation stage

4. Surface relief modulation of polymeric films

We have obtained very good surface relief gratings (SRG) with submicron pitch on commercial photoresist or PMMA foil (Apostol et al., 2006). Gratings inscription was realized in a single step (no developing stage) under the action of an exciter laser with emission at 193 nm wavelength, pulse length 7 ns. The obtained SRG were characterized through various investigations means like optical microscopy, atomic force microscopy (Fig. 7, 8), and also with scatterometry. The characterization of the gratings is necessary not only for insight in the creation process of gratings but also for checking the reproducibility and uniformity. The diffraction efficiencies of the various orders diffracted by the grating were measured and they were fitted to theoretical predictions corresponding to various structure models of the gratings. The fitting procedure was used to provide the parameters of the gratings, such as the width, the grating height, the pitch or shape factors, such as the wall angles for a trapezoidal structure (Logofatu et al., 2008).

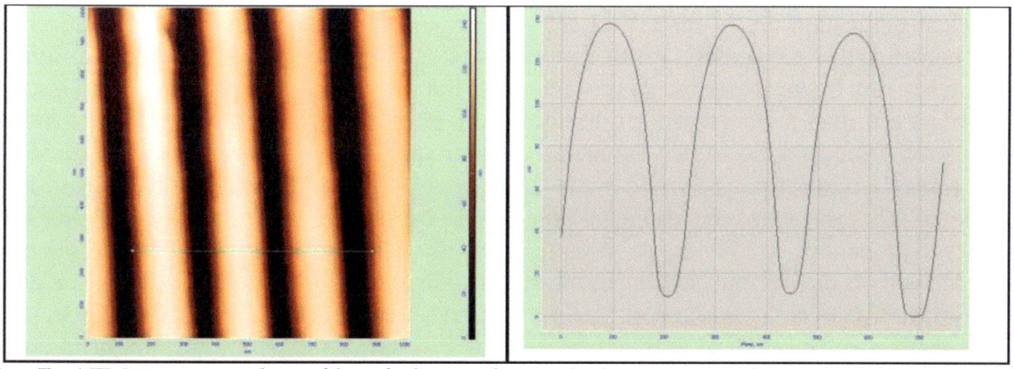

Fig. 7. AFM image and profile of the surface relief grating with the pitch of 250 nm, modulation depth 130 nm, induced on commercial photoresist under the action of an interferometric field at the wavelength of 193 nm.

The modulation depth of the SRGs obtained in the process of the surface relief modulation under the action of a structured light field depends on the incident fluence/intensity but also on the subsequent irradiation pulses number. As a consequence the modulation depth was possible to be varied between 10 nm to hundreds of nanometers. The smooth profiles evidenced by the AFM investigation of the obtained SRG evidenced that the surface modulation process is not induced by laser ablation. This fact suggests that an inner reorganization process of the polymeric chains take place and is responsible for the surface modulation. Generally the value of the incident irradiation light field intensity was between 10^5 - 10^6 W/cm^2. The complexity of the photoresist material makes difficult to analyse the mechanism of the surface relief modulation under the action of a light field.

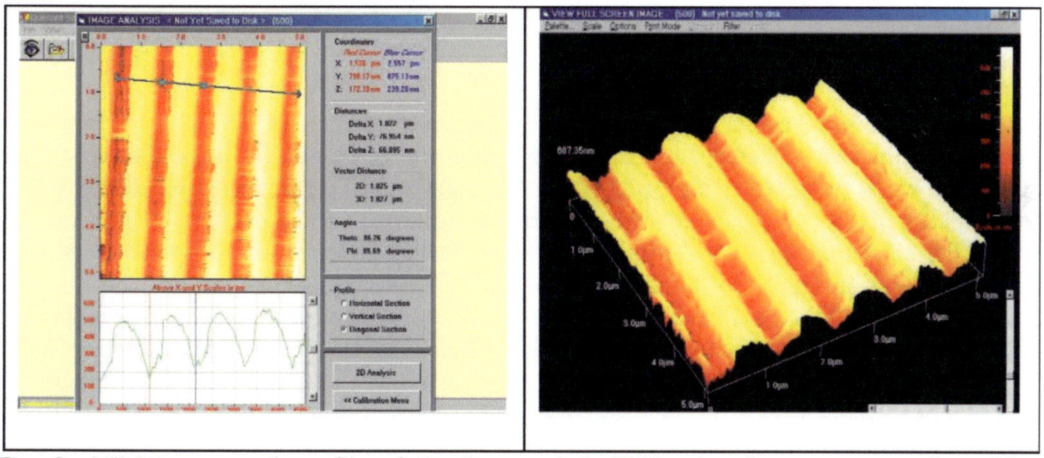

Fig. 8. AFM image and profile of the 1 μm pitch SRG printed on commercial positive photoresist.

In order to have a better control on the photochromic properties of the studied materials but also because of their specified applications we have studied single step surface relief modulation of azopolymeric materials. Based on the structural changes induced in the polymeric materials by UV light we have studied the possibility to induce tridimensional surface structuration under the UV laser radiation in azopolymeric films. Also analyses of the photochromic properties of the azopolymers evidenced an absorption maximum at around 350 nm. In our experiments we have used the third harmonic of the Nd:YAG laser, at 355 nm. A group of azopolymers studied from the point of view of the possibility to induce surface relief modulation under the action of laser radiation in a under ablation threshold regime are the azo-azopolysiloxane modified with nucleobases (Enea et al., 2008) Irradiation geometry schemes are given in Fig. 5 and 6. The irradiation beam is the Nd: YAG laser harmonic of 266 nm or 355 nm, pulse length 5 ns. By using a phase mask of 1 μm pitch we have formed on the sample surface an interference image with the period of the same order of magnitude as that of the phase mask. In the same set up it is possible to go to a lower pitch of the grating in the hundred of nanometer range. The surface relief gratings induced on the azopolisiloxane films were investigated with a Zeiss AXIO Imager microscope and an AFM. Laser induced effects on the material surface was studied as a function of incident laser fluence and number of laser pulses subsequent on the same place.

Fig. 9. Evolution of the surface relief formation for a fixed incident laser fluence (8.4 mJ/cm^2) for 10 pulses (*left*) and 100 pulses (*right*) and AFM profile (right) on for an azo-polysiloxane modified with thymine units film.

In Fig. 9 the evolution of the surface relief formation is evidenced for fixed incident laser fluence and for two values of the number of incident laser pulses. It can be seen that for a lower number of pulses the structuration which results in SRG formation is not complete and for 100 subsequent pulses a very uniform line structure was obtained. The low fluence for which the structuration was observed evidences that the mechanism responsible for the surface relief gratings formation is an inner material reorganization and not the material ablation. In Fig. 10 the evolution of the relief surface structuration as a function of incident laser fluence for a fixed number of incident laser pulses is considered. For the fluence of 35 mJ/cm^2 and 15 pulses material organization is not complete and the SRG is not yet completely formed. It can be seen also that using high irradiation fluence (196 mJ/cm^2) situated near but under the laser ablation threshold a good SRG is obtained, but also some damages can appear on the surface.

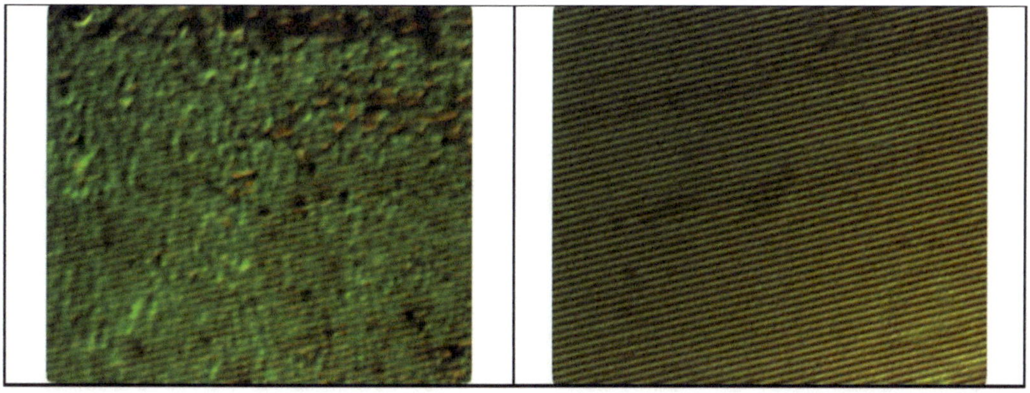

Fig. 10. SRG formation for 15 incident laser pulses with incident fluencies of 35 mJ/cm^2 (*left*) and 196 mJ/cm2 (*right*) on azo-polysiloxane modified with thymine units film

For surface relief gratings formation understanding and also for technical applications the time evolution and stability of the induced surface structure is an important parameter and is a problem to be discussed. In case of single step surface relief formation on polymeric materials the process is connected with the photo chromic behavior of the materials. As it is known the UV light induced effects on the material structure are reversible under the action of visible light, but with different speeds. The time evolution of the SRG obtained on polymeric films under the action of the UV light has to be analyzed from two points of view (Apostol et al., 2009):

- surface relief evolution during the light treatment process;
- surface relief evolution in time after structuration (time stability)

Generally the studies about the trans-cis isomerisation degree were realized under the action of the classical UV lamps and the isomerisation time is of the order of hundreds of seconds. Irradiation of the sample surface with the interference field, having as a light source the high power pulsed laser beam, induces effects in a period of the order of magnitude of the laser pulse duration. In our case the laser pulse has 5 ns at the FWHM (full width half maximum) and we have obtained effects of surface structuration even at 1 irradiation pulse. The laser irradiation intensity is lower than the material ablation threshold, in order to avoid damages. Surface induced structure analyses were realised, as we have mentioned, with high resolution optical microscopy and atomic force microscopy (AFM). Both techniques are not adapted for analyses of the effects induced under the action of laser treatment in-situ and during the patterning process. To be able to analyse the time evolution (or the evolution with the number of irradiation pulses) of the surface induced relief it was used a sequential technique. Based on the property of high reproducibility of the effects induced under the action of laser radiation (if the laser-matter interaction is realised in reproducible conditions) there were realised a great number of samples (irradiations) in known conditions (incident laser fluence, interference field structure, film surface quality as composition and cleaning). Sequences of irradiations for a constant incident fluence and different number of irradiation pulses give us the possibility to analyse the time evolution of the surface structuring process. Samples were analyzed immediately after laser treatment and after a known time with optical microscopy and higher time delays with AFM.

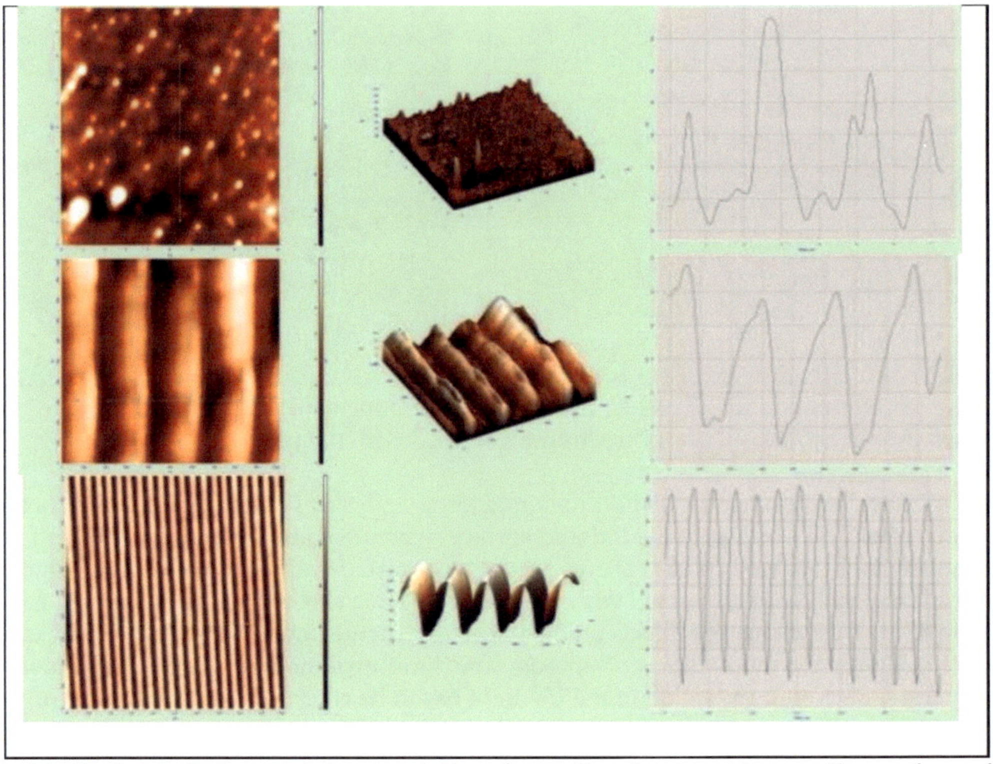

Fig. 11. Surface relief modulation evolution from disorder to order on the surface of a photoresist film: sequence of AFM images and AFM profiles of the surface relief formation for an irradiation fluence of 4 mJ/cm^2 (I = 0.8 x 105 W/cm^2) as a function of the irradiation pulses number (from up to down): for 20 pulses: surface structuration apprears; for 50-100 pulses the modulation depth is 10-15 nm and a more regular organization of the surface relief and for 400 pulses the modulation depth is about 90 nm and SRG is formed

The induced surface profile time evolution during the multipulse irradiation has evidenced the evolution of the surface structuration from disorder to order, up to a sinusoidal profile corresponding to the surface grating formation (Fig. 11). If we compare the profile of the interference pattern calculated for the phase mask we have used in the experiments with the induced surface relief grating profile visualised with an AFM they are completely similar. (Apostol et al., 2009).

The surface structuration evolution is similar for the photoresist films and for the azopolymer films. The structuration time depends also on the incident laser fluence/intensity from 1 pulse (5 ns) to up to 500 pulses of 5 ns each. If we consider that the surface volume structuration effect is due to the trans-cis isomerisation effect we can consider that under the action of laser radiation the isomerisation time is much less than for the classical UV lamps. Taking into account the fast response of the material at the UV irradiation it can be considered that the surface relief formation is due to a spontaneous reorientation of the molecules due to the conformational changes as a result of the isomerisation process. It is considered that in this case the surface relief formation effect is reversible under the action of visible light or at the temperatures higher than the vitrification temperatures of the material. The relaxation time reported is generally of the same order of magnitude with the isomerisation time. But in case of surface structuring under the action of laser radiation we have obtained for some of the studied materials very good time stability. In case of the photoresist irradiation we have obtained gratings lasting in good conditions as modulation depth and pitch for more than two years. Only if mechanically damages like scratches are produced the grating is damaged.

Microscope image after 15 min. from the irradiation moment	Microscope image after 15 min. + 24 h from the irradiation moment
	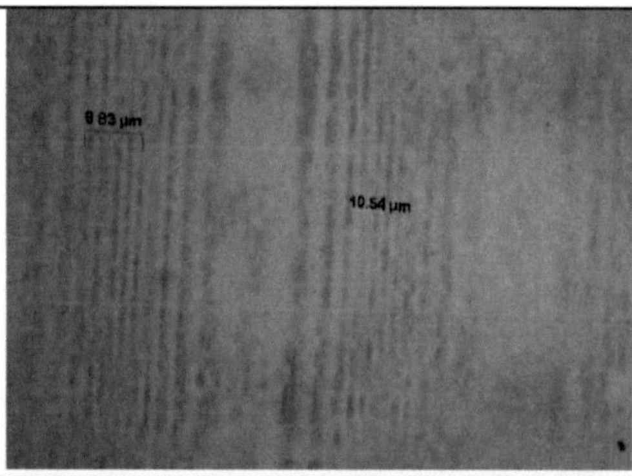

Fig. 12. Surface relief grating relaxation, after 24 h from the irradiation of the film of azopolysiloxane modified with azophenol (95-98)%; Irradiation conditions: Fluence = 17 mJ/cm2, Intensity = 3.5 x 107 W/cm2

Also in case of an azopolymer film surface structuration it is possible to observe the surface modulation evolution from disorder (Fig. 10, left) to order (Fig. 10, right). In case of surface relief structuration of azopolymer films the stability of the induced SRG depends on the type of polymer. To analyze the time stability of the induced structures on the surface of azo-polysiloxane modified with thymine units films the samples were analyzed also after a month, taking into account that the *cist-trans* relaxation curves under the visible light and in dark indicate relaxation times from 500 s to hours. The samples were kept at the normal ambient (summer) temperature. The microscope analyses evidenced the same structure without damage, so their time stability can be reported (Enea et al., 2008). In case of a sample of polysiloxane modified with cu azophenol (substitution degree 95-98%) the time evolution of the structured surface was monitored up to 30 hours from the irradiation time. In Fig. 12 can be seen the microscope images of the grating induced under the action of laser radiation at 355 nm at 15 min. after irradiation moment and after 24h (Apostol et al. 2009). A sequence of microscope images is presenting the evolution of the decay of the contrast in a grating which is disappearing from the surface in about 24 hours (Fig.11.). The host material is also polysiloxane modified with cu azophenol. It was selected in photos a region with small defects, to have a spatial reference to recognize the analyzed region. The sample was kept at the normal room temperature (about 23-26°C). It can be observed that the line contrast is reduced up to the complete disappearance of the lines after 27 hours (Fig.13.). In case of films of azo-polyimide, with rigid main chain and azo-polysiloxane modified with thymine with flexible main chain the surface structure was induced under the action of 1laser pulse (5 ns) up to 500 pulses. The microscope image was realized after 15 minutes from irradiation and the AFM analyses after more than 3 month (Fig. 12.).

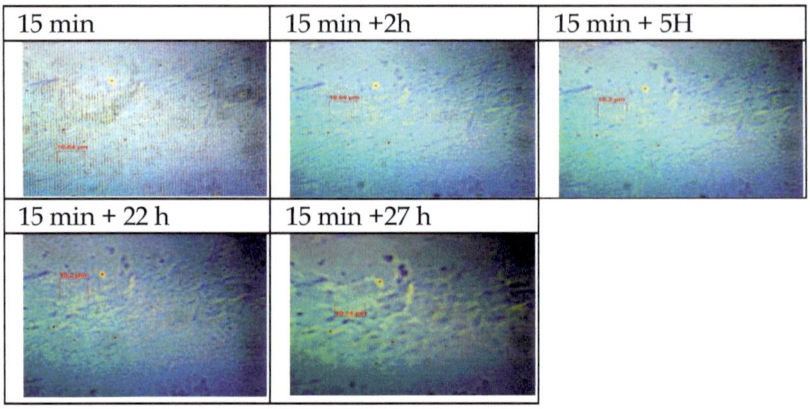

Fig. 13. Time decay of the surface relief grating in a film of azopolysiloxane modified with azophenol (95-98) %; Irradiation conditions: Fluence = 17 mJ/cm2, Intensity = 3.5 x 107 W/cm2

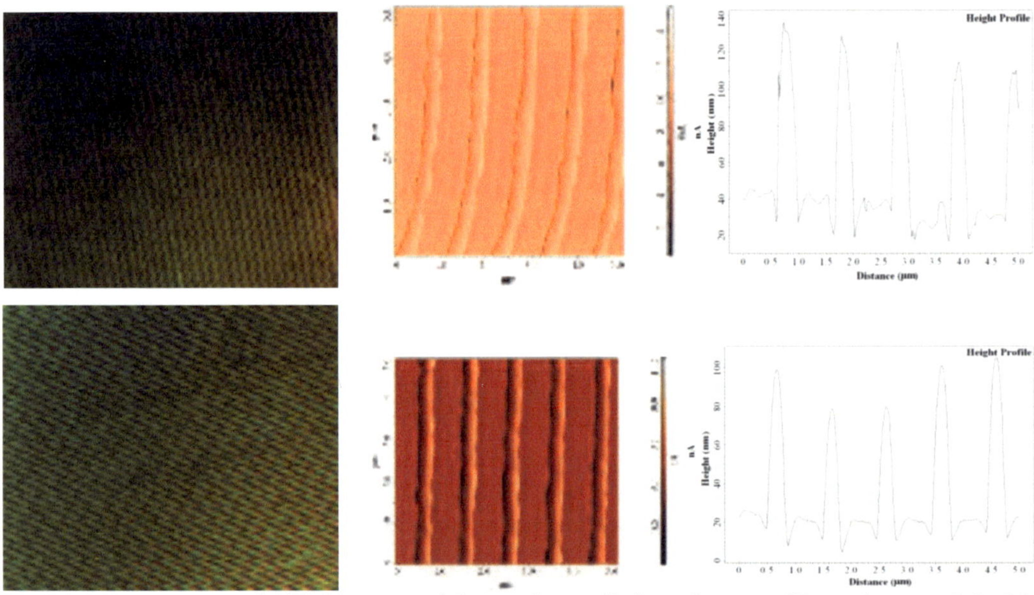

Fig. 14. Microscope and AFM images of the surface relief gratings on films of azo-polyimide (upper row) and azo-polysiloxane modified with thymine (lower row). The microscope images are registered 15 min. after irradiation time, the AFM images and profiles are registered after more than three month after irradiation time; irradiation conditions: fluence = 8.4 mJ/cm2 and 100 irradiation pulses.

AFM profiles of the surface relief induced under the action of an interference field with a medium fluence of 8.4 mJ/cm2 and 100 subsequent laser pulses are similar for both azo-polymers, with rigid and flexible main chain (Sava et al. 2008). The depth of the induced structure is about 90 nm for the azo-polymer film and 100 - 110 nm for the azoplyimide film (Fig. 14.). The difference is made by the evolution of the structure with the number of incident laser pulses, respectively irradiation time. After only 10 irradiation pulses the height of the "hills" formed on the surface of azo-polyimide was half from the height of the profiles induced on the azo-polysiloxane films. This fact could be the result of the rigid main chain of the azo-polyimide for which the molecular reorganization is slower. For both azo-polymers the AFM analyse was realized at about 3 month from the irradiation moment. The samples were preserved during this time at ambient temperatures between 23 - 35 °C at daily light. This indicates that the surface structuration was stable for a rather long time.

5. Conclusions

Two classes of polymeric films were analyzed from the point of view of the capability to induce single step surface relief modulation in the form of SRGs under the action of a UV interference field having as a light source pulsed laser radiation at 193 nm or 355 nm wavelength: photoresists and azopolymers. The incident laser fluence was lower than the ablation threshold of the material and the transversal profile of the induced structures has a continuous shape, without phase changes.

There were obtained SRGs with a pitch of 250 nm and 1 μm, depending on the irradiation set-up. The modulation depth was between 10 nm and 800 nm, depending on the incident fluence/intensity and the number of subsequent incident pulses. The surface relief modulation time is of the order of laser pulse duration (5 -7 ns). There were obtained surface relief gratings with sinusoidal profile on photoresist films. The obtained surface relief gratings had very good time stability from the point of view of the pitch and modulation depth. In case of the azopolymers the time stability of the SRG depends on the specific composition. For azopolysiloxane modified with azophenol (95-98) % the surface induced gratings begins to decay after 1-2 hours from the irradiation moment up to a complete loss of the structuration after 24 hour. A stable structure was obtained on the surface of films of azo-polyimide and azo-polysiloxane modified with thymine films. The surface structuration was monitored 3 month after irradiation and a good contrast of the surface relief structuration was observed. In case of azopolymers the single step surface relief modulation under the action of a light field is considered to be the consequences of the photo-induced conformational changes in the molecular chain. More generally the property of a polymeric material to have different configurations as a function of external stimuli (laser light in this case) offers the possibility to obtain surface relief structures in functional surface coatings with applications in biophysics, pharmaceutics, electronics and optoelectronics.

6. References

Apostol, I; Castex, M.C.; Logofatu, P.C.; Damian, V; Savu, B; Stanciu, G, Iordache, I; Garoi, F; M.-C. Castex, Apostol, I; P.C. Logofatu, V. Damian, B. Savu, G. Stanciu, I. Iordache, F. Garoi, (2006), Production and analyses of surface relief gratings with submicron period, Workshop on Laser Interface Interaction and Laser Cleanig, LIILAC 2006

Apostol, I.; Apostol, D.; Damian, V.; Iordache, I.; Hurduc, N.; Sava, I.; Sacarescu, L.; Stoica, I.; (2009), UV radiation induced surface modulation time evolution in polymeric materials, Proc. of SPIE Vol. 7366 73661U-1–8

Bolle, M.; Lazare, S.; Le Blanc, M.; Vilmes, A.; Submicron periodic structures produced on polymer surfaces with polarized excimer laser ultraviolet radiation, Appl. Phys. Lett. 60, 674 (1992)

Castex, M.C.; Oliveiro, C.; Fischer, A.; Mousel S.; Michelon, J.; Ades, D.; Siove, A.; (2002) Polycarbazoles microcavities: towards plastic blue lasers , *Appl. Surf. Sci*, 197-198, 822-825

Castex, M.C.; Fischer, A.; Simeonov, D.; Ades, D.; Siove, A. (2003), Réalisation de réseaux sur polymères par laser UV, *J. de Physique IV*, 108 173-177

Dyer, P.E.; Farley, R.J.; Giedl, R. (1996), Analysis and application of a 0/1 order Talbot interferometer for 193nm laser grating formation, *Optics Communications*, 129, 98-108.

Enea, R.; Apostol, I.; Damian, V.; Hurduc, N; Iordache I. (2008) a, Photo-sensible (thymine containing) azo-polysiloxanes: synthesis and light induced effects, IOP:Conf. Ser., vol. 100, 012022

Enea, R; Hurduc, N.; Apostol, I.; Damian, V.; Iordache, I.; Apostol, D.; (2008) b, The capacity of nucleobases azopolysiloxanes to generate a surface relief grating, JOAM, 10 (3), 541-545

Hiraoka, H. & Sendova, M., (1994), Laser induced sub-half-micrometer periodic structure on polymer surfaces, Appl. Phys. Lett. 64 (5), 31

Hurduc, N.; Enea, R.; Scutaru, D.; Sacarescu, L.; Donose, B.C.; Nguyen, A.V., Nucleobases modified azo-polysiloxanes, materials with potential application in biomolecules nanomanipulation -Journal of Polymer Science Part A: Polymer Chemistry, 45, Issue 18, 4240-4248, 2007)

Logofatu, P.C.; Apostol, I.; Castex, M.C.; Damian, V; Iordacche, I.; Bojan, M.; Apostol, D.; (2008) Proc. Of SPIE, Vol. 6617 -661717, 1-12

Naydenova, I.; Mihailova, E.; Martin, S.; Toal, V.; (2005), Holographic patterning of acrylamide based photopolymer surface, Optics Express, Vol. 13, No. 13, 4878

Pelissier, S.; Blancc, D.; Andrews, M.P.; Najafi, S. I.; Najafi, A.V..; Tishenko, A.V.; Parriaux, O.; (1999), Single step UV recording of a sinusoidal surface gratings in hybrid solgel glasses, Appl. Opt. 38, 6744-6748

Rochon, P.; Batalla, E.; Natansohn, A.; (1995), Optically induced surface gratings on azoaromatic polymer filme, Appl. Phys. Let., 66(2), 1995

Sava, I.; Sacarescu, L.; Stoica, I.; Apostol,I.; Damian, V.; Hurduc, N., (2008), Photocromic properties of polymide and polysiloxane azopolymers, Polym. Int. 58, 163 -170

Shishido, A.; Tsutsumi, O.; Kanazawa, A.; Shiono, T.; Ikeda, T.; Tamai, N.; (1997), Distinct Photochemical Phase-Transition Behavior of Azobenzene Liquid-Crystals Evaluated by Reflection-Mode Analysis, Journal of Physical Chemistry B 1997, 101, 2806-2810.

9

Unconventional Layer-by-Layer Assembly for Functional Organic Thin Films

Guanglu Wu and Xi Zhang
Tsinghua University
China

1. Introduction

The layer-by-layer (LbL) assembly is a powerful technique for fabricating multilayer thin films with controlled architecture and functions (Zhang & Shen, 1999; Decher & Schlenoff, 2002; Hammond, 2004). Although the research could be traced back to pioneering work of Iler in 1966 (Iler, 1966), this important work did not become public until it was rediscovered by Decher and Hong in the beginning of 1990s (Decher & Hong, 1991a, 1991b; Decher et al., 1992). Since then, the field of LbL has gained rapid progress. Besides electrostatic driven LbL assembly (Decher, 1997), many different intermolecular interactions, such as hydrogen bonding (Wang et al., 1997; Stockton & Rubner, 1997), charge transfer interaction (Shimazaki et al., 1997; Shimazaki et al., 1998), molecular recognition (Hong et al., 1993; Decher et al., 1994; Bourdillon et al., 1994; Lvov et al., 1995; Anzai et al., 1999), coordination interactions (Xiong et al., 1998), have been used as driving force for the multilayer buildup. In addition, layer-by-layer reactions have been also employed to construct robust multilayer thin films (Kohli et al., 1998; Major & Blanchard, 2001; Chan et al., 2002; Zhang et al., 2005; Such et al., 2006). Diversified building blocks have been used to construct LbL multilayer thin films, including polyelectrolytes (Kleinfeld & Ferguson, 1994), colloid and nanoparticles (Gao et al., 1994; Rogach et al., 2000; Fu et al., 2002a), dyes (Zhang et al.,1994; Sun et al., 1996), dendrimers (Zhang et al., 2003; Huo et al., 2003), clay minerals (Wei et al., 2007), carbon materials (Olek et al., 2004; Correa-Duarte et al., 2005), enzymes and proteins (Kong et al., 1994; Lvov & Moehwald, 2000; Sun et al., 2001), DNA (Lvov et al., 1993; Shchukin et al., 2004), viruses (Lvov et al., 1994) and so on. These building blocks can be fabricated into multilayer thin films simply by alternating deposition at liquid-solid interface, so-called conventional LbL assembly.

In order to fabricate single charged or water-insoluble building blocks, a series of unconventional LbL methods have been proposed. The key idea of these approaches includes more than one step in the assembly process, as shown in Figure 1. For example, the building blocks can self-assemble in solution to form molecular assemblies, and the molecular assemblies can be used as one of the building blocks subsequently for LbL assembly at liquid-solid interface. In this way, those building blocks which can not fabricated by conventional LbL assembly can be assembled by this unconventional LbL assembly. In addition, the unconventional LbL assembly can not only bring new structures but also endow the multilayer thin films with new functions (Zhang et al., 2007).

This chapter is to summarize different methods of unconventional LbL assembly, including electrostatic complex formation, hydrogen-bonded complex, block copolymer micelles and polymer-assisted complex. It will be noted that single charged or water-insoluble building blocks become self-assembling after these treatments in solution. When fabricating into multilayer thin films, this unconventional LbL assembly leads to development of new concept of surface imprinting, nanocontainers and nanoreactors.

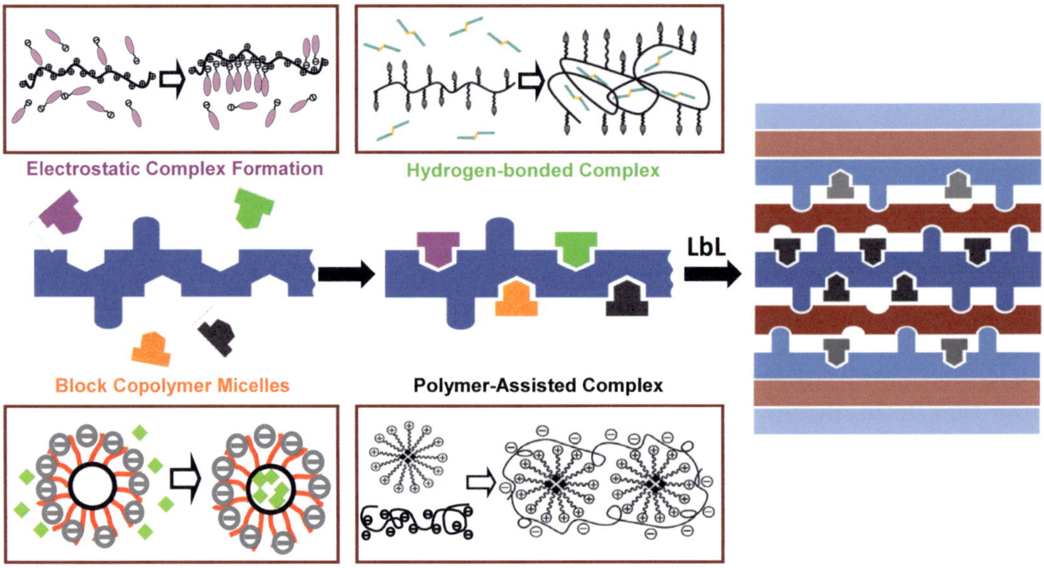

Fig. 1. Schematic illustration of unconventional LbL assembly.

2. Electrostatic complex formation

An electrostatic complex for the fabrication of LbL films can be described as follows. First, polyelectrolytes are mixed with counter-charged molecules in aqueous solution to form electrostatic complex; second, the complex are deposited alternatively with counter-polyelectrolyte to form LbL films. Electrostatic complex formation is a convenient way to fabricate LbL films with embedded charged organic molecules, including single-charged or oligo-charged (Fabianowski et al., 1998; Chang-Yen et al., 2002; Das and Pal, 2002; Nicol et al., 2003; Chen et al., 2005).

One typical example is the incorporation of single charged molecules, e.g. sodium 9-anthracenepropionate (SANP) into LbL films (Chen et al., 2005), as shown in Figure 2. This negatively charged moiety is used to form a macromolecular complex with positively charged poly(diallyldimethylammonium chloride) (PDDA), PDDA-SANP in short, and multilayer films are fabricated by alternating deposition of the PDDA-SANP complex with poly(4-styrenesulfonate) (PSS) at the liquid-solid interface. It is well known that small molecules, such as SANP, can diffuse into conventional LbL films of PDDA/PSS. However, the amount of SANP assembled in this method is much larger than that in the diffusion method, and moreover, a controllable amount of SANP can be incorporated by adjusting the initial concentration of SANP in the PDDA-SANP complex solution.

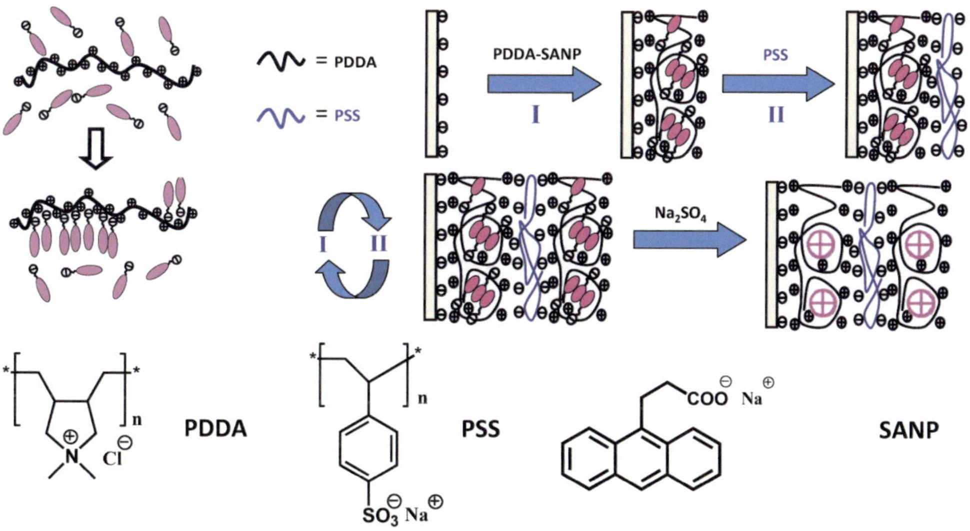

Fig. 2. Schematic illustration of the incorporation of single charged SANP into LbL film.

Can LbL films act as a nanoreactor? To answer this question, the LbL film of PDDA-SANP/PSS is a nice model system, since anthracene moiety in SANP can undergo photo-cycloaddition under UV irradiation. As shown in Figure 3, the characteristic absorbance of anthracene between 250 and 425 nm decreases with UV irradiation, at the same time the absorbance of benzene around 205 nm increases, which indicates that SANP moieties incorporated in the LbL film undergo photocycloaddition to produce a photocyclomer. Interestingly, the quantum yield of photocycloaddition is about four times higher than that in the solution. The reason such photocycloaddition occurs with an enhanced quantum yield should be correlated with the aggregations of SANP in the LbL films which facilitates the reaction.

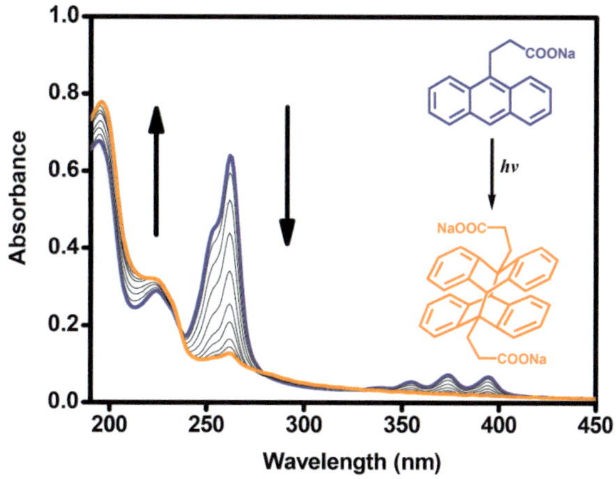

Fig. 3. Absorption spectra of PDDA-SANP/PSS multilayer film under UV irradiation for different times. Arrows indicate the transformation of the spectrum with increasing irradiation time.

It should be noted that the combination of macromolecular complexes and LbL deposition allows not only for incorporation of single charged moieties into LbL films, but also for controlled release of them from LbL films. For example, when immersing an LbL film of PDDA-SANP/PSS into an aqueous solution of Na_2SO_4, the SANP can be released from the film quickly depending on the ionic strength of the solution. An interesting finding is that after releasing SANP, the LbL film has been endowed the property of charge selectivity. That is to say, the as-prepared LbL film can readsorb only negatively charged moieties, whereas it repels positively charged moieties. As control experiment, small molecules can diffuse into normal LbL films of PDDA/PSS, however, either positively charged or negatively charged species can be equally incorporated, indicative of no charge selectivity. In addition, the loading capacity of SANP in a PDDA-SANP/PSS film is seven times higher than that in a PDDA/PSS film. Therefore, the LbL films fabricated by this unconventional LbL method can be used as materials of permselectivity.

We are wondering if the above unconventional LbL method can be extended to incorporate positive charged building blocks and to fabricate films that are able to readsorb only positively charged moieties, whereas it repels negatively charged moieties. For this purpose, 1-pyrenemethylamine hydrochloride (PMAH) is chosen as a positive charged moiety (Chen et al., 2007). Similar to the previous discussion on SANP, PMAH can be incorporated into LbL films by the unconventional LbL method that involves the electrostatic complex formation of PMAH and PSS in solution and alternating deposition between the complex and PDDA at liquid-solid interface. When immersing the LbL films of $(PDDA/PSS\text{-}PMAH)_{10}$ into Na_2SO_4 aqueous solution of varying concentration, PMAH can be released from the LbL films and the releasing rate depends on the concentration of Na_2SO_4 solution. At a high Na_2SO_4 concentration of 0.62 mol/L, PMAH can be released completely in about 90 s. However, at a low concentration of 6.2×10^{-3} mol/L, it takes nearly 500s for the completely release of PMAH. Notably, the LbL films after releasing PMAH can selectively readsorb positively charged moiety while repelling the opposite.

Not all small molecules are suitable templates for fabrication of LbL films that can trap ion of one sign of charge while repelling the opposite. We have tried different cations and anions and realized that single-charged molecules bearing condensed aromatic structures are good candidates. The reasons are listed as following: (1) Single-charged molecules can form complexes with polyelectrolytes and also unbind easily, which is an important factor for successful incorporation into LbL films as we have mentioned above. Molecules with two or more charges can hardly unbind from the polyelectrolytes. (2) The small molecules we used in our experiment have a hydrophilic group and a hydrophobic group with condensed aromatic moiety. When forming a complex in aqueous solution, the aromatic hydrophobic groups might get together due to hydrophobic interaction as well as the π-π stacking interaction.

3. Hydrogen bonding complex

Hydrogen-bonded LbL assembly was first demonstrated by Rubner and our group simultaneously in 1997 (Stockton & Rubner, 1997; Wang et al., 1997; Wang et al., 2000). Since then, various building blocks have been fabricated into thin film materials on the basis of hydrogen bonding (Fu et al., 2002b; Zhang et al., 2003; Zhang et al., 2004; Zhang et al., 2007). This method is suitable for building blocks with hydrogen donors and acceptors, and it can be feasible not only in the environment of aqueous solution but also in suitable organic solvent. Considering that hydrogen bonding is sensitive to environmental conditions, such as pH, the hydrogen-bonded LbL films can be erasable (Sukhishvili & Granick, 2000; Sukhishvili & Granick, 2002).

Inspired by the concept of unconventional LbL assembly, we attempt to develop unconventional method of LbL assembly on the basis of hydrogen bonding. It involves hydrogen-bonding complexation in solution and hydrogen-bonded LbL assembly at liquid-solid interface. The solvent used could be organic, which favors the formation of hydrogen-bonding. In this way, some water-insoluble small organic molecules can be loaded into multilayer thin films.

One of the examples of hydrogen-bonded unconventional LbL assembly is shown in Figure 4 (Zeng et al., 2007). First, a small organic molecule, bis-triazine (DTA) is mixed with poly(acrylic acid) (PAA) in methanol to form a hydrogen-bonding complex (PAA-DTA); second, LbL assembly is performed between the methanol solutions of PAA-DTA and diazo-resin (DAR), driven by hydrogen-bonding. In this way, DTA is loaded into the LbL film in a convenient and well-controlled manner. Since DAR is a photoreactive polycation, one can irradiate the film with UV light to convert the hydrogen bonding into covalent bond, therefore forming a stable multilayer film (Sun et al., 1998, 1999, 2000; Zhang et al., 2002).

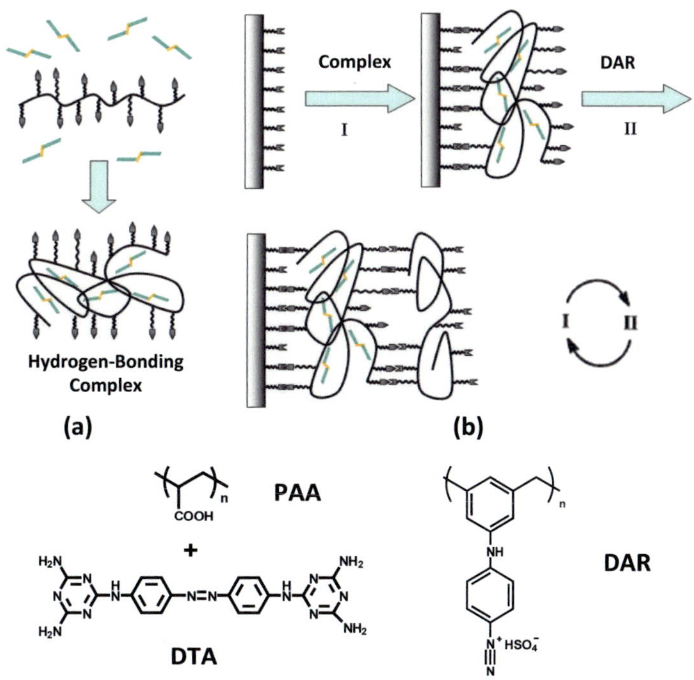

Fig. 4. Schematic illustration of hydrogen-bonded unconventional LbL assembly: Step 1, formation of hydrogen-bonding PAA-DTA complexes (a); Step 2, LbL assembly of PAA-DTA and DAR (b).

We have applied this method to a series of structurally related molecules with an increasing number of hydrogen bond donors and acceptors to find out the structural demand of the method. Our conclusion is only the molecules that can form multiple and strong hydrogen bonds with PAA are suitable for our method. One simple technique to test if molecules can interact with PAA strongly is described below: when mixing the molecules with PAA in solution, it means that there exist a strong interaction between the molecule and PAA if a floccule is formed. Therefore, those molecules are usually suitable for this unconventional LbL assembly.

4. Block copolymer micelles

Amphiphilic block copolymers are able to self-assemble into core–shell micellar structures in selective solvent. In order to take advantage of hydrophobic cores of the block copolymer micelles, we have incorporated water-insoluble molecules, e.g. pyrene, into the hydrophobic micellar cores of poly(styrene-b-acrylic acid) and then employed the loaded block copolymer micelles as building blocks for LbL assembly (Ma et al., 2005). As shown in Figure 5, the block copolymer micelles of poly(styrene-b-acrylic acid) with acrylic acid on the shell functioned as polyanions, allowing for LbL assembly by alternating deposition with polycations. This is certainly another unconventional LbL assembly that involves micellar formation in solution and use of loaded micelles for LbL deposition at liquid-solid interface. In this way, small water-insoluble molecules can be fabricated.

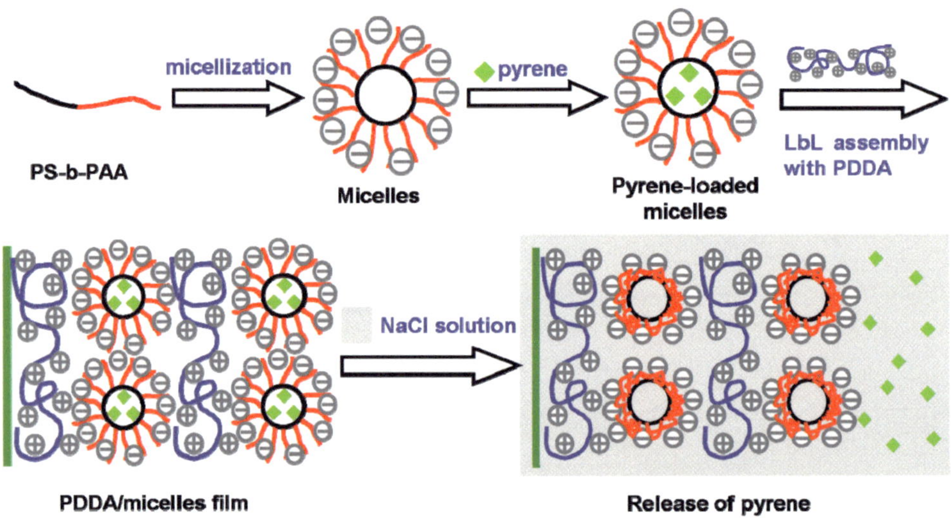

Fig. 5. Schematic illustration of the incorporation of pyrene into block copolymer micelles, LbL deposition of loaded micelles with PDDA, and the release of pyrene from the multilayer thin film.

The same concept can be extended to incorporate different water-insoluble molecules, such as azobenzene, for LbL assembly (Ma et al., 2006, 2007). It is well known that azobenzene can undergo a reversible photoisomerization under UV irradiation, but the rate of photoisomerization is faster in solution than in solid films. For a multilayer film of azobenzene loaded poly(styrene-b-acrylic acid) micelles and PDDA, we have found, interestingly, that the photoisomerization of the azobenzene in the multilayer film needs only several minutes, which is much faster than in normal solid films, but similar to that in dilute solutions, suggesting a way for enhancing the photophysical properties in the LbL films.

The above discussion concerns LbL films of block micelles when micelles are used to replace just one of the polyelectrolyte layers. The preparation of micelle-only multilayer is also possible. For this purpose, positively and negatively charged block copolymer micelles are needed as building blocks (Qi et al., 2006; Cho et al., 2006). For example, Block copolymer micelle/micelle multilayer films can be fabricated by alternating deposition of protonated poly(styrene-b-4-vinylpyrinde) and anionic poly(styrene-b-acrylic acid), as shown in Figure 6. The film growth is governed by electrostatic and hydrogen-bonding interactions between the block copolymer micelles. Multilayer films with antireflective and photochromic properties are obtained by incorporating water-insoluble photochromic (spiropyran) into the hydrophobic core (Cho et al., 2006). In addition, the micelle-only multilayer can be prepared not only on planar substrates but also on colloidal particulate substrates (Biggs et al., 2007).

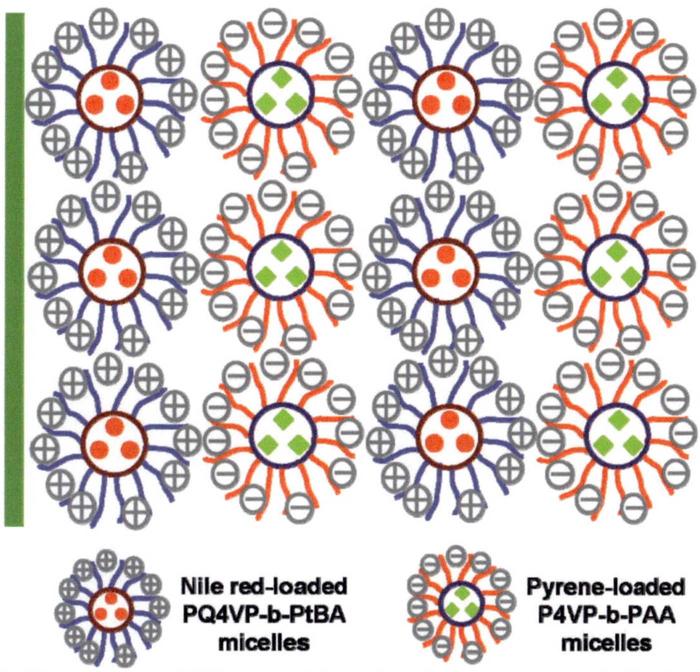

Fig. 6. Schematic illustration of LBL assembly of block copolymer micelle/micelle multilayer films with encapsulated guests.

The stability of micelles formed by low molecular weight surfactant is lower than block copolymer micelles, which usually cannot be used for LbL deposition. To improve the stability of micelles, a strategy is put forward that involves the use of polyelectrolyte to stabilize the micelles, which will be discussed in the following section.

5. Polymer-assisted complex

Polymer-assisted complex can be formed by the complexation of polymer with organic or inorganic components in solution through weak interaction such as electrostatic interactions, hydrogen-bonds, coordination interactions, guest-host interactions and so on. It has been demonstrated that diversified polymer-assisted complexes can be used as building blocks for the unconventional LbL assembly of multilayer thin films with well-tailored structures and functionalities, including polyelectrolyte-stabilized surfactant (Liu et al., 2008), polymeric complexes (Zhang & Sun, 2009; Liu et al., 2009; Guo et al., 2009), organic/inorganic hybrid complexes (Zhang et al., 2008).

Instead of using block copolymer micelles mentioned above as containers, Sun and co-workers found that the inexpensive polyelectrolyte-stabilized surfactant could be used as containers for noncharged species. For instance, they used this unconventional LbL assembly to realize the incorporation of noncharged pyrene molecules into multilayer films (Liu et al., 2008). First, noncharged pyrene molecules were encapsulated into the hydrophobic cores of the commonly used micelles formed by cetyltrimethylammonium bromide (CTAB); Second, the pyrene-loaded CTAB micelles were complexed with poly(acrylic acid) to obtain PAA-stabilized CTAB micelles, noted as PAA-(Py@CTAB), as shown in Figure 7; Then PAA-(Py@CTAB) were alternately deposited with PDDA through electrostatic interaction to produce PAA-(Py@CTAB)/PDDA multilayer thin film. As a consequence, pyrene molecules were firmly incorporated in the PAA-(Py@CTAB)/PDDA films with a high loading capacity. The assisted polymer plays an important role in stabilizing the micelles because CTAB micelles without assisted polymer can disassemble during the LbL deposition process. Considering that the surfactant micelles and polyelectrolytes are easily available, it is anticipated that this method can be extended to a wide range of polyelectroyte-stabilized surfactant micelles and will open a general and cost-effective avenue for the fabrication of advanced film materials containing noncharged species, such as organic molecules, nanoparticles and so forth by using LbL assembly technique.

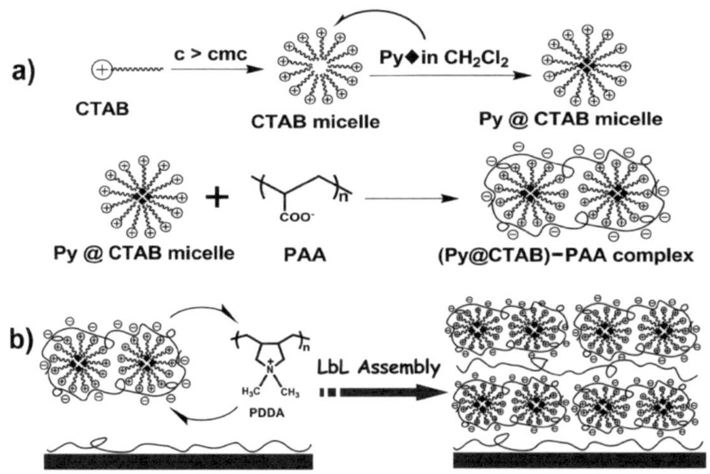

Fig. 7. (a) Preparative process of PAA-stabilized Py@CTAB micelles. (b) LbL deposition process for fabrication of PAA-(Py@CTAB)/PDDA multilayer films.

LbL assembled porous films could be hardly fabricated through conventional LbL assembly by directly alternate deposition of oppositely charged polyelectrolytes because of the flexibility of polyelectrolytes, which tends to close up any pre-designed pores and produce thin and compact films. However, by firstly preparing the polyelectrolyte complexes of negatively charged PAA and DAR (noted as PAA-DAR) and positively charged DAR and PSS (noted as DAR-PSS) as building blocks for further LbL assembly, a robust macroporous foam coating could be rapidly fabricated by direct LbL deposition of PAA-DAR and DAR-PSS complexes combined with subsequent photocross-linking (Zhang & Sun, 2009). These macroporous PAA-DAR/DAR-PSS foam coatings have a high loading capacity toward cationic dyes and can be used for dye removal from wastewater because of the large surface area and the abundance of negatively charged carboxylate and sulfonate groups provided by the foam coatings.

In addition of electrostatic interaction, hydrogen-bonded interaction could be also employed to form the polymer-assisted complex. For instance, poly(vinylpyrrolidione) (PVPON) and PAA could pe-assemble to polymeric complex through hydrogen-bonding interaction (denoted PVPON-PAA). Then, the pre-assembly complex could fabricate with poly(methacrylic acid) (PMAA) to a micrometre-thick PVPON-PAA/PMAA film with hierarchical micro- and nanostructures. After chemical vapor deposition of a layer of fluoroalkylsilane on top of the as-prepared multilayer thin film, superhydrophobic coatings were conveniently fabricated (Liu et al., 2009). The structure of the as-prepared PVPON-PAA/PMAA films could be well tailored by the mixing ratio of the PVPON-PAA complexes and the film preparative process. A non-drying LbL deposition process is critically important to realize the rapid fabrication of PVPON-PAA/PMAA films with hierarchical structures because the spherical structure of the PVPON&PAA complexes can be well preserved during film fabrication. In contrast, A N_2 drying step during LbL deposition process can produce a lateral shearing force, which produces thin and smooth films because of the spread and flattening of the PVPON-PAA complexes.

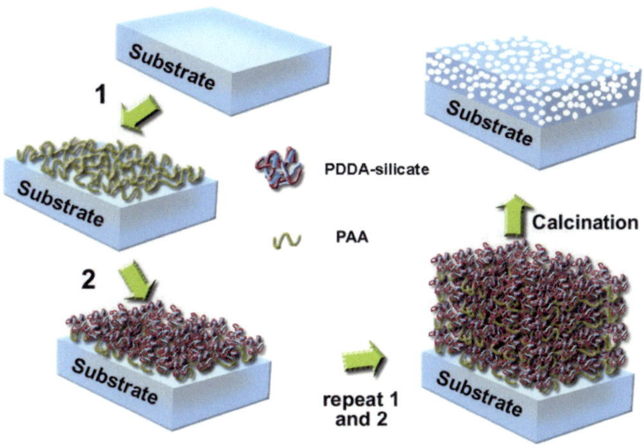

Fig. 8. Schematic illustration of the LbL deposition of PDDA-silicate complexes and PAA for fabrication of antireflection and antifogging coatings.

Besides polymeric complexes, polymer-assisted organic/inorganic hybrid complexes can be also assembled with counter species through unconventional electrostatic LbL assembly to fabricate functional film materials. As shown in Figure 8, complexes of PDDA and sodium silicate (PDDA-silicate) were alternately deposited with PAA to fabricate PAA/PDDA-silicate multilayer thin films (Zhang et al., 2008). The removal of the organic components in the PAA/PDDA-silicate mulilayer films through calcination produces highly porous silica coatings with excellent mechanical stability and good adhesion to substrates. Quartz substrates covered with such porous silica coatings exhibit both antireflection and antifogging properties because of the reduced refractive index and superhydrophilicity of the resultant films. A maximum transmittance of 99.8% in the visible spectral range is achieved for the calcinated PAA/PDDA-silicate films deposited on quartz substrates. The use of PDDA-assisted silicate complexes instead of simplex sodium silicate can largely increase the ratio of the organic components in the LbL-deposited organic/inorganic hybrid films and therefore enhance the porosity of the calcinated films, which favors the fabrication of antireflection and antifogging coatings with enhanced performance.

6. Surface imprinting LbL film

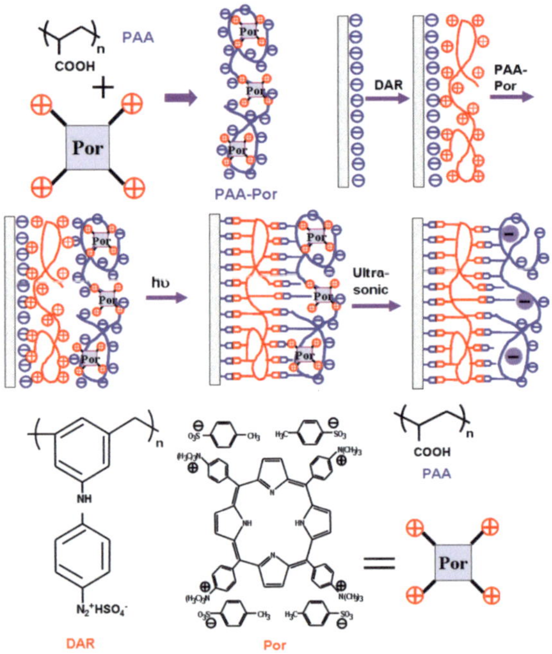

Fig. 9. Schematic illustration of the formation of the imprinting complexes (PAA-Por) and the experimental procedure for the formation of surface imprints of multilayer thin film.

The unconventional LbL assembly is not only used for assembling building blocks which cannot be assembled by conventional method, but also for introducing new functions. Among them, surface imprinting in LbL nanostructured films is one typical example. It is well-known that molecularly imprinting polymers provide a general means to generate specific binding sites in polymer matrices (Wulff & Sarhan, 1972; Vlatakis et al., 1993; Zimmerman et al., 2002; Haupt & Mosbach, 2000; Wulff, 2002; Komiyama et al., 2002). However, they suffer from basic limitations associated with the limited concentration of imprinted sites, and the bulk volume of the polymer matrices that requires long diffusion paths of the imprinted host molecules. In fact, a few previous reports have addressed the possibility of imprinting molecular-recognition sites in monolayer systems (Kempe et al., 1995; Lee et al. 1998; Lahav et al., 1999), but the effectiveness of these systems and their utility are limited. Surface imprinting LbL film can provide a solution to solve these problems, therefore opening a new avenue for surface imprinting with enhanced efficiency.

A general procedure for preparation of surface imprinting LbL films includes four steps. Taking the generation of imprinted sites for the porphyrin derivative (Por) as an example, as shown in Figure 9, firstly, an electrostatically stabilized complex between the positively charged porphyrin and PAA is formed in aqueous solution. Secondly, multilayer thin film is fabricated by alternating deposition of the complex PAA-Por and photoactive DAR. Thirdly, the layered structure is photo cross-linked to yield the covalent bridging of the layers by UV irradiation. In the final step, the template porphyrin molecules are washed off from the film to yield the surface imprinted matrix (Shi et al., 2007).

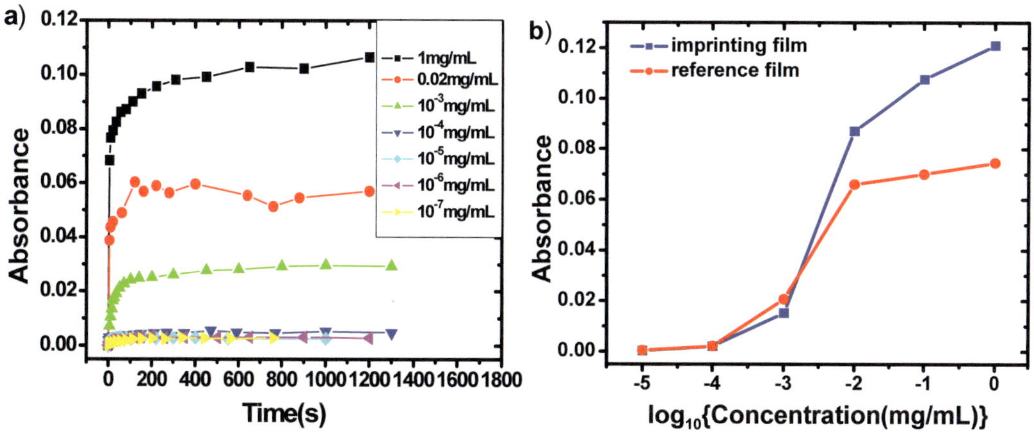

Fig. 10. a) Time-dependent absorbance of the Por association to the imprinted film at different bulk concentration of Por; b) Isothermal absorption curve of the Por relative to the imprinted film (blue curve) and the reference film (red curve).

The surface imprinting LbL films have advantage over other method in terms of thermodynamics and kinetics. As shown in Figure 10(a), the rate of binding of Por to the imprinted film is very fast, and the loading process reaches a saturation value at less than two min. Moreover, the absorbance of the saturation value of Por increases upon elevating the bulk concentration of Por, which indicates that the binding of Por is concentration-dependent. Furthermore, the formation of different saturation values for the absorbance of Por bound to the polymer at different bulk concentrations of Por implies that the association of Por is an equilibrium process. Knowing the saturation value of bound Por at different bulk concentrations, the binding constant of Por to the imprinted site is estimated to be 2×10^5 M^{-1}. The isothermal absorption is shown in Figure 10(b) by relating the absorbance of Por at 30 min against the concentration of Por. When the concentration of Por is lower than 10^{-2} mg mL^{-1}, the absorption of Por is similar in the imprinted film and the reference film. However, upon increasing the concentration of Por to 10^{-2} mg mL^{-1}, we observe that the imprinted film absorbs substantially more Por than the reference film.

Further support that the film binds the Por substrate to specific imprinted sites, rather than by sole electrostatic interactions, is obtained by the electrochemical probing of the association of the positively charged $Ru(NH_3)_6^{3+}$ label to the LbL film before and after exclusion of the template, Por, from the polymer. No redox response is observed for the $Ru(NH_3)_6^{3+}$ label when the polymeric film is loaded with the template Por, implying that the film insulates the interfacial electron transfer to the redox label, Figure 11(curve 1). Exclusion of the template results in the electrical response of $Ru(NH_3)_6^{3+}$, Figure 11(curve 2). That is, after exclusion of the template, the film is permeable to the redox label and the positively charged units bind to the negatively charged empty sites from which the Por is removed. Interaction of the $Ru(NH_3)_6^{3+}$ loaded polymer with the imprinted substrate, Por, results in a decrease in the electrical response of $Ru(NH_3)_6^{3+}$, implying that the redox label is competitively displaced by Por. That is, Por exhibits a substantially higher affinity for the polymer as compared to $Ru(NH_3)_6^{3+}$, Figure 11(curves 3 and 4). This is attributed to the fact that the imprinted sites in the film can sterically accommodate Por, in addition to its electrostatic stabilization by the negative charges associated with the polymer.

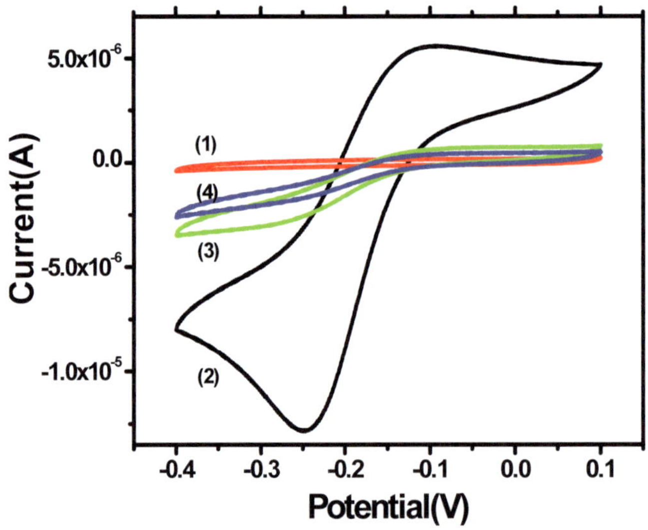

Fig. 11. Cyclic voltammograms of the gold electrodes modified by $(DAR/PAA\text{-}Por)_5DAR$ multilayer films in the presence of an electrolyte solution consisting of 3 mM $Ru(NH_3)_6^{3+}$ and 0.1 M KCl; 1) after UV irradiation and in the presence of bound Por; 2) removal of Por from the film by ultrasonic agitation in the ternary solution; 3, 4) after immersing the unloaded film in a mixed solution of 0.2 $mgmL^{-1}$ Por, 2.4 mM $Ru(NH_3)_6^{3+}$,and 0.08 M KCl for 2 min and 20 min, respectively.

To improve the selectivity of surface imprinting LbL films, we have attempted to introduce the cooperativity of various specific interactions within the binding sites. We choose theophylline derivatives as the model template molecules to investigate the feasibility of our method in the fabrication of surface imprinting LbL films. Theophylline-7-acetic acid (THAA) is covalently conjugated to polyelectrolyte PAA with a cystamine bridge by amide linkage to form precursor assemblies PAAtheo15, which is a PAA with 15% of its carboxylic acid grafted of THAA. The disulfide bond moiety of the cystamine bridge can introduce extra recognition sites as the thiol group obtained from reduction of disulfide is able to form hydrogen-bonding interactions with hydroxyl groups of the guest molecules. This additional mercapto recognition site, combined with the other hydrogen-bonding interactions established through template incorporation and film construction, renders selectivity for the nanostructured binding cavities in the LbL film (Niu et al., 2007; Niu et al., 2008).

7. Conclusions

The LbL assembly has experienced several stages of development: extension of various building blocks, LbL method driven by different driving forces, combination of layered nanoarchitectures and functional assemblies, and the unconventional LbL assembly as summarized and discussed in this chapter. In general, the unconventional LbL assembly includes supramolecular assembly in solution and LbL deposition at liquid-solid interface. Therefore, it can be regarded as one of the multi-level assembly. As you can see, the unconventional LbL method brings not only new supramolecular structures but also functions. However, no matter whether the conventional or the unconventional LbL method is employed, each method has its own scope of applications as well as limitations. The combination of different methods may facilitate the assembly of thin film materials with complex and elaborate structures for the integration of functionalities.

8. References

Anzai, J.;Kobayashi, Y.; Nakamura, N.; Nishimura, M. & Hoshi, T. (1999). Layer-by-Layer construction of multilayer thin films composed of avidin and biotin-labeled Poly(amine)s. *Langmuir,* 1999, 15, 221-226.

Biggs, S.; Sakai, K.; Addison, T.; Schmid, A.; Armes, S. P.; Vamvakaki, M.; Bütün, V. & Webber, G. (2007). Layer-by-Layer formation of smart particle coatings using oppositely charged block copolymer micelles. *Adv. Mater.,* 2007, 19, 247–250.

Bourdillon, C.; Demaille, C.; Moiroux, J. & Savéant, J. M. (1994). Step-by-Step immunological construction of a fully active multilayer enzyme electrode. *J. Am. Chem. Soc.,* 1994, 116, 10328-10329.

Chan, E. W. L.; Lee, D. C.; Ng, M. K.; Wu, G. H.; Lee, K. Y. C. & Yu, L. P. (2002). A novel Layer-by-Layer approach to immobilization of polymers and nanoclusters. *J. Am. Chem. Soc.,* 2002, 124, 12238-12243.

Chang-Yen, A.; Lvov, Y.; McShane, M. J. & Gale, B. K. (2002). Electrostatic self-assembly of a ruthenium-based oxygen sensitive dye using polyion–dye interpolyelectrolyte formation. *Sens. Actuators, B,* 2002, 87, 336-345.

Chen, H.; Zeng, G. H.; Wang, Z. Q.; Zhang, X.; Peng, M. L.; Wu, L. Z. & Tung, C. H. (2005). To combine precursor assembly and Layer-by-layer deposition for incorporation of single-charged species: nanocontainers with charge-selectivity and nanoreactors. *Chem. Mater.,* 2005, 17, 6679-6685.

Chen, H.; Zeng, G. H.; Wang, Z. Q. & Zhang, X. (2007). To construct "ion traps" for enhancing the permselectivity and permeability of polyelectrolyte multilayer films. *Macromolecules,* 2007, 40, 653-660.

Cho, J.; Hong, J.; Char, K. & Caruso, F. (2006). Nanoporous block copolymer micelle/micelle multilayer films with dual optical properties. *J. Am. Chem. Soc.*, 2006, 128, 9935–9942.

Correa-Duarte, M. A.; Kosiorek, A.; Kandulski, W.; Giersig, M. &,Liz-Marzán, L. M. (2005). Layer-by-Layer assembly of multiwall carbon nanotubes on spherical colloids. *Chem. Mater.*, 2005, 17, 3268-3272.

Das, S. & Pal, A. J. (2002). Layer-by-Layer self-assembling of a low molecular weight organic material by different electrostatic adsorption processes. *Langmuir*, 2002, 18, 458-461.

Decher, G. & Hong, J. D. (1991a). Buildup of ultrathin multilayer films by a self-assembly process: I. Consecutive adsorption of anionic and cationic bipolar amphiphiles. *Makromol. Chem., Macromol. Symp.*, 1991, 46, 321-327.

Decher, G. & Hong, J. D. (1991b). Buildup of ultrathin multilayer films by a self-assembly process: II. Consecutive adsorption of anionic and cationic bipolar amphiphiles and polyelectrolytes on charged surfaces. *Ber. Bunsenges. Phys. Chem.*, 1991, 95, 1430-1434.

Decher, G.; Hong, J. D. & Schmitt, J. (1992). Buildup of ultrathin multilayer films by a self-assembly process: III. Consecutively alternating adsorption of anionic and cationic polyelectrolytes on charged surfaces. *Thin Solid Films*, 1992, 210/211, 831-835.

Decher, G.; Lehr, B.; Lowack, K.; Lvov, Y. & Schmitt, J. (1994). New nanocomposite films for biosensors: layer-by-layer adsorbed films of polyelectrolytes, proteins or DNA. *Biosens. Bioelectron.*, 1994, 9, 677-684.

Decher, G. (1997). Fuzzy nanoassemblies: toward layered polymeric multicomposites. *Science*, 1997, 277, 1232-1237.

Decher, G. & Schlenoff, J. B. (2002). *Multilayer Thin Films – Sequential Assembly of Nanocomposite Materials.* Wiley-VCH, Weinheim, Germany 2002.

Fabianowski, W.; Roszko, M. & Brodziñska, W. (1998). Optical sensor with active matrix built from polyelectrolytes–smart molecules mixture. *Thin Solid Films*, 1998, 327–329, 743- 747.

Fu, Y.; Xu, H.; Bai, S. L.; Qiu, D. L.; Sun, J. Q.; Wang, Z. Q. & Zhang, X. (2002a). Fabrication of a stable polyelectrolyte/Au nanoparticles multilayer film. *Macromol. Rapid Commun.*, 2002, 23, 256-259.

Fu, Y.; Chen, H.; Qiu, D. L.; Wang, Z. Q. & Zhang, X. (2002b) Multilayer assemblies of poly(4-vinylpyridine) and poly(acrylic acid) bearing photoisomeric spironaphthoxazine via hydrogen bonding. *Langmuir*, 2002, 18, 4989-4995.

Gao, M. Y.; Gao, M. L.; Zhang, X.; Yang, Y.; Yang, B. & Shen, J. C. (1994). Constructing PbI_2 nanoparticles into a multilayer structure using the molecular deposition (MD) method. *J. Chem. Soc., Chem. Commun.*, 1994, 2777-2778.

Guo, Y. M.; Geng, W. & Sun, J. Q. (2009). Layer-by-layer deposition of polyelectrolyte-polyelectrolyte complexes for multilayer film fabrication. *Langmuir*, 2009, 25, 1004-1010.

Hammond, P. T. (2004). Form and function in multilayer assembly: new applications at the nanoscale. *Adv. Mater.*, 2004, 16, 1271–1293.

Haupt, K. & Mosbach, K. (2000). Molecularly imprinted polymers and their use in biomimetic sensors. *Chem. Rev.*, 2000, 100, 2495–2504.

Hong, J. D.; Lowack, K.; Schmitt, J. & Decher G. (1993). Layer-by-layer deposited multilayer assemblies of polyelectrolytes and proteins: From ultrathin films to protein arrays. *Prog. Colloid Polym. Sci.*, 1993, 93, 98-102.

Huo, F. W.; Xu, H. P.; Zhang, L.; Fu, Y.; Wang, Z. Q. & Zhang, X. (2003). Hydrogen-bonding based multilayer assemblies by self-deposition of dendrimer. *Chem. Commun.*, 2003, 874-875.

Iler, R. K. (1966). Multilayers of colloidal particles. *J. Colloid Interface Sci.*, 1966, 21(6), 569-594.

Kempe, M.; Glad, M. & Mosbach, K. (1995). An approach towards surface imprinting using the enzyme ribonuclease A. *J. Mol. Recognition*, 1995, 8, 35-39.

Kleinfeld, E. R. and Ferguson, G. S. (1994). Stepwise formation of multilayered nanostructural films from macromolecular precursors. *Science*, 1994, 265, 370-373.

Kohli, P.; Taylor, K. K.; Harris, J. J. & Blanchard, G. J. (1998). Assembly of covalently-coupled disulfide multilayers on gold. *J. Am. Chem. Soc.*, 1998, 120, 11962-11968.

Komiyama, M.; Takeuchi, T.; Mukawa, T. & Asanuma, H. (2002). *Molecular Imprinting*. Wiley-VCH, Weinheim, Germany 2002.

Kong, W.; Zhang, X.; Gao, M. L.; Zhou, H.; Li, W. & Shen, J. C. (1994). A new kind of immobilized enzyme multilayer based on cationic and anionic interaction. *Macromol. Rapid Commun.*, 1994, 15, 405-409.

Lahav, M.; Katz, E.; Doron, A.; Patolsky, F. & Willner, I. (1999). Photochemical imprint of molecular recognition sites in monolayers assembled on Au electrodes. *J. Am. Chem. Soc.*, 1999, 121, 862–863.

Lee, S. W.; Ichinose, I. & Kunitake, T. (1998). Molecular imprinting of azobenzene carboxylic acid on a TiO_2 ultrathin film by the surface sol-gel process. *Langmuir*, 1998, 14, 2857-2863.

Liu, X. K.; Zhou, L.; Geng, W. & Sun, J. Q. (2008). Layer-by-layer-assembled multilayer films of polyelectrolyte-stabilized surfactant micelles for the incorporation of noncharged organic dyes. *Langmuir*, 2008, 24, 12986-12989.

Liu, X. K.; Dai, B. Y.; Zhou, L. & Sun, J. Q. (2009). Polymeric complexes as building blocks for rapid fabrication of layer-by-layer assembled multilayer films and their application as superhydrophobic coatings. *J. Mater. Chem.*, 2009, 19, 497–504.

Lvov, Y.; Decher, G. & Sukhorukov, G. (1993). Assembly of thin films by means of successive deposition of alternate layers of DNA and poly(allylamine). *Macromolecules*, 1993, 26, 5396–5399.

Lvov, Y.; Haas, H.; Decher, G. & Möhwald, H. (1994). Successive deposition of alternate layers of polyelectrolytes and a charged virus. *Langmuir*, 1994, 10, 4232-4236.

Lvov, Y.; Ariga, K.; Ichinose, I. & Kunitake, T. (1995). Layer-by-layer architectures of concanavalin A by means of electrostatic and biospecific interactions. *J. Chem. Soc., Chem. Commun.*, 1995, 2313-2314.

Lvov, Y. & Moehwald, H. (2000). *Protein Architectures: Interfacing Molecular Assemblies and Immobilization Biotechnology*, Marcel Dekker, New York 2000.

Ma, N.; Zhang, H. Y.; Song, B.; Wang, Z. Q. & Zhang, X. (2005). Polymer micelles as building blocks for Layer-by-Layer assembly: an approach for incorporation and controlled release of water-insoluble dyes. *Chem. Mater.*, 2005, 17, 5065-5069.

Ma, N.; Wang, Y. P.; Wang, Z. Q. & Zhang, X. (2006). Polymer micelles as building blocks for the incorporation of azobenzene: enhancing the photochromic properties in Layer-by-Layer films. *Langmuir*, 2006, 22, 3906-3909.

Ma, N.; Wang, Y. P.; Wang, B. Y.; Wang, Z. Q.; Zhang, X.; Wang, G. & Zhao, Y. (2007). Interaction between block copolymer micelles and azobenzene-containing surfactants: from coassembly in water to Layer-by-Layer assembly at the interface. *Langmuir*, 2007, 23, 2874-2878

Major, J. S. & Blanchard, G. J. (2001). Covalently bound polymer multilayers for efficient metal ion sorption. *Langmuir*, 2001, 17, 1163-1168.

Nicol, E.; Habib-Jiwan, J. L. & Jonas, A. M. (2003). Polyelectrolyte multilayers as nanocontainers for functional hydrophilic molecules. *Langmuir*, 2003, 19, 6178-6186.

Niu, J.; Shi, F.; Liu, Z.; Wang, Z. Q. & Zhang, X. (2007). Reversible disulfide cross-linking in Layer-by-Layer films: preassembly enhanced loading and pH/reductant dually controllable release. *Langmuir*, 2007, 23, 6377-6384.

Niu, J.; Liu, Z. H.; Fu, L.; Shi, F.; Ma, H. W.; Ozaki, Y. & Zhang, X. (2008). Surface imprinted nanostructured layer-by-layer film for molecular recognition of theophylline–derivatives. *Langmuir*, 2008, 24, 11988–11994.

Olek, M.; Ostrander, J.; Jurga, S,; Moehwald, H.; Kotov, N.; Kempa, K. & Giersig, M. (2004). Layer-by-layer assembled composites from multiwall carbon nanotubes with different morphologies. *Nano Lett.*, 2004, 4, 1889-1895.

Qi, B.; Tong, X. & Zhao, Y. (2006). Layer-by-Layer assembly of two different polymer micelles with polycation and polyanion coronas. *Macromolecules*, 2006, 39, 5714-5719.

Rogach, A. L.; Koktysh, D. S.; Harrison, M. & Kotov, N. A. (2000). Layer-by-Layer assembled films of HgTe nanocrystals with strong infrared emission. *Chem. Mater.*, 2000, 12, 1526–1528.

Shchukin, D. G.; Patel, A. A.; Sukhorukov, G. B. & Lvov, Y. M. (2004). Nanoassembly of biodegradable microcapsules for DNA encasing. *J. Am. Chem. Soc.*, 2004, 126, 3374-3375.

Shi, F.; Liu, Z.; Wu, G. L.; Zhang, M.; Chen, H.; Wang, Z. Q. & Zhang, X. (2007). Surface imprinting in Layer-by-Layer nanostructured films. *Adv. Funct. Mater.* 2007, 17, 1821–1827.

Shimazaki, Y.; Mitsuishi, M.; Ito, S. & Yamamoto, M. (1997). Preparation of the Layer-by-Layer deposited ultrathin film based on the charge-transfer interaction. *Langmuir*, 1997, 13, 1385-1387.

Shimazaki, Y.; Mitsuishi, M.; Ito, S. & Yamamoto, M. (1998). Preparation and characterization of the Layer-by-Layer deposited ultrathin film based on the charge-transfer interaction in organic solvents. *Langmuir*, 1998, 14, 2768-2773.

Stockton, W. B. & Rubner, M. F. (1997). Molecular-level processing of conjugated polymers. 4. Layer-by-Layer manipulation of polyaniline via hydrogen-bonding interactions. *Macromolecules*, 1997, 30, 2717-2725.

Such, G. K.; Quinn, J. F.; Quinn, A.; Tjipto, E. & Caruso, F. (2006). Assembly of ultrathin polymer multilayer films by click chemistry. *J. Am. Chem. Soc.*, 2006, 128, 9318-9319.

Sukhishvili, S. A. & Granick, S. (2000). Layered, erasable, ultrathin polymer films. *J. Am. Chem. Soc.*, 2000, 122, 9550–9551.

Sukhishvili, S. A. & Granick, S. (2002). Layered, erasable polymer multilayers formed by hydrogen-bonded sequential self-assembly. *Macromolecules,* 2002, 35, 301-310.
Sun, Y. P.; Zhang, X.; Sun, C. Q.; Wang, Z. Q.; Shen, J. C.; Wang, D. J. & Li, T. L. (1996). Supramolecular assembly of alternating porphyrin and phthalocyanine layers based on electrostatic interactions. *Chem. Commun.*, 1996, 2379-2380.
Sun, J. Q.; Wu, T.; Sun, Y. P.; Wang, Z. Q.; Zhang, X.; Shen, J. C. & Cao, W. X. (1998). Fabrication of a covalently attached multilayer via photolysis of layer-by-layer self-assembled films containing diazo-resins. *Chem. Commun.*, 1998, 1853-1854.
Sun, J. Q.; Wang, Z. Q.; Sun, Y. P.; Zhang, X. & Shen, J. C. (1999). Covalently attached multilayer assemblies of diazo-resins and porphyrins. *Chem. Commun.*, 1999, 693-694.
Sun, J. Q.; Cheng, L.; Liu, F.; Dong, S. J.; Wang, Z. Q.; Zhang, X. & Shen, J. C. (2000). Covalently attached multilayer assemblies containing photoreactive diazo-resins and conducting polyaniline. *Colloids Surf., A,* 2000, 169, 209-217.
Sun, J. Q.; Sun, Y. P.; Wang, Z. Q.; Sun, C. Q.; Wang, Y.; Zhang, X. & Shen, J. C. (2001). Ionic self-assembly of glucose oxidase with polycation bearing Os complex. *Macromol. Chem. Phys.*, 2001, 202, 111-116.
Vlatakis, G.; Andersson, L. I.; Müller, R. & Mosbach, K. (1993). Drug assay using antibody mimics made by molecular imprinting. *Nature,* 1993, 361, 645 - 647.
Wang, L. Y.; Wang, Z. Q.; Zhang, X. & Shen, J. C. (1997). A new approach for the fabrication of an alternating multilayer film of poly(4-vinylpyridine) and poly(acry1ic acid) based on hydrogen bonding. *Macromol. Rapid Commun.,* 1997, 18, 509-514.
Wang, L. Y.; Cui, S. X.; Wang, Z. Q.; Zhang, X.; Jiang, M.; Chi, L. F. & Fuchs, H. (2000). Multilayer assemblies of copolymer PSOH and PVP on the basis of hydrogen bonding. *Langmuir,* 2000, 16, 10490-10494.
Wei, H.; Ma, N.; Shi, F. Wang, Z. Q. & Zhang, X. (2007). Artificial nacre by alternating preparation of Layer-by-Layer polymer films and $CaCO_3$ strata. *Chem. Mater.*, 2007, 19, 1974-1978.
Wulff, G & Sarhan, A. (1972). Use of polymers with enzyme-analgous structures for the resolution of raeemates. *Angew. Chem. Int. Ed. Engl,* 1972, 11, 341-344.
Wulff G. (2002). Enzyme-like catalysis by molecularly imprinted polymers. *Chem. Rev.,* 2002, 102, 1–28.
Xiong, H. M.; Cheng, M. H.; Zhou, Z.; Zhang, X. & Shen, J. C. (1998). A new approach to the fabrication of a self-organizing film of heterostructured polymer/Cu_2S nanoparticles. *Adv. Mater.,* 1998, 10, 529-532.
Zeng, G. H.; Gao, J.; Chen, S. L.; Chen, H.; Wang, Z. Q. & Zhang, X. (2007). Combining hydrogen-bonding complexation in solution and hydrogen-bonding-directed Layer-by-Layer assembly for the controlled loading of a small organic molecule into multilayer films. *Langmuir,* 2007, 23, 11631-11636.
Zhang, X.; Gao, M. L.; Kong, X. X.; Sun, Y. P. & Shen, J. C. (1994). Build-up of a new type of ultrathin film of porphyrin and phthalocyanine based on cationic and anionic electrostatic attraction. *J. Chem. Soc. Chem. Commun.*, 1994, 1055-1056.
Zhang, X. & Shen, J.C. (1999). Self-Assembled ultrathin films: from layered nanoarchitectures to functional assemblies. *Adv. Mater.*, 1999, 11, 1139–1143.

Zhang, X.; Wu, T.; Sun, J. Q. & Shen, J. C. (2002). Ways for fabricating stable layer-by-layer self-assemblies:combined ionic self-assembly and post chemical reaction. *Colloids Surf., A*, 2002, 198–200, 439-442.

Zhang, H. Y.; Fu, Y.; Wang, D.; Wang, L. Y.; Wang, Z. Q. & Zhang, X. (2003). Hydrogen-bonding-directed Layer-by-Layer assembly of dendrimer and poly(4-vinylpyridine) and micropore formation by post-base treatment. *Langmuir,* 2003, 19, 8497-8502.

Zhang, H. Y.; Wang, Z. Q.; Zhang, Y. Q. & Zhang, X. (2004). Hydrogen-bonding-directed Layer-by-Layer assembly of poly(4-vinylpyridine) and poly(4-vinylphenol): effect of solvent composition on multilayer buildup. *Langmuir,* 2004, 20, 9366-9370.

Zhang, F.; Jia, Z. & Srinivasan, M. P. (2005). Application of direct covalent molecular assembly in the fabrication of polyimide ultrathin films. *Langmuir*, 2005, 21, 3389-3395.

Zhang, X.; Chen, H & Zhang, H. Y. (2007). Layer-by-layer assembly: from conventional to unconventional methods. *Chem. Commun.*, 2007, 1395-1405.

Zhang, L. B.; Yang, L.; Sun, J. Q. & Shen, J. C. (2008). Mechanically stable antireflection and antifogging coatings fabricated by the layer-by-layer deposition process and postcalcination. *Langmuir*, 2008, 24, 10851-10857.

Zhang, L. & Sun, J. Q. (2009). Layer-by-layer deposition of polyelectrolyte complexes for the fabrication of foam coatings with high loading capacity. *Chem. Commun.,* 2009, 3901–3903.

Zimmerman, S. C.; Wendland, M. S.; Rakow, N. A.; Zharov, I. & Suslick, K. S. (2002). Synthetic hosts by monomolecular imprinting inside dendrimers. *Nature*, 2002, 418, 399-403.

10

Surface Wetting Characteristics of Rubbed Polyimide Thin Films

Wenjun Zheng
National Sun Yat-Sen University
Kaohsiung 80424
Taiwan RoC

1. Introduction

Amide and imide based polymers are a catalogue of versatile materials that have a wide range of applications from scientific interests to commercial products because of their great thermal stability, excellent electric properties, highly mechanical strength, and superior chemical resistance (Sroog, 1976). In a thin film form, polyimides have been found to have many important uses in optoelectronic and photonic applications. One of the most successful examples in industrial applications is the use of polyimide thin films as molecular alignment layers in liquid crystal displays. By unidirectionally rubbing a thin layer of polyimide coated on a substrate a template with some form of anisotropy can be created. When a liquid crystal is put in contact with the rubbed polyimide film, the interactions between the surface and the liquid crystal molecules degenerate into actions with orientational features. As a result, liquid crystal molecules are driven to orient in a desired direction. Because of its outstanding ability and reliability in molecular alignment, the easiness in processing and cost effective, rubbing polyimide becomes the standard liquid crystal alignment technique, and rubbed polyimide thin films as efficient alignment layers are, up to date, still irreplaceable components in modern LCDs.

On the other hand, a surface process will cause changes in chemicophysical and physcochemical properties at outmost surface of a polymer, and these changes, in turn, will induce many interesting surface phenomena, and impose a number of interesting aspects for scientific research and may lead to engineering applications. In many circumstances, a comprehensive knowledge of surface properties of polyimide thin films is of prime importance for elucidating mechanisms behind surface phenomena. A number of experimental results have shown that rubbing causes polymer chains to become oriented unidirectionally along the rubbing direction (Sawa et al., 1994; Sakamot et al., 1994; Hirosawa, 1996; Arafune, 1997), and the anisotropy in the distribution of the polymer chains is considered to be the main factor responsible for liquid crystal alignment. Wetting characteristics of a polymer surface are remarkably sensitive to chemical compositions and morphology of the outmost surface, and can provide a wide range of information on physical properties of the surface. The changes in surface characteristics of polyimide thin films due to mechanical rubbing must be reflected by surface wettability of the polymer films. There are some reports on the influence of rubbing on the surface energy of the polyimide (Lee et al., 1996; Ban & Kim, 1999). However, very little work has been done on the anisotropic surface wettability of rubbed polymers.

In this chapter, attention has been concentrated on the influence of mechanical rubbing on the surface wettability of polyimide thin films. As unidirectional rubbing creates a preferential direction on polyimide surface, how a liquid wets the surface about this direction is an interesting aspect. An insight into the effects of mechanical rubbing on surface wetting characteristics of polyimide will allow us to reveal some key correlations between inter- and intra-molecular interactions at the interface.

2. Surface energetic characteristics of polymers

2.1 Surface energy and surface free energy of continua

For a continuum, in either a solid state or a liquid form, in thermal equilibrium state, all interactions that act upon each molecule in the bulk are balanced. When a surface is created, the molecules at the surface loss the balance, which they initially possessed in the bulk and extra forces are required to maintain the molecules at the surface in the stable state. The unbalance forces for the molecules at the surface lead to additional energy at the surface, and this additional energy at the surface is known as surface energy. Microscopically, surface energy of a solid state matter is the reversible work per unit area required for the creation of a new surface, and quantifies the disruption of intermolecular bonds that occurs when the surface is created. In nature, the surface energy originates from a break in the physicochemical uniformity in the bulk. The surface energy may therefore be referred to the excess energy at the surface of a material compared to that in the bulk.

A surface is a physical boundary that separates the two continua. The two continua can either be different materials or the same material in different phases. At the surface, molecules are in relatively stable state maintained by various intermolecular forces. When a flat membrane of a continuum is stretched, the force, F, involved in stretching the membrane is

$$F = \gamma L, \tag{1}$$

where γ is the surface tension. Surface tension is therefore a measure, in forces per unit length with a dimension of N/m, the extra force stored at surface to balance the difference between the interactions in the bulk and at the surface respectively.

The same issue can be approached based on the thermodynamic consideration. In order to increase the surface area of a continuum by an amount, dA, the amount of work, dW, is needed. This work can be thought to be the potential energy stored at the surface. When the surface is stretched by dx, the work, dW, involved in increasing the surface by the length is,

$$dW = F\,\mathrm{dx} = \gamma L\,\mathrm{dx} = \gamma\, dA = \sigma. \tag{2}$$

The surface free energy is then defined as

$$\sigma = \frac{dW}{dA}, \tag{3}$$

whith a dimension of J/m^2.

Comparing Eq. 3 to Eq. 2, it can be found that surface free energy has the same value as surface tension does, but with different dimensions. The two physical quantities are interchangeable in terms of numerical value. However, the physical meanings behind the two quantities are different: one represents the energy stored at the surface, while another measures force stored in unit length at the surface.

Regarding surface energy of continua, one must distinguish between solids and liquids. For liquids, constituent atoms can move from the surface with the higher level of energy into the bulk of liquid with the lower energy, so that the area of the free surface can be significantly changed, and the surface free energy can be determined by connecting the energy with the area of contact between phases. In the case of solids, the geometry of the solid and the mechanical state of the solid may affect the apparent value of the surface energy. In particular, the surface energy inferred from the creation of a finite surface by peeling or cleavage is not necessarily equal to that of exposed surface when an infinite solid is cut along a plane and the resulting half-spaces are drawn apart to infinity, with their surfaces kept parallel at all times (Yudin & Hughes, 1994). For solids, usually surface tension does not equal to the surface free energy in value (Vanfleet & Mochel, 1995; Yu & Stroud, 1997).

2.2 Extra aspects in surface energy of polyimide

In general, polymers are the sort of uniform media in which those periodically spatial arrangements of molecules or molecular groups usually seen in an inorganic solid disappear. However, the physical origin of the surface energy remains the same: it arises from a break in the continuity at the surface.

The surface free energy of polyimides is related to chemistry of the surface, and significantly influenced by the nature of the functional group packing at the surface. For instance, Fluorination of polymers causes dramatic changes in their surface characteristics with respect to the corresponding fully hydrogenated materials. Perfluorinated polymers show low intra- and inter-molecular interactions and exhibit low surface free energy (Smart, 1994). While fluorine atoms lower the surface energy of polymer, oxygen raises the surface energy of most polymers. The technique most widely used to oxidization of polymer surface is to bombard polymer surface using oxygen plasma. The oxygen plasma bombardment of polyimide film can cause some atoms to be sputtered away and substituted by oxygen atoms. This substitution produces highly polar groups at the surface by breaking the imide and benzene rings and forming new polar species of carbon-oxygen and carbon-nitrogen-oxygen , and raises the surface energy (Naddaf et al., 2004).

The surface free energy of polyimides may be modified by polymerization of the precursors. For polyimide, the ratios of different functional groups vary with the degree of imidization (Zuo et al., 1998). Thus the degree of imidization can affect the surface energetic state of the resultant polyimide (Flitsch & Shih, 1990; Sacher, 1978; Inagaki et al., 1992). With the development of the imidization, more polar functional groups such as amide and acid become less polar imid groups, and this leads to a polyimide film with lower surface free energy. For thermal set polyimides, the degree of imidization is dependent on curing temperature and the duration the amide acid agent is kept at the temperature. Therefore, a proper curing temerature is crucial for imidization of amide acid. The curing temperature, depending upon the type of amic acid precursor, can be between 180 ~ 400°C, and the duration for thermal curing is normally one hour.

2.3 Wet a polymer surface

As long as surface wetting is concerned, at least one liquid and one solid surface are involved. Wetting a solid surface by a liquid is a surface phenomenon in which the liquid spreads on the surface and tends to cover it. Surface wetting has been thought to be a thermodynamic process which ends at equilibrium state of the system. According to their chemical activities, wetting of solid surfaces can be classified into two categories: non-reactive wetting, in which a liquid spreads on a substrate with no chemical reaction or absorption, and reactive wetting which is influenced by chemical reactions between spreading liquid and substrate material. Depending upon its basis - how the process is initiated and driven, wetting can be classified into two types: spontaneous spreading, which is defined as the spreading of a liquid on a solid by itself without any external interference; and driven spreading which is initiated and driven by some kind of external actions. Within the frame of this chapter, the discussions are focused on non-reactive spontaneous wetting.

2.3.1 Static contact angle

For thermal dynamic system, if the space is filled up with one continuum, the assembly of all co-contact points at which two thermal dynamic phases join together forms a surface; the assembly of all co-contact points at which three phases join together can form a line; the co-contact points for four phases joining together cannot contact each other in the real space. Therefore, topologically, the spatial boundary that separates two thermal dynamic phases is a two dimensional surface; when one more phase joins in, the boundary that separates the three phases degenerates to one dimensional line, and the boundary that separates four phases becomes isolated dimensionless points. There will be no real boundary that can connect more than four thermal dynamic phases in a real space.

When a small amount of a liquid is put in contact with a flat polymer surface, the tri-phase boundary that separates the three phases, i.e. solid state (S) of the substrate, liquid state (L) of the liquid droplet and vapour state (V), is known as the contact line (c.f. Fig. 1). If the substrate is chemical homogeneous and the surface is uniform, the contact line is a circle. The plane containing the normal of the solid surface and cutting through the apex of the liquid droplet is known as the meridian plane. The contact angle is defined as the angle between the solid surface and the tangent of the liquid at the tri-phase contact point in the meridian plane, through the liquid phase.

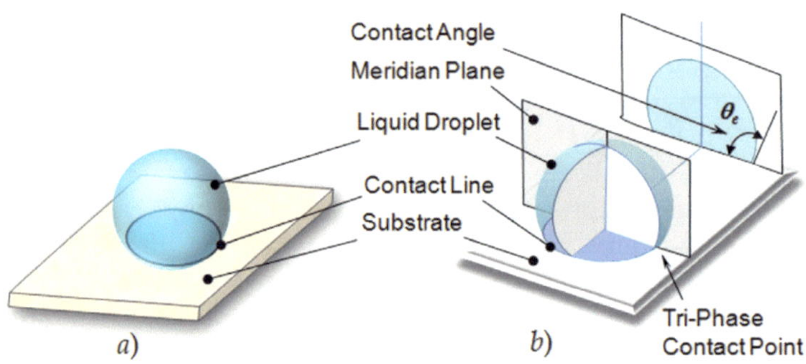

Fig. 1. a) A liquid droplet is put in contact with a solid surface, and b) the main features of the liquid droplet.

2.3.2 Contact angle hysteresis

Contact angle measurement must be carried out on an ideal solid surface, which is smooth, homogeneous, chemically and physically inert with respect to the probe liquid. Actually, no real surface exists that entirely satisfied to these exigencies. For dynamic liquid droplets on polymer surfaces, a range of contact angles appear along the contact line. Among all observed contact angles for a liquid droplet on a polymer surface, the largest one is advancing contact angle θ_a which is the contact angle measured while the volume of the liquid droplet is increasing and the contact line is moving outwards, whereas the smallest one is receding contact angle which is the one measured while the volume of the liquid droplet is decreasing and the contact line moving inwards (Fig. 2). The phenomenon of existence of multiple contact angles for the same probe liquid is known as hysteresis. The difference between advancing and receding contact angles is defined as contact angle hysteresis

$$h = \Delta\theta = \theta_a - \theta_r \, . \tag{4}$$

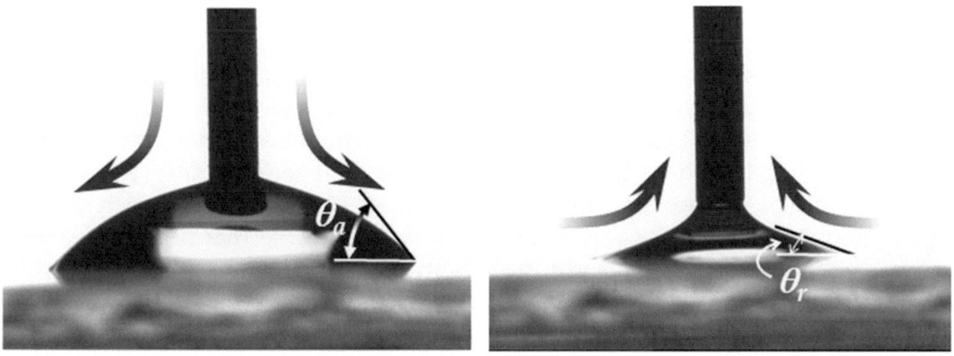

(a) (b)

Fig. 2. Dynamical profiles of a liquid droplet on a JASL-9800 polyimide surface during (a) the advancing cause in which extra amount of liquid is added on, and (b) the receding cause in which liquid is withdrawn from the droplet, respectively. θ_a and θ_r are contact angles measured during the advancing and the receding causes, respectively.

The contact angle hysteresis could be due to substrate surface roughness and heterogeneity, impurities adsorbing on to the surface, rearrangement or attraction of the surface by the solvent, etc. It is generally observed that cleaner the surface, smaller the contact angle hysteresis. For a clean and chemically homogeneous surface, it is thought that roughness and chemical heterogeneity of the surface are major factors that cause the contact angle hysteresis (Li, 1996; Chibowski & Gonzalez-Caballero, 1993). Busscher et al. showed that surface roughening tends to increase the observed contact angle as far as the contact angle on the smooth is above 86°, whereas contact angle decreases if on a smooth surface the angle becomes 60° (Busscher et al., 1984). For polymer surfaces, the surface swelling may become an important factor that contributes to contact angle hysteresis.

In wetting a rough and chemically homogeneous solid, two different effects may be observed (Kamusewitz et al., 1999): (i) the barrier effect, in which the contact angle hysteresis increases with growing roughness, and (ii) the capillary attraction/depression. In the case of a pure barrier effect, advancing contact angle increases by the same amount as receding contact angle decreases with growing roughness. Thus the equilibrium contact angle θ_e can be given by: $\theta_e = (\theta_a + \theta_r)/2$. Hence the relationship between static wetting and the dynamic one can be expressed as

$$\begin{cases} \theta_a = \theta_e + \dfrac{\Delta\theta}{2} \\ \theta_r = \theta_e - \dfrac{\Delta\theta}{2} \end{cases} . \tag{5}$$

As a result of capillary attraction or depression of grooves in the surface, for $\theta_e < 90°$, wettability will be worse on a rough surface than on a corresponding smooth surface. It is reported that, capillary effect causes an increase in both advancing and receding contact angles with growing roughness for $\theta_e < 90°$ and an opposite effect is observed if $\theta_e > 90°$. Only at $\theta_e = 90$, capillary has no effect.

2.3.3 Wettability

In wetting a polymer surface with a liquid, one of the following phenomena may take place: the liquid spread a little or may not spread at all, a case of non-wetting; the liquid spreads continuously and covers the entire substrate with a thin film of the liquid, the case is known as complete wetting; the liquid droplet spreads partially to some extent – a case generally referred as partial or incomplete spreading. Each of these phenomena depicts the degrees that a polymer surface may be wetted by a liquid. The degree that a polymer surface is wetted by a liquid is defined as the wettability of the surface wetted by the liquid. Wettability describes the tendency for a liquid to spread on a polymer surface, i.e. the degree of intimate contact between a liquid and the polymer surface.

There is no direct measure of wettability. In practice, the wettability of a polymer surface is evaluated by examining the profile of a probing liquid droplet which is put in contact with the polymer, and characterized by contact angle. For example, the two distinct extreme equilibrium regimes may be characterized by the value of contact angle as: complete wetting with the contact angle $\theta = 0$, or absolute non-wetting with the contact angle $\theta \rightarrow 180°$. When the contact angle is measured with a finite value $0 < \theta < 180°$, the surface is then partial wetted by the liquid.

In reality, a complete non-wetting is rarely seen, and most surfaces are partially wettable. In engineering, the wettability of a solid is classified as

$$\begin{cases} \theta > 90^\circ : unwettable \\ 0 < \theta < 90^\circ : partially \quad wettable \\ \theta = 0 : completely \quad wettable \end{cases} . \tag{6}$$

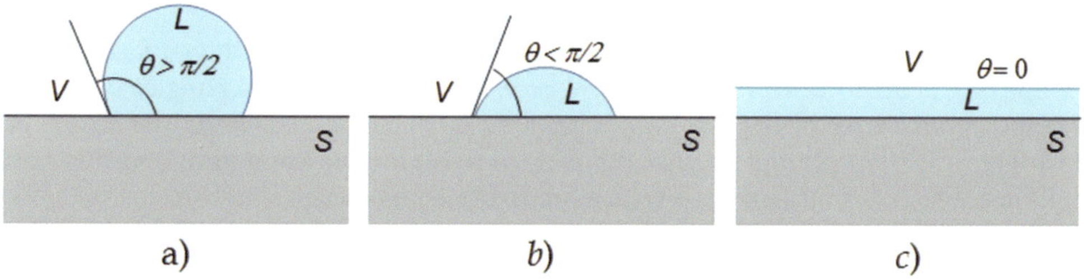

Fig. 3. When wetting a solid surface, three cases of the spreading of a wetting liquid are normally seen: a) non-wetting ($\theta > \pi/2$), b) partial wetting ($\theta < \pi/2$), and c) complete wetting ($\theta = 0$).

If the probing liquid is water, a wettable surface is known as a hydrophilic (or lyophilic) surface; whereas an unwettable surface is referred to as a hydrophobic (or lyophobic) surface.

2.4 Evaluation of wetting characteristics of polymer surface

2.4.1 Measurement of surface free energy

The driving force for the spreading a wetting liquid on a solid surface can be written as:

$$F_d(t) = \gamma_S - \gamma_{SL} - \gamma_L \cos\theta(t), \tag{7}$$

where θ is contact angle, γ_S, γ_{SL} and γ_L are interfacial tensions in solid-vapour, solid-liquid and liquid-vapour interfaces, respectively. Eq. 7 is also known as the equation of state. γ_{SL} is a parameter that connects the properties of the solid and probing liquid. At thermodynamic equilibrium, the energy of the system must be stationary and the dynamic driving force is cancelled out, i.e. $F_d = 0$, due to a balance between all interactions at the surface, and as a result, the spreading of the liquid droplet comes to rest. These conditions lead to the famous Young's equation

$$\gamma_S - \gamma_{SL} = \gamma_{LV} \cos\theta . \tag{8}$$

Eq. 8 shows that contact angle θ is defined and is decided by the surface and interfacial energies. This indicates the importance of surface energetic states on determining the surface wetting characteristics. Therefore, the measurement of surface free energy forms an important part of the evaluation of surface wetting properties of a polymer surface.
Although it draws the basic principles for surface characterization, Young's equation cannot be solved straight away. Usually, $\gamma_{LV} \equiv \gamma$ can be obtained by separate measurements. Thus we are left with two unknown variables γ_{SL} and γ_S with only one datum θ.
A number of thermodynamic approaches have been proposed to determine γ_S and γ_{SL}. Detailed descriptions about these approaches can be found in literature (de Gennes P G, 1985; Gindl et al., 2001; Kumar & Prabhu, 2007). We adopt geometric mean approach for this study.

Zisman (Zisman, 1963) introduced the concept of critical surface free energy γ_c, which is defined as the surface tension of a probing liquid which fully wets the surface ($\cos\theta = 1$). The value of γ_c is determined from empirical investigations, and contact angles of the liquids of a homologous series of organic compounds on a solid are measured. The cosine of the contact angels is then plotted against the surface tension γ_L of the liquid, and this forms a straight line which can be described with a following relationship,

$$\cos\theta = 1 - b(\gamma_L - \gamma_c), \tag{9}$$

where b is the slope of the regression line. Extrapolation of this line to the point of $\cos\theta = 1$ yields the value of $\gamma_L = \gamma_c$ at the point. Combining Eq. 8 with Eq. 9, one can obtain

$$\gamma_S = \frac{(b\gamma_c + 1)^2}{4b}. \tag{10}$$

Zisman's method is the geometric mean approach.

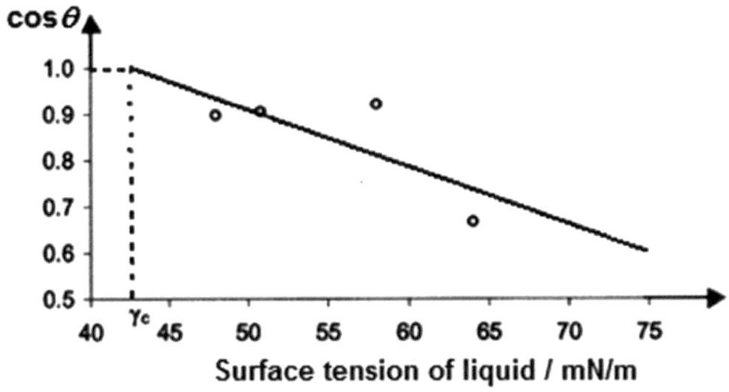

Fig. 4. A Zisman plot for estimating surface tension of a liquid.

Later an idea to partition of surface free energy into individual components includes the assumption that the quantity γ_{SL} is determined by various interfacial interactions that depend on the properties of both the measuring liquid and the solid-liquid of the studied solid. In his pioneer work, Fowkes assumed that the surface free energy of a surface is a sum of independent components, associated with specific interactions:

$$\gamma_S = \gamma_S^d + \gamma_S^p + \gamma_S^h + \gamma_S^i + \gamma_S^{ab} + \dots, \tag{11}$$

Where γ_S^d, γ_S^p, γ_S^h, γ_S^i, and γ_S^{ab} are the dispersion, polar, hydrogen bond, induction, and acid-base components, respectively. According to Fowkes, the dispersion component of the surface free energy is connected with the London interactions, arising from the electron dipole fluctuations. These interactions occur commonly in the matter and result from the attraction between adjacent atoms and molecules. The London forces depend on the kind of mutually attracting elements of the matter and are independent of other types of interactions. The remaining van der Waals interactions have been considered by Fowkes as a part of the induction interactions. This method is not widely accepted due to its complex.

With the consideration of the idea of the surface free energy partition, Owens and Wendt improved Zisman's fundamental work and developed a new method which has been widely accepted for measurement of contact angle for evaluation of surface free energy measurement (Owens & Wendt, 1969). In the Owens-Wendt method, it has been assumed that the sum of all the components occurring on the right-hand side of Eq. 6 except γ_S^d, can be considered as associated with the polar interaction (γ_S^p), and the equation of state can be written as

$$\gamma_{Sl} = \gamma_S + \gamma_L - 2\left(\sqrt{\gamma_S^d \gamma_L^d} + \sqrt{\gamma_S^p \gamma_L^p}\right). \tag{12}$$

The combination of Eq. 8 and Eq. 12 leads to

$$\frac{(1+\cos\theta)\gamma_L}{2\sqrt{\gamma_L^d}} = \sqrt{\gamma_S^p}\sqrt{\frac{\gamma_L^p}{\gamma_L^d}} + \sqrt{\gamma_S^d}, \tag{13}$$

The form of the Eq. 13 is of the type $y = bx + m$. For a certain solid, the surface free energy is assumed to be constant without varying with different probing liquids. One can graph $(\gamma_L^p)^{1/2} / (\gamma_L^d)^{1/2}$ vs $\gamma_L(1+ \cos\theta) / (\gamma_L^d)^{1/2}$. The slope will be $(\gamma_S^p)^{1/2}$ and the y-intercept will be $(\gamma_S^d)^{1/2}$. The total free surface energy is merely the sum of its two component forces.

2.4.2 Experimental determination of surface free energy

Young's equation explains theoretically the necessary conditions for a liquid drop to reside on a surface statically. The measurement of contact angle is then a practical way to obtain surface free energy. Depending upon how the probe liquid wets the surface to be tested, two different approaches are commonly used for the measurement of contact angles, goniometry and tensiometry. Tensiometry involves measuring the forces of interaction as a solid is contacted with a probe liquid whose surface tension is known. This technique is particularly suitable for the porous surfaces which may absorb the wetting liquid. Goniometry involves the observation of a sessile drop of test liquid on a solid substrate. Analysis of the shape of a drop of test liquid placed on a solid is the basis for goniometry, and this is particularly useful for evaluation of contact angle hysteresis. Goniometry is the technique we used to observe the wetting characteristics of rubbed polyimide films.

The equipment used for goniometrical measurement contact angles is a DSA100 which is commercially available from Krüss. During measurement, droplets of about 2 μl of test liquids are dispensed onto the polymer surface to be tested, and monitored with a charge-coupled device (CCD) camera. The images of test liquid captured are then analyzed with computer software which is written based on Owens-Wendt model (described by Eq. 13).

In order to detect unusual features created due to rubbing of polyimide films, the surface tension meter has been modified to have a stage, which can be rotated azimuthally, mounted.

3. Breaking down surface uniformity of polyimide thin films due to rubbing

3.1 Preparation of polyimide thin films

3.1.1 Coating a polymer precursor on to substrate

Several techniques are available for coating polyimide resin onto a surface. The most popular and reliable one is the spin-coating technique, which is also the one we used to prepare polyimide thin films for our studies. Spin coating provides uniform, pinhole free coating polymer layer on a substrate. Any standard photoresist spin coating technique can be used for the coating of polyimide. The factors which affect the thickness uniformity and overall quality of the final coating can be listed as following:

- substrate preparation (cleaning)
- Volume of solution dispensed
- Substrate acceleration
- Final spin speed
- Spin time
- environment conditions (e.g. Temperature, humidity, exhaust air flow rate, etc.)

Coating thickness for a solution with a particular concentration will vary as a function of spin speed and spin time. A spin speed of at least 1000 rpm and a spin time of at least 30sec are recommended for applications in which surface uniformity is of primary concern. If the packed resin is thinned, the diluted solution should be left still for de-bubbling. All dispensing should be as close as possible to avoid bubble formation. Tiny bubbles in the solution will cause comet-like defect in the coated film (cf. Fig. 5). The volume of solution dispensed should remain constant for each substrate to insure substrate to substrate uniformity.

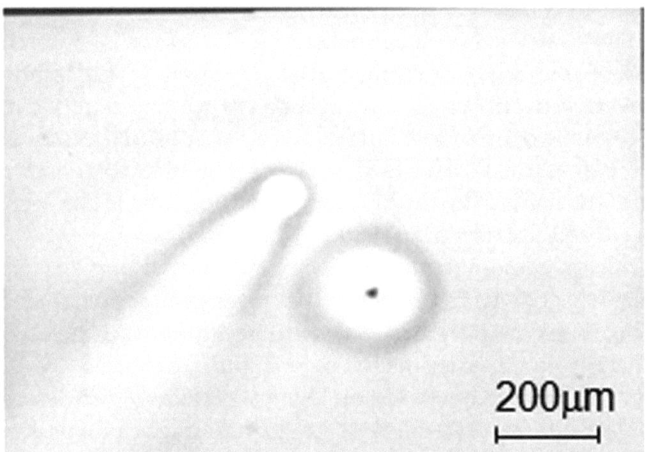

Fig. 5. A 'comet' defect in polyimide coating film due to a micro-bubble in the resin solution, and a defect resulted from a solid particle on the substrate.

3.1.2 Imidization

Before thermal imidization, the amic acid solution coated on the substrate is soft bake to remove the residual solvent. The soft baking process also provides the precursor with sufficient chemical resistance and adhesion so that the coating will not be attacked.
The soft baking of precursor is carried out by putting the coated substrates on a hot plate at a temperature in a range of 60°C to 105°C for 30 - 60 min. The substrates should remain in a horizontal position during this process to avoid the reflow of the coated solution. An insufficient drying can result in the attack of the coating by some contaminants, such as residual thinner and some organic solvent, causing defects on the coating surface and/or the formation of pinholes. A too high temperature soft-baking can initiate partial crosslinking and /or imidization.
The minimum final cure temperature is dependent upon the type of amid acid resin used. For most polyimide precursors, imidization can occur when temperature exceeds 100°C, and the curing temperature for imidization can be within a wide range from 150°C to 300°C. To achieve a good imidization, amid acid is usually cured at 200°C for a period of 1 hour. The curing temperature can affect the surface free energy of the final polyimide film because of the correlation of the degree of imidization to the curing temperature. It has been shown that the degree of imidization increases with curing temperature (Lee et al., 1996; Zuo et al., 1998). The effect of the degree of imidization on the dispersed part of free energy, which relates to the long range molecular interactions, is small and can be ignored. However, the polar part of the surface free energy is strongly influenced by the degree of imidization. With the development of the imidization, more polar functional groups such as amid acid become less polar imid groups, and this causes a significate decrease in the strength of the polar part free energy. As a result, the surface free energy of the resultant polyimide film is reduced.

3.1.3. General features of polyimide thin films coated on Indium-Tin-Oxide glass substrates

During cure, a net weight loss up to 50% may occur to the coating film accompanied by a corresponding decrease in coating thickness. With this imidization induced film shrinkage being taken into account, the thickness of the final polyimide films is thought to be decided by the viscosity of the amid acid solution and the spin speed of the substrates. Figure 6 shows the thickness of polyimide films prepared from a commercial 5 wt% amide-acid solution JALS-9800 (JSR, Japan) against spin speed. The curing temperature for imidization was set at 240°C.

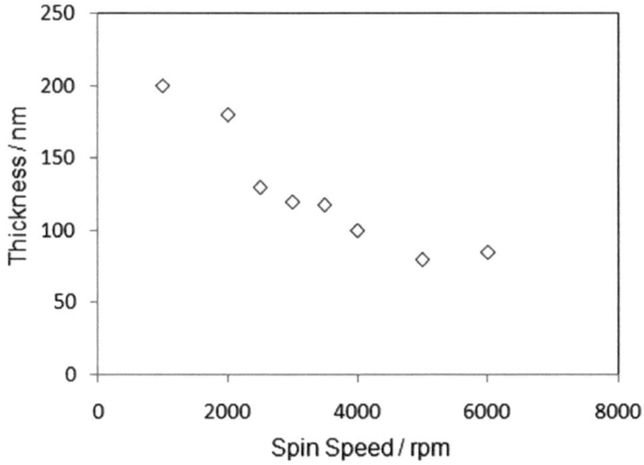

Fig. 6. Thickness of polyimide films vs spin speed of the substrate.

The atomic force microscopy (AFM) examination of polyimide films coated on the ITO glass substrates reveals that the surface of the polymer films are flat and smooth. As far as the surface characteristics of a thin polymer film coating on a solid surface is concerned, it is necessary to learn whether the measured results are distorted by the effects of the material beneath the polymer film. Experimental results reveal that surface free energy of the polyimide films is rather stable when the thickness of the polymer films is within the range from 80 - 150 nm (Fig. 7). These polyimide films were produced by coating the amic acid solution onto substrates which were spinning at speed ranged from 2000 to 4000 rpm (c.f. Fig. 6). We preferentially set the spin speed of the coater at 4000 rpm, and the polyimide thin films produced are 100 nm thick. The surface free energy of the films before further process is measured to be 45.532 (± 2.794) mJ/m^2.

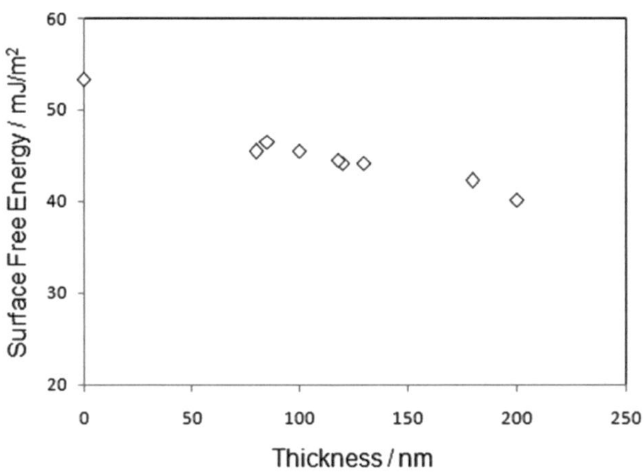

Fig. 7. Surface free energy of polyimide films vs film thickness.

3.2 Rubbing process

The mechanical rubbing of polyimide films is carried out using a rubbing machine. An in house made rubbing machine is schematically illustrated in Fig. 8. The rubbing machine consists of a rotating drum which is wrapped with a piece of velvet textile. The sample holder is mounted on a translationally movable flat stage. The rubbing strength is the most important parameter for the rubbing process. It is a measure of the strength of the interaction between the rubbing textile and the polymer thin film, and depends on many factors such as the pressure of the rubbing textile applying to the surface, the hardness of the fibre of the velvet etc. A satisfactory method to determine mechanical rubbing strength is yet to be developed. In engineering, the rubbing strength is evaluated using following equation

$$RS = N \cdot \Lambda \left(\frac{2\pi R\omega}{v} \pm 1 \right), \tag{14}$$

where N is the number of rubbing cycles, Λ is the pile impression of the velvet fibres, ω is the rotaion speed of the drum, R is the radius of the drum, and v is the translational speed of the sample holder. The sign before the factor of 1 indicating the relative moving direction between the sample and the rubbing volvet: " – " means the sample moving against rubbing volvet, whereas "+" means both the sample and the rubbing volvet moving in the same direction. The RS calculated using Eq. 14 is also known as specific rubbing length because it has a dimension of length.

Before rubbing the polyimide films are rather flat and smooth. The average roughness of the polyimide film, measured using AFM, is 0.33 nm. Mechanical rubbing is a crude process during which large quantities of polymer material in some regions may be excavated leading to considerable damage to a polymer surface. A macroscopic effect in a microscopic scale of the mechanical rubbing is the formation of microgrooves on the polymer surface.

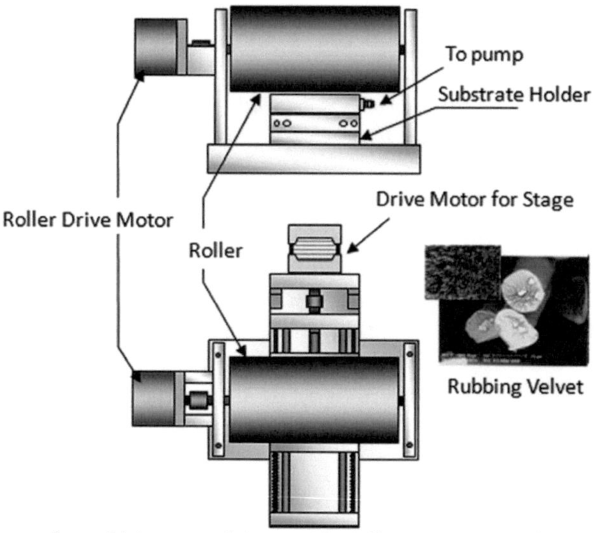

Fig. 8. An in house made rubbing machine with following main features: the radius of the drum R = 30 mm, the rotation speed of the drum ω = 135 rpm, the translational speed of the stage for the sample holder v = 30 mm/min, average length of fibre of velvet = 1.8 mm.

For a unidirection rubbing, the microgrooves, which can be clearly seen in an AFM image (Fig. 9), are parallel to the rubbing direction. The geometric dimension of the grooves and the density of the groove on the surface are determined by the phyical characteristics, such as the length, the elastidity, the surface features etc., of the rubbing velvet, and the number of rubbings (Zheng et al., 2004).The surface roughness increases with rubbing strength.

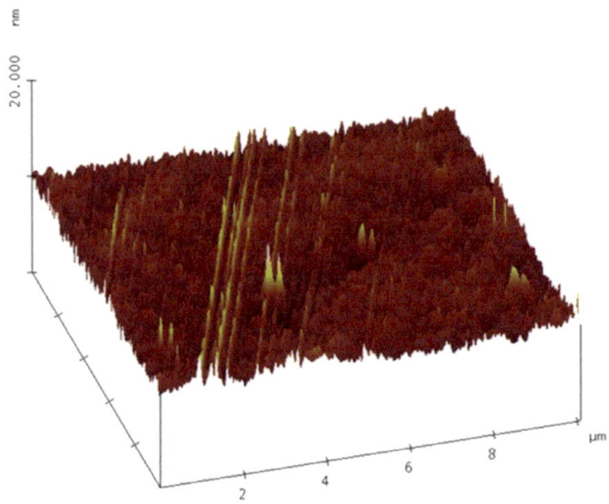

Fig. 9. Atomic force microscopic image of a rubbed polyimide surface. The polymer film was rubbed 4 times by a volvet with a pile impression of 0.3 mm.

The changes in the surface roughness of the polyimide film due to rubbing may not be significant (Zheng et al., 2009). For JASL-9800, with the pile impression of rubbing velvet being set at 0.3 mm, the average surface roughness of the polymer films, which are rubbed up to seven times, is below 1.0 nm (Fig. 10). A restruction in surface topography has been observed. The surface roughness increases with the first two rubbing cycles, and drops when the film is rubbed three times; then increases as the rubbing continues and peaks at the completion of the fifth rubbing, then drops again when the polymer film is further rubbed. The surface roughness increases and decreases alternatedly with the rubbing cycle. The topographic reconstruction can be explained as follows. Rubbing causes the formation of grooves at the surface of the polyimide film. Although the grooved surface will lead to only a small variation in pile impression, and hence rubbingstrength, across the surface, the peaks in the corrugated surface suffer higher abrasion rates than thoughs leading to a reduction in surface roughness. Subsequent rubbings will cause more polyimide material to be excavated from the surface leading to a rougher surface. As rubbing continuing, a new course of flatness is started. It seems that with the polyimide (JALS-9800) used for the observation the repeating period in the variation of surface roughness with rubbing is three rubbing cycles.

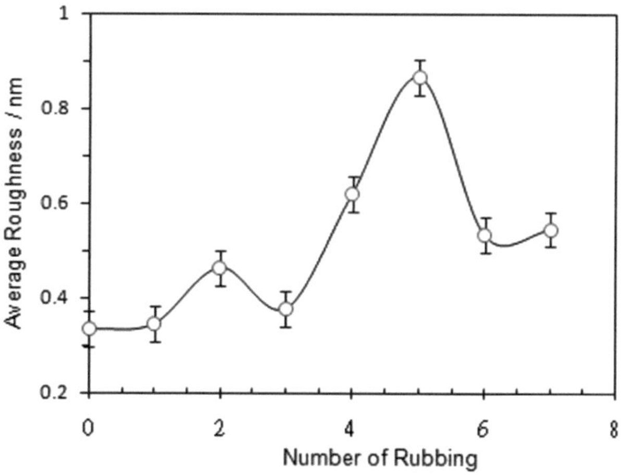

Fig. 10. Surface roughness of rubbed polyimide films against rubbing strength.

The mechanical rubbing can force polar groups to reorient at the surface and thus leads to changes in polar strength of the polyimide surface (Lee et al., 1996). The way the polar strength changes depends on the chemical properties of the polyimide materials. For the polyimides whose surface polar strength can be enhanced by rubbing, the surface free energy will increase with rubbing strength (Ban & Kim, 1999). For polyimide thin films produced using JALS-9800, increasing rubbing strength, as illustrated in Fig. 11, results in a decrease in the surface free energy.

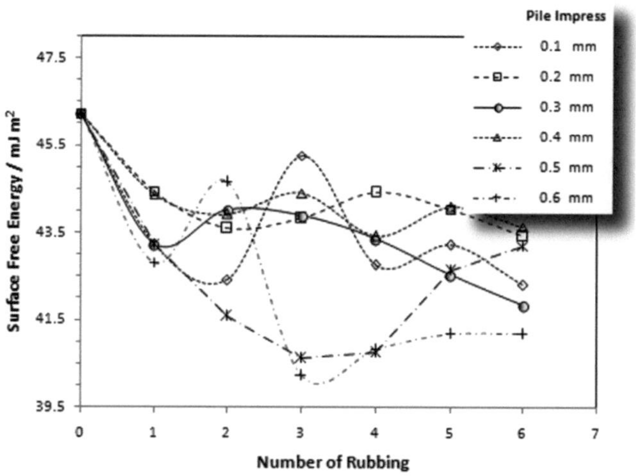

Fig. 11. The surface free energy of polyimide thin films against the number of rubbing cycles for different pile impression of the rubbing velvet.

3.3 Anisotropic wettability of rubbed polyimide films

The formation of the grooved surface clearly indicates that the topographical uniformity of the surfaces of the polyimide films has been broken, and anisotropy in surface topography has been created. As the topgraphic uniformity of the surface is broken, the two dimensional uniformity in many physical properties at the surface may be lost or changed. The unidirectional rubbing produces grooves, which are parallel to the rubbing direction, on the polyimide surface. So the rubbing creates a preferential direction, which is parallel to the rubbing direction, on the surface. It is nature to take the rubbing direction as reference direction for the study of surface anisotropy.

For a solid state surface, several phenomena, such as the surface roughness, chemical heterogeneities, surface restructuring, material swelling and dissolution etc., can contribute to the contact angle. In many cases, the surface roughness and chemical heterogeneities are considered as major factors that affect the contact angle. However, for rubbed polyimide films, surface restructure may have significant effect on contact angle. The effect of the orientation of the polymer chains and rearrangement of polar groups at the surface due to rubbing should be reflected by wetting characteristics of the polymer films. The modified surface tension analyzer DSA100, equipped with a rotating state, enable us to carry out the observation. A static water droplet on the rubbed polyimide films exhibits a different contact angle in different viewing direction. Fig. 12a shows azimuthal variation in the contact angel of a deionized water droplet resided statically on the rubbed polyimide surfaces. The amplitude of the contact angles varies with rubbing strength. This indicates that wettability of the polyimide can be changed by mechanical rubbing. However, the profiles of the water droplet, evaluated by the curve of the contact angle, on polyimide films rubbed with different rubbing strength are similar. Therefore the anisotropy in wettability of rubbed polyimide about the rubbing direction is evident. The difference in contact angles measured respectively towards and against the rubbing direction is marked (cf. Fig. 12b).

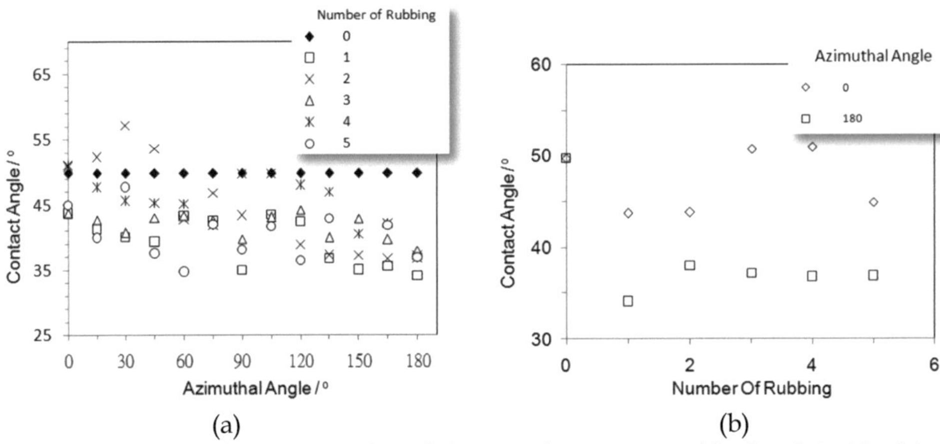

Fig. 12. (a) Variation of contact angles of deionized water on rubbed polyimide thin films against azimuthal angle for different rubbing strength. (b) Variation of contact angle of deionized water with rubbing strength measured at tri-phase points towards and against rubbing direction respectively.

It was suggested that the effect of surface roughness on the contact angle can be omitted when the surface roughness is not greater than 100nm (Morra et al., 1990). Heng et al. (Heng et al., 2006) confirmed that the effect of surface roughness on contact angle hysteresis for single crystalline paracetamol was negligible. For rubbed JASL-9800 films displayed here, the surface roughness is less than 1 nm. The effect of the surface roughness on the contact angle can be omitted. The evidence that support this argument can be found from the experimental results. For the parallel grooved surface, maximum surface roughness appears in the direction perpendicular to the grooves. The observed results, which show that the maximum contact angle, as illustrated in Fig. 12a, does not necessarily appear in the direction perpendicular to the rubbing direction, demonstrate that surface roughness of the rubbed PI under studying is not a decisive factor that determines the contact angle, and the anisotropic wettability of rubbed polyimide is resulted from other mechanisms rather than the geometrical surface topography.

3.3 Anisotropy in contact angle hysteresis

In principle, the measurement of static contact angle provides an effective means to evaluate wettability of a solid surface. In practical, however, it is often difficult to measure the static contact angle since the tri-phase system can hardly reach thermodynamic equilibrium in a laboratory environment, thus the volume of the probe liquid is changing all the time. In many cases, a dynamic analysis, in which the contact angle hysteresis is examined, can provide results which are more closed to the true wetting characteristics of a surface.

In order to examine dynamical wetting characteristics of the rubbed polyimide films, a drop of 2 μl deionized water was initially dispensed onto the polyimide surface, then extra deionized water is added to the droplet at a rate of 1 μl/min and the advancing contact angle is measured during the contact line of the deionized water at the surface was moving outwards, whereas the receding contact angle was determined during the deionized water is withdrawn from the droplet and the contact line of the water moving inwards. For an unrubbed JSAL-9800 film, the contact angle hysteresis is measured to be 34.0°, with an advancing contact angle 86.8°. The contact angle hysteresis is independent of azimuthal angle indicating that the wettability of the polyimide film is symmetric. This is expected as there is no preferential direction on a uniform polyimide surface.

The profiles of a water droplet on a rubbed polyimide in both advancing and receding curses are asymmetric. In the advancing course, as illustrated in Fig. 13a, more water accumulated on the side of hte droplet that the contact line move against rubbing direction, while in the receding course (cf. Fig. 13b), on the side of the droplet the contact line moves against the rubbing direction, the movement of the contact line is hindered and the droplet was stretched and elongated.

In the case of rubbed polyimide films, in addition to the movement of the contact line, the moving direction of the contact line to the rubbing direction must also be taken into account when evaluating contact angle hysteresis. The parallel contact angle hysteresis h_p is determined by substracting the parallel receding contact angle θ_r^p measured at the tri-phase point, which is moving towards rubbing direction in the receding curse, from the parallel advancing contact angle θ_a^p measured at the tri-phase point which is moving towards the rubbing direction in the advancing curse, whereas the anti-parallel contact angle hysteresis h_{ap} is given as the difference between the anti-parallel advancing contact angle θ_a^{ap} measured at the tri-phase point which is moving against the rubbing direction and anti-parallel receding contact angle θ_r^{ap} measured at the tri-phase points which are moving against the rubbing direction during receding curse. Therefore, parallel contact angle hysteresis h_p and anti-parallel contact angle hysteresis h_{ap} are defined as

$$\begin{cases} h_p = \theta_a^p - \theta_r^p \\ h_{ap} = \theta_a^{ap} - \theta_r^{ap} \end{cases}. \tag{14}$$

Notice that θ_a^p and θ_r^p (and also the θ_a^{ap} and θ_r^{ap} pare) are not at the same side of the droplet. This is due to the reversal of the moving direction of the tri-phase points with reference to the rubbing direction with the dynamic liquid droplet being switched over between the advancing and the receding courses.

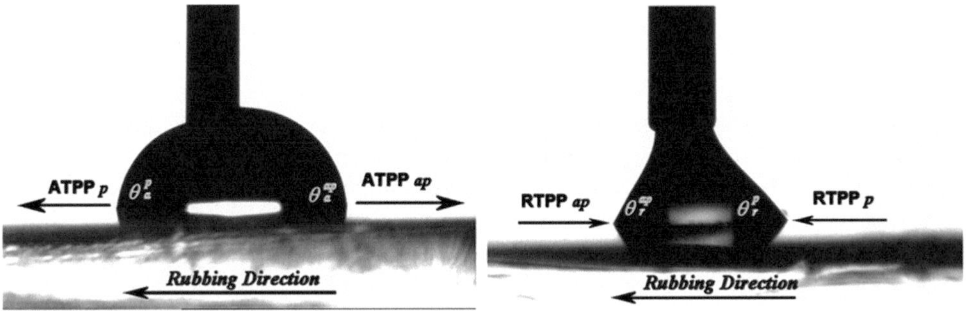

(a) (b)

Fig. 13. Images of profiles of deionized water droplet on the surface of a rubbed polyimide film in (a) advancing course and (b) receding course, respectively. θ_a^p and θ_r^p are advancing and receding contact angles measured whent the trip-phase contact point $ATPP_p$ and $RTPP_p$ are moving in the rubbing direction in the advancing and receding courses, respetively, whereas θ_a^{ap} and θ_r^{ap} are advancing and receding contact angles obtained when the tri-phase points $ATPP_{ap}$ and $RTPP_{ap}$ are moving against the rubbing direction in the advancing and the receding courses, respectively.

The contact angle hysteresis varies with the rubbing strength. For JASL-9800, an increase in rubbing strength causes both parallel contact angle hysteresis and anti-parallel contact angle hysteresis to decrease. In general, a small contact angle hysteresis corresponds to a less polar surface, i.e. a more hydrophobic surface. The variation in wetting characteristics with rubbing strenght is a well known phenomenon which has been observed by other researchers. What is interesting here is that the parallel contact angle hysteresis is different from the anti-parallel one indicating the anisotropy in wettability of rubbed polyimide films. This anisotropy in wettability can be clearly seen in Fig. 14.

A macroscopical effect of a mechanical rubbing in a microscopical scale is the formation of grooves on the surface of polyimide films. The surface with parallel grooves exhibits anisotropy in surface topography with a preferential direction that is parallel to the grooves. The anisotropy created due to a unidirectional geometrical structure is known as form anisotropy. The form anisotropy was once thought to be the main cause that was responsible for some interfacial phenomena such as a unidirectional alignment of liquid crystal molecules (Berreman, 1972). Roughness and chemical heterogeneity of the surface are usually considered as two major factors that determine the contact angle hysteresis. Chemical heterogeneity is a more complecated issue. It has been shown that the effect of surface topography on the contact angle hysteresis is negligible when the surface roughness

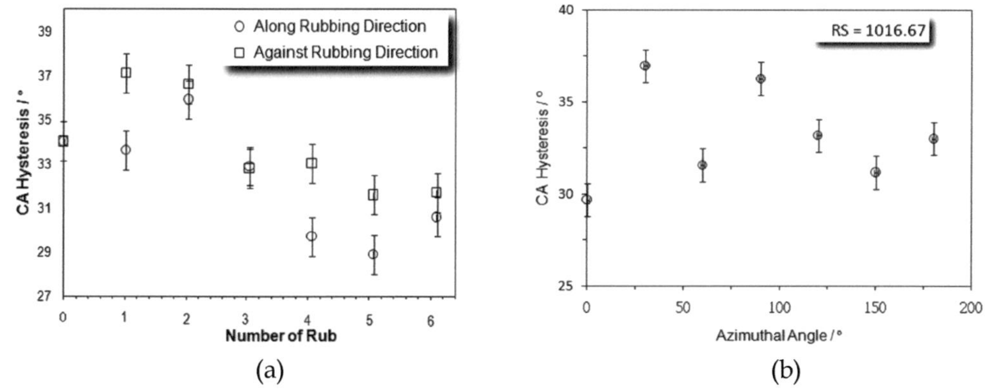

Fig. 14. (a) Contact angle hysteresis of deionized water on rubbed polyimide thin films against number of rubbing. (b) Variation of contact angle hysteresis with azimuthal angle against the rubbing direction. The polyimide films is rubbed with a rubbing strength of 1016.67 mm.

is not greater than 100 nm. As demonstrated in a previous section, the surface roughness can be controlled to be well below this amplitude with proper rubbing conditions. Furthermore, it has been revealed that even on molecularly smooth surfaces contact angle hysteresis can be quite significant (Chibowski, 2003; Lam et al., 2002). It seems that surface roughness becomes a less important factor when it is small enough (e.g. < 100 nm). This also suggestes that the form anisotropy may not be the decisive factor for the ainsotropic wettability of the rubbed polymimide films. Molecular scale topography at outmost surface might be the key to elucidate contact angle hysteresis.

3.4 Anisotropy in surface free energy

Rubbing does not always produce a observable geometrical structure on the surface, thus the surface anisotropy does not necessarily result from form anisotropy. Stöhr et al. (Stöhr et al., 1998) proposed a model to describe how rubbing pulls the polymer chains orienting them in one direction. According to Stöhr's model, at the rubbed polyimide surface, the polymer chains are pulled by the velvet fibres to align themselves with the rubbing direction, and there is a preferential out-of-plane tilt of phenyl rings. It has been shown by many researchers that rubbing can induce a reorientation of polymer chains (Arafune et al., 1997; 1998) and the rearrangement of functional groups in the polymer (Lee et al., 1997). The changes in surface energy, and consequently in surface wettability, have be attributed to the variation and rearrangement of polar and/or non-polar groups at the surface due to rubbing. However, how these changes in surface restruction act on a probe liquid has remained unknown. Surface free energy can be thought to be the total sum of the effects of all interactions at the surface. Owing to the close relation between surface free energy and contact angle, the anisotropy in contact angle hysteresis, in turn, indicates that the surface free energy may have an asymmetric pattern.

The surface free energy of the rubbed JASL-9800 polyimide thin films, as shown in Fig. 15, is anisotropic: for a rubbed polyimide film, the surface free energy towards the rubbing direction, e.g. for JASL-9800, is higher than that in the direction against rubbing direction.

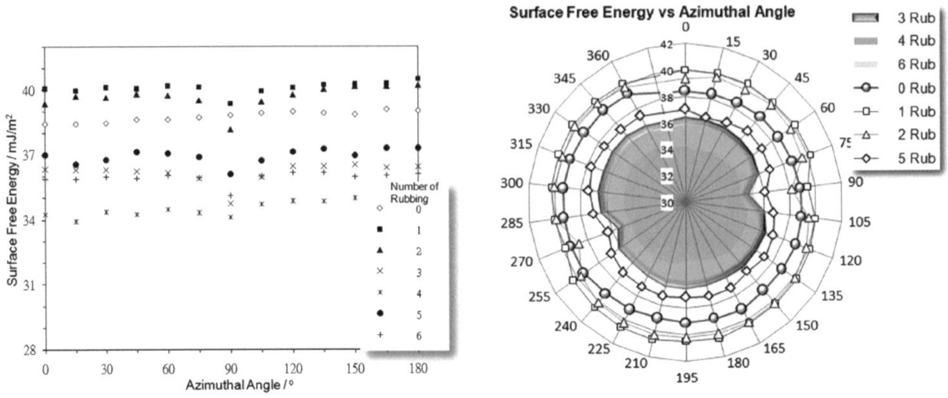

(a) (b)

Fig. 15. Azimuthal variation in surface free energy of rubbed JASL-9800 polymer films (a) within 180° range, and (b) for a full circle. The azimuthal angle is the angle the meridian plane of the water drop made against the rubbing direction.

The link between the orientation of the polymer chains and surface free energy is still missing. An imprical model based on the experimental observations is proposed as follows (Zheng et al., 2008). The overall anisotropy in the surface free energy of the rubbed polyimide films can be attributed to the macroscopic orientational order of the polymer chains at the surface, whereas the difference in the respective values measured parallel and antiparallel to the rubbing direction may be due to the microscopic orientation of functional groups in the polymer chains. The rubbing also has significant effects on the wettability of the rubbed polyimide.

It is widely accepted that the distribution of polar groups at the polyimide surface would determine the surface energetic state. We evaluated surface polarity of polyimide thin films using polar part of surface free energy. Fig. 16 shows the variation in the polar part of surface free energies as a function of rubbing strength. The surface polarity of polyimide increases with rubbing strength. The polarity in the rubbing direction is smaller than that against the rubbing direction. It is well known that the contact angle is very sensitive to the surface polarity, and a surface with a larger polarity exhibits lower hysrophobicity (Lee, K. W. Et al. 1997). The increase in the surface polarity of rubbed polyimide is considered to result from an outwards reorientation of polar groups at the polymer surface (Lee et al., 1996; 1997). Considering the orientation in polymer backbones induced by rubbing, a possible mechanism for the appearance of the anisotropy in the contact angle hysteresis is inferred as follows. The overall anisotropy in the contact angle hysteresis on the rubbed PI thin films may result from the anisotropic dispersion surface tension, which originates from a unidirectional orientation of the polymer backbonds, whereas the local orientation of the polar groups at outmost surface owing to the rubbing may be responsible for the difference in contact angle hysteresis measured in and against the rubbing direction, respectively.

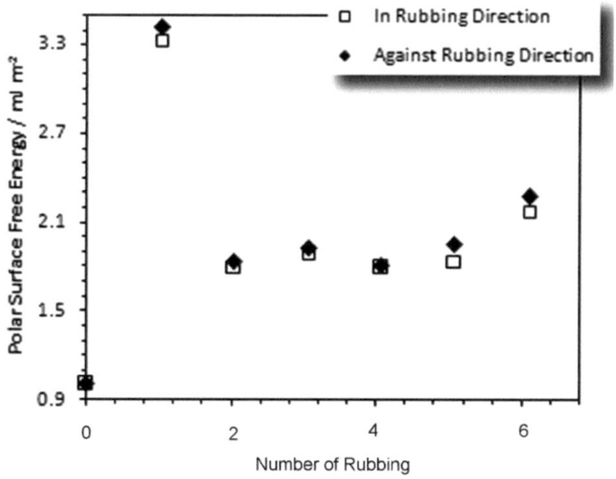

Fig. 16. The variation of polar part surface free energy of polyimide thin films with rubbing strength. The hollow squares are data for the polar surface free energy measured towards rubbing direction, whereas the solid diamond spots are data for the polar surface free energy measured against rubbing direction.

4. Conclusion

Mechanically rubbing polyimide thin film is a simple process. It, however, imposes some interesting surface phenomena. Mechanical rubbing breaks the two-dimensional topographical uniformity of the polyimide surface and causes changes in the surface energy of the polyimide thin films. The wettability of rubbed polyimide films is anisotropic. Water spreading on the rubbed polyimide thin films exhibits an asymmetric behaviour. For the rubbed polyimide thin films, the hydrophilicity of the surface towards the rubbing direction is different from that in the direction against the rubbing direction. The surface anisotropy in the rubbed polyimide surface is thought to be created due to an orientational arrangement of polymer chains at the surface. However, the evidence for this argument still remains unclear.

To find out links between thermodynamic phenomena and interactions in the interface at molecular level will be helpful for elucidating the mechanisms behind surface wetting phenomena.

5. References

Arafune, R.; Sakamoto, K. & Ushioda, S. (1997). Correlation between the pretilt angle of liquid crystal and the inclination angle of the polyimide backbone structure, *Appl. Phys. Lett.*, Vol. 71 (19), 2755-2757. ISSN 0003-6951

Arafune, R.; Sakamoto, K.; Ushioda, S.; Tanioka, S. & Murata, S. (1998). Importance of rubbing-induced inclination of polyimide backbone structures for determination of teh pretilt angle of liquid crystals, *Phys. Rev. E*, Vol. 58 (5), 5914-5918. ISSN 1539-3755

Ban, B. S. & Kim, Y. B. (1999). Surface free energy and pretilt angle on rubbed polyimide surfaces, *J. Appl. Polym. Sci.*, Vol. 74 (2), 267-271. ISSN 1079-4628

Berreman, D.W. (1972). Solid surface shape and the alignment of an adjacent nematic liquid crystal, *Phys. Rev. Lett.* Vol. 28 (26), 1683-1686. ISSN 1539-3755

Busscher H. J.; Van Pelt A.W. J.; de Boer P.; de Jong H. P.; & Arends J., (1984), The effect of surface roughening of polymers on measured contact angle of liquid, *Colloid. Surf.* Vol. 9 (4), 319-331. ISSN 0927-7757

Chibowski, E. (2003), Surface free energy of a solid from contact angle hysteresis, Adv Colloid Interface Sci. Vol. 103 (2), 149-172. ISSN 0001-8686

Chibowski, E. Gonzalez-Caballero F. (1993), Interpretation of contact angle hysteresis, *J. Adhes. Sci. Technol.* Vol. 7, 1195-1205. ISSN 0169-4243

de Gennes P G, (1985), Wetting: statics and dynamics, *Rev. Mod Phys.* Vol. 57 (3), 827-863. ISSN 0034-6861

Flitsch R.; Shih D.Y.; (1990), A study of modified polyimide surfaces as related to adhesion, *J. Vac. Sci. Technol.* A, Vol. 8 (3), 2376-2381. ISSN 0734-2101

Gindl M, Sinn G, Gindl W, Geiterer A, Tschegg S, (2001), A comparison of different methods to calculate the surface free energy of wood using contact angle measurements, *Colloids and surfaces A*, Vol. 181 (1-3), 279-287. ISSN 0927-7757

Heng, J.Y.Y.; Bismarck, A.; Lee, A.F.; Wilson, D.; & Williams, D.R. (2006). Anisotropic surface energetics and wettability of macroscopic form I paracetamol crystals, *Langmuir*, Vol. 22 (60), 2760-2769. ISSN 0743-7463

Hirosawa, I. (1996). Method of characterizing rubbed polyimide film for liquid crystal display devices using reflection ellipsometry, *Jpn. J. Appl. Phys.* Vol. 35 (11), 5873-5875. ISSN: 0021-4922

Inagaki, N.; Tasaka, S.; & Hibi, K. (1992), Surface modification of Kapton film by plasma treatments, *J. Polym. Sci. A: Polym Chem.*, 30 (7), 1425-14311. ISSN 0887-624X

Kamusewitz, H.; Possart, W. & Paul, D. (1999). The relation between Young's equilibrium contact angle and the hysteresis on rough paraffin wax surfaces, *Colloids Surf A*, Vol. 156 (1-3), 271-279. ISSN 0927-7757

Kumar, G. & Prabhu, K. N. (2007). Review of non-reactive and reactive wetting of liquids on surfaces, *Adv. Colloid Interface Sci.* Vol 133 (2), 61-89. ISSN 0001-8686

Lam, C. N. C.; Wu, R.; Li, D.; Hair, M. L. & Neumann, A. W. (2002), Study of the advancing and receding contact angles: liquid sorption as a cause of contact angel hysteresis, *Adv Colloid Interface Sci.* Vol. 96 (1-3), 169-191. ISSN 0001-8686

Lee, K. W.; Lien, A.; Paek, S. H.; During, C. J. & Fukuro, H. (1996). Microscopic molecular reorientation of alignment layer polymer surfaces induced by rubbing and its effect on LC pretilt angles, *Macromolecules,* Vol. 29 (27), 8894-8899. ISSN 0024-9297

Lee, K. W.; Lien, A.; Stathis, J. H. & Peak, S. H. (1997), Control and modification of nematic liquid crystal pretilt angles on polyimides, *Jpn. J. Appl. Phys.* Vol. 36(6A), 3591-3597. ISSN 0021-4922

Li, D. (1996). Drop size dependence of contact angles and line tensions of solid-liquid systems, *Colloid. Surf. A*, Vol. 116 (1-2), 1-23. ISSN 0927-7757

Morra, M.; Occhiello, E. & Garbassi, F. (1990), Knowledge about polymer surfaces from contact angle measurements, *Adv. Colloid Interface Sci.* Vol. 32 (1), 79116. ISSN 0001-8686

Naddaf, M.; Balasubramanian, C.; Alegaonkar, P. S.; Bhoraskar, V. N.; Mandle, A. B.; Ganeshan, V. & Bhoraskar, S. V. (2004). Surface interaction of polyimide with oxygen ECR plasma, *Nucl. Instr. Meth. Phys. Res. B*, Vol. 222 (1-2), 135-144. ISSN 0168-583X

Owens, D. K. & Wendt, R. C. (1969). Estimation of the surface free energy of polymers, *J. Appl. Polym. Sci.* Vol. 13 (8), 1741-1747. ISSN 1079-4628

Sakamoto, K.; Arafune, R.; Ito, N.; Ushioda, S.; Suzuki, Y. & Morokawa, S. (1994). Molecular orientation of rubbed and unrubbed polyimide films determined by polarized infrared absorption, *Jpn. J. Appl. Phys.* Vol. 33 (9B), L1323-L1326. ISSN: 0021-4922

Sawa, K.; Sumiyoshi, K.; Hirai, Y.; Tateishi, K. & Kamejima, T. (1994). Molecular orientation of polyimide films for liquid crystal alignment studied by infrared dichroism, *Jap. J. Appl. Phys.* Vol. 33 (11), 6273-6276. ISSN: 0021-4922

Sacher, E. (1978), The effect of paracrystal formation on the surface tension of annealed polyimide, *J. Appl. Polym. Sci.*, 22 (8), 2137-2139. ISSN 1079-4628

Smart, B. E. (1994), in Organofluorine Chemistry: Principles and Commercial Applications, Banks, R. E.; Smart, B.E. & Tatlow, J. C. Ed, Plenum Press: New York. ISBN: 978-0-306-44610-8

Sroog, C. E. (1976). Polyimides, *J. Polym. Sci. Macromol. Rev*, Vol. 11 (1), 161-208.

Stohr J.; Samant M. G.; Cossy-Favre A.; Diaz J.; Momoi Y.; Odahara S.; Nagata T. (1998), *Macromolecules*, Vol. 31, 1942. ISSN 0024-9297

Vanfleet, R. R. & Mochel, J. M. (1995). Thermodynamics of melting and freezing in small particles, *Surf. Sci.* Vol. 341 (1-2), 40-50. ISSN 0039-6028

Wu, H. A. (2006). Molecular dynamics study of the mechanics of metal nanowires at finit temperature, *Eur. J. Mech. A/Solid* , Vol. 25(2), 370-377. ISSN 0997-7358

Yu, W. & Stroud, D. (1997). Molecular-dynamics study of surface segregation in liquid semiconductor alloys, *Phys. Rew. B*, Vol. 56(19), 12243-12249. ISSN 1098-0121

Yudin, M. & Hughes, B. D. (1994). Surface energy of solids, *Phys Rev. B*, Vol. 49 (8), 5638-5642. ISSN 1098-0121

Zheng, W.; Underwood, I.; Macdonald, B. F. & Cole, R. J. (2004). The effect of rubbing strength on the formation of zigzag defects in surface stabilised ferroelectric liquid crystals, *Mol. Cryst. Liq. Cryst.* Vol. 412(1), 237-245. ISSN: 1542-1406

Zheng, W.; Lu, C. H. & Ye, Y. C. (2008). Effects of mechanical rubbing on surface tension of polyimide thin films, Jpn. J. Appl. Phys. Vol 47 (3), 1651-1656. ISSN: 0021-4922

Zheng, W. J.; Wang, C. C. & Lu, C. H. (2009). Preparation of zigzag-free ferroelectric liquid crystal between rubbed polyimide thin films, *J. Phys. D: Appl. Phys.* Vol. 42 (4), 045402. ISSN 0022-3727

Zisman, W. A. (1963). Influence of constitution on adhesion, *Ind. Eng. Chem.* Vol. 55 (10), 18-38.

Zuo, M.; Takeichi, T.; Matsumoto, A. & Tsutsumi, K. (1998). Surface characterization of polyimide films, *Colloid Polym. Sci.* Vol. 276(7), 555-564. ISSN 0303-402X

11

Cryochemistry of nanometals

Tatyana I. Shabatina and Gleb B. Sergeev
Department of Chemistry, M.V. Lomonosov Moscow State University
Leninskie Gori 1/3, 119991 Moscow, Russia

1. Introduction

The development of nanotechnologies is one of the most promising prospective of nowadays (Poole&Owens, 2003). Nanoscience has been established as a new interscience field of research. It can be defined as a whole knowledge on fundamental properties of nanosized objects. The results of nanoscience are realized in nanotechnology as new materials and functional facilities. At present time nanochemistry becomes one of the main growing directions of nanoscience objects (Sergeev, 2006; Ozin&Arsenault, 2005). Understanding the peculiarities controlling the size, shape and self-organization of nano- and subnanosized particles and the properties of materials including such particles is the main problem of nanochemistry. Another problem is connected with the existence of size effect. As size effect we can define the qualitative changes in physical and chemical properties and chemical activity depending on the number of atoms and molecules in nanosized particle. The existence of such dependences is a particular feature of nanochemistry.

The use of low temperatures temperature technique (4-100 K) enlarges the possibilities of nanochemistry and opens new prospects in creation of bulk and film materials with new conducting, protecting and sensor properties. Low temperatures and matrix isolation methods are used for stabilization of highly energetic and very active metal species as atoms, clusters and nanoparticles. Using metal atoms, clusters and nanoparticles the effect of reacting particles size (number of atoms) on their chemical activity and properties of reaction products can be revealed. This effect is the intrinsic feature of nanochemistry, which is the base of production of new compounds and materials with unusual properties. (Sergeev, 2001; Sergeev&Shabatina, 2002; Shabatina&Sergeev, 2003; Shabatina&Sergeev, 2007).

The main scope of this work is to combine the unique properties of metal atoms, clusters and nanoparticles with different organic and inorganic substances, particularly liquid crystals and polymers using methods of cryochemistry. The joint and separate condensation of metals (Ag, Mg, Cu, Pb, Sm and Eu) and different active and inert organic and inorganic compounds on the cooled surfaces and in cooled liquids have been made. The problems of stabilization and of activity and selectivity in competitive reactions of metal species are discussed.

2. Size effects in reactions of metal atoms and clusters stabilized in matrices of noble gases and hydrocarbons

Low temperatures and matrix isolation methods are used for stabilization of highly energetic and very active metal species as atoms, clusters and nanoparticles. The scheme of our cryochemical synthesis and some methods of characterisation of film samples obtained are presented in Fig.1. Low temperatures can be used also for study of unusual chemical reactions of metal species. The effect of reacting particle's size (size-effect) on their chemical activity and properties of reaction products is of great interest. As size effect we can define the qualitative changes in physical and chemical properties and chemical activity depending on the number of atoms and molecules in nanosized particle. For metal species of several nanometers in size, containing up to 10 nm in size (10-1000 atoms) the dependence of the reaction rate possesses not monotonous character (Sergeev, 2003). The existence of such dependencies is a particular feature of nanochemistry at low temperatures and is based on changing of electronic and geometry structure of metal species by raising the number of atoms formed the particle. Analyzing of such dependencies is of great importance for understanding of the nature of size effects, which can be considered as structural-size effects.

Metal atoms and small metal particles possess high chemical activity (Klabunde, 1994). The interaction between separated metal atoms and ligand molecules can be described by the following scheme including parallel and consecutive reactions, where M is metal atom and L is ligand molecule [5]. Aggregation of metal atoms (the reaction pathway 1) and their reactions with ligand molecules (the reaction pathway2):

The problem of producing of the exact compound can be solved for the reactions of naked clusters in the gas phase using double mass-spectral selection method under molecular beam conditions. For metal species of several nanometers in size, containing up to 100 atoms the dependence of the reaction rate ordinary possesses not monotonous character.

It is important to note, that physical and chemical properties of metal atoms and small clusters in the gas phase and stabilized in inert gas matrices, for example, argon under low temperatures are practically the same. This fact allowed us to discover the size effects in chemical reactions of metal atoms and clusters in condensed phase. One of the exiting examples is the interaction of magnesium atoms and clusters with carbon tetrachloride in low temperature film co-condensates of different metal concentration. It is important that the reaction doesn't occur for bulk metal at ambient temperatures. In low temperature co-condensates according to the results:

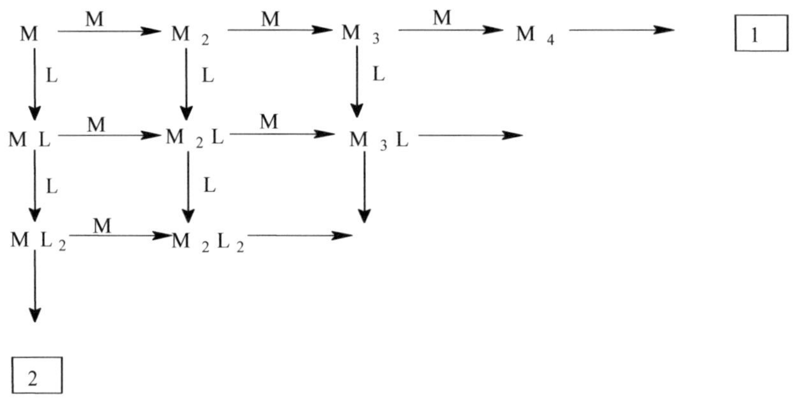

The spectra of different magnesium species stabilized in argon at 10 K are presented in Fig.2a (Mikhalev et.al, 2004). The changes in UV-VIS spectra in presence of carbon tetrachloride are shown in Fig.2b. The data presented allowed us to compare the changes in relative activity of magnesium species and it made possible to assume that the activity of magnesium particles in reaction with carbon tetrachloride decreases in the series $Mg_2>Mg_3>Mg_n>Mg$. According to the results of IR-spectroscopic study of Mg-CCl_4-Ar co-condensate system hexachloroethane and tetrachloroethylene are the main products of the reaction (Rogov et.al, 2004). The experimental data and theoretical quantum chemistry calculations allowed us to propose the following scheme of chemical transformations in this case:

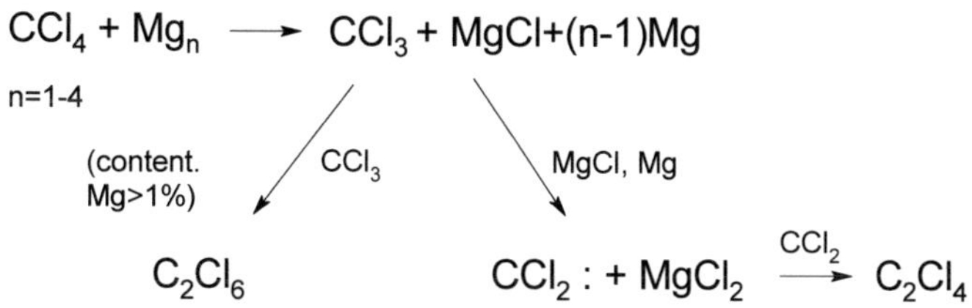

Some interesting results have been obtained for Sm/CO_2 co-condensate system. The spectroscopic study of low temperature co-condensates of samarium with carbon dioxide in argon matrix allowed us to propose the following reaction scheme (Sergeev, 2001):

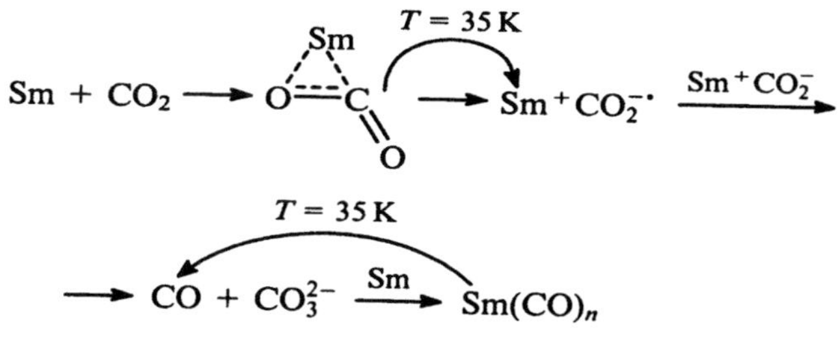

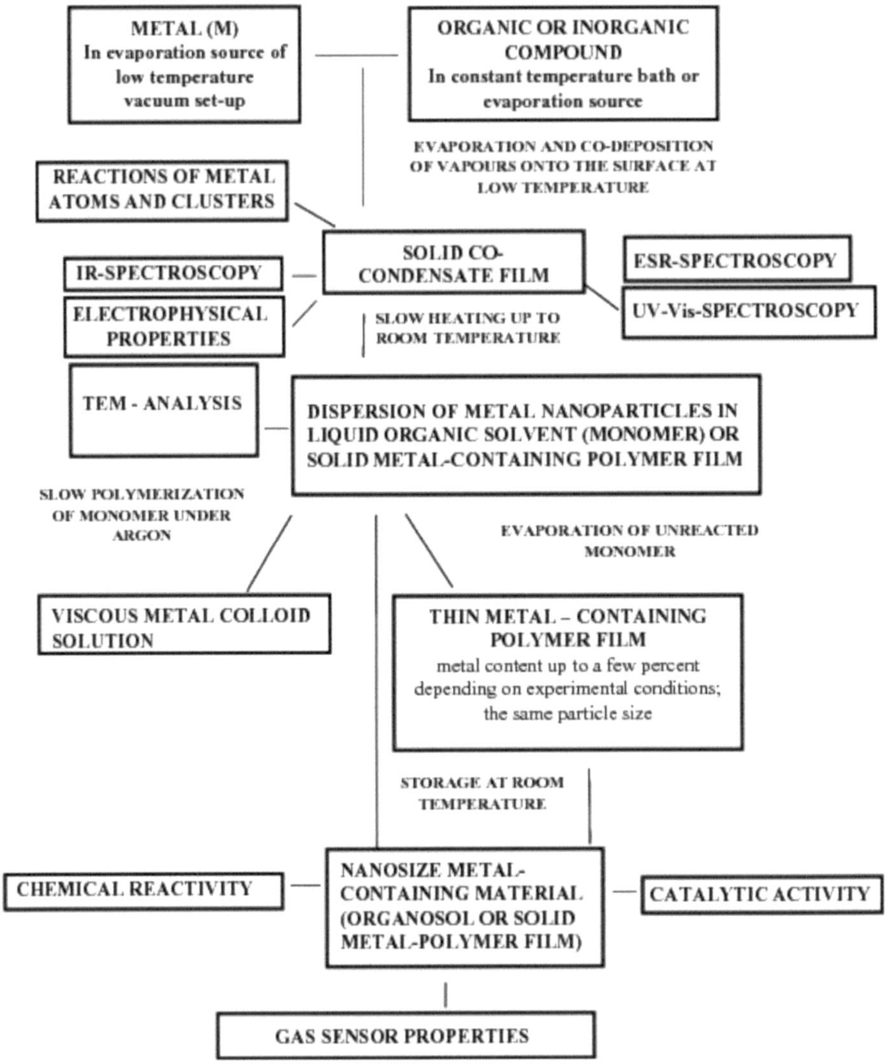

Fig. 1. Cryochemical synthesis of nanosize materials encapsulated into inorganic, organic and polymeric matrices (Sergeev & Shabatina, 2008)

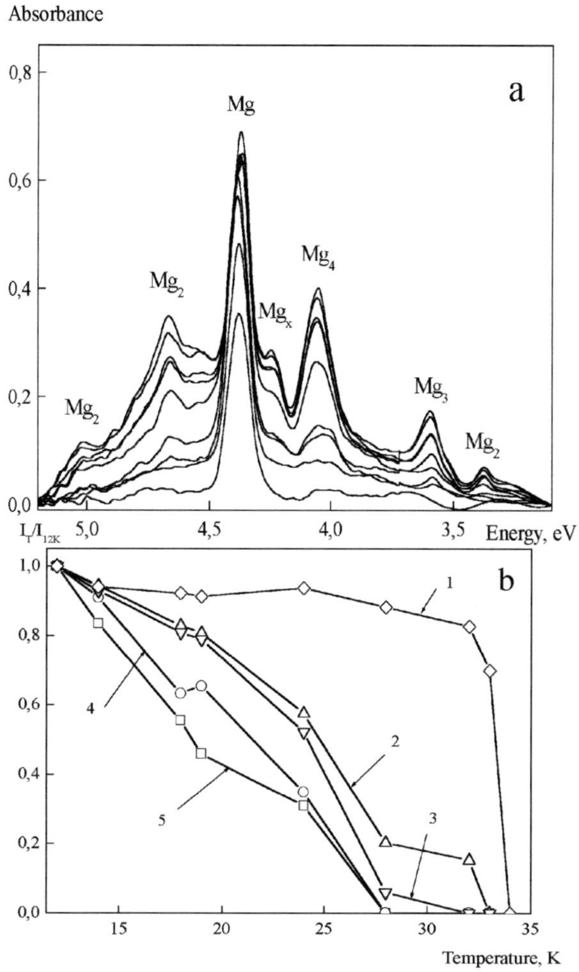

Fig. 2. The changes in UV-visible spectra in the temperature range 12–35 K in the system Mg : CCl_4 : Ar = 1 : 100 : 1000 (a) and normalized integral intensity of magnesium particles absorption at different temperatures: (*1*) Mg, (*2*) Mg_4, (*3*) Mg_x, (*4*) Mg_3, (*5*) Mg_2. (*b*) (Mikhalev et.al, 2004)

The relative activity of samarium species in reaction with carbon dioxide at different temperatures show the higher activity of samarium clusters as compared with samarium atoms.

Evidently the effect of metal particle size on its reactivity is of primary importance for the development of nanochemistry. From our viewpoint of no less importance is to compare the chemical activity of uni-size particles of different chemical nature, of different substances. The data of Table 1 allowed us to compare the activity of atoms and clusters of two different metals - magnesium and samarium and for each metal particles of the same size with two different ligands CO_2 and C_2H_4. It was shown that in double systems: Mg-CO_2 and Mg-C_2H_4, Sm-CO_2 and Sm-C_2H_4 both metals react with both ligands CO_2 and C_2H_4, but in triple systems Mg-CO_2-C_2H_4 and Sm-CO_2-C_2H_4 both metals react performable with carbon dioxide molecules. These results reveal the problem of activity and selectivity in competitive reactions of metal species (Sergeev, 2001).

Metal		Ligand	
specie	CO_2	C_2H_4, C_2D_4	CH_3X, X=Cl, Br
Mg	$Mg^+CO_2^-$ at matrix annealing		CH_3MgX at irradiation (λ=280 nm)
$Mg_{2\text{-}4}$	$Mg^+CO_2^-$ at co-condensation	cycle $Mg(C_2H_4)_2$ at matrix annealing	CH_3MgX at irradiation
Mg_x	$Mg^+CO_2^-$ at matrix annealing		(λ>300 nm)
Sm	$Sm^+CO_2^-$, CO, $SmCO_3$ at matrix annealing	complexes $Sm(C_2H_4)$–(C_2D_4) and $Sm(C_2H_4)_2$–$(C_2D_4)_2$	methane
Sm_2	$Sm^+CO_2^-$, CO, $SmCO_3$		at co-condensation
Sm_x	co-condensation		

Table 1. Reaction products of magnesium and samarium with ligands at 10–40 K

3. Competitive interactions of metal species with organic molecules in low temperature co-condensates.

Cryochemical synthesis allowed us to obtained new compounds and complexes for series of d- and f-metals (Shabatina, 2007). The competitive interactions of atoms and dimers of europium and samarium with alkylcyanobiphenyls and cyanophenylpyridines and metastable complexes have been obtained and characterized by FTIR, UV-Vis and ESR-spectroscopy (Shabatina et.al, 2005, Vlasov 2005). The formation of two sandwich-like complex structures with different stoichiometric metal to ligand ratio ML_2 and M_2L_2 were shown by combination of spectral data with the results of DFT-B3LYP modeling of the system. The solid state transformation of mononuclear lanthanide complex to the binuclear one was established by heating of the system up to 173-243 K.

The reactions of silver and copper atoms and clusters were studied using mesogenic alkylcyanobiphenyl compounds as stabilizing matrix and carbon tetrachloride as the third active reagent (Shabatina, 2003; Timoshenko, 2005). In framework of these investigations the following tasks were solved:to carry out the cryosynthesis of new metal atom and cluster complexes and to study their thermal stability; to establish the competitive reactions and relative chemical activity of metal species of different size with the reagent molecules added to the system; to establish of intermediate reaction products and propose the reaction mechanism. It was shown the competition in reactions of silver atoms and clusters with carbon tetrachloride and complexation with cyanobiphenyl molecules. The triple Ag-CCl_4-5CB co-condensate systems of different reagent's ratios from 1:1:100 to 1:10:1000 have been studied by ESR-technique in temperature range 77-350 K. There are three reaction pathways for silver atoms and small clusters existing in cocondensate samples at different temperstures. It was established the competitive formation of silver atoms complexes of π-type with two cyanobiphenyl molecules, formation of σ-type complexes with two different ligands 5CB and CCl_4, and aggregation of the silver atoms and small clusters resulting in formation of silver nanoparticles. The preferred reaction channel was dependent upon the reagent ratio, and, first of all, on carbon tetrachloride contents in the system. By heating up of the co-condensate system silver complexes undergo decomposition and freed silver atoms aggregated with formation of additional amount of metal nanoparticles. The formation of $AgCl{*}CCl_3^{\bullet}$ complex was shown also for triple Ag-CCl_4-5CB system and also for double Ag-CCl_4 system. Depending on carbon tetrachloride content in the system and temperature we can obtain performable stabilized in matrix metal clusters or paramagnetic chloro-containing products (Fig.3).

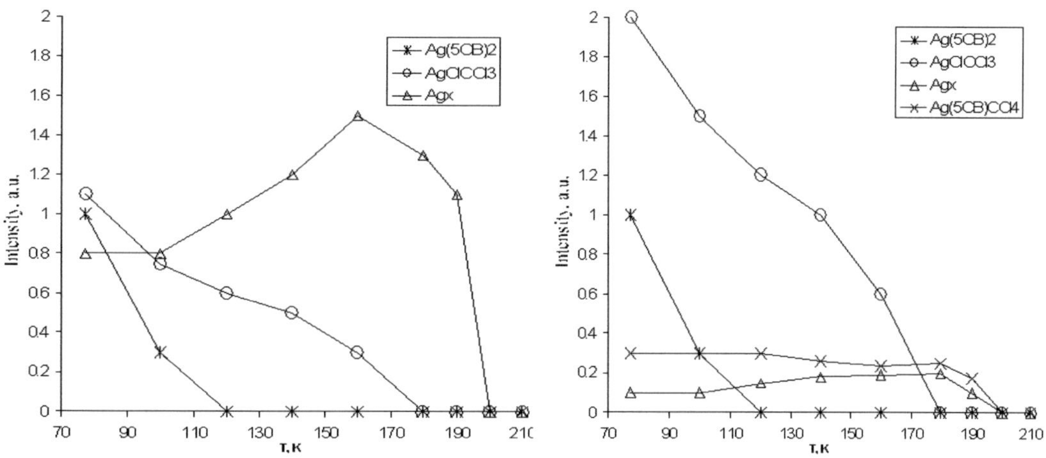

Fig. 3. Temperature dependence of the relative yields of paramagnetic products of metal atoms and clusters reactions in Ag/5CB/CCl_4 co-condensate. Δ- Agx, ④-AgCl-CCl_3; ⊠ - $Ag(5CB)_2$; x-$Ag(5CB)CCl_4$.

4. Encapsulation of nanometals in thin polymer films

Low temperature technique was used also for encapsulation of metal particles of definite size in polymer films. Metal atom aggregation was prevented by sterical hindrance of polymer matrix cage (Sergeev, 2005). In this case we have used different monomers which can polymerize just at low temperatures. One of them is highly reactive para-xylylene monomer, obtained via pyrolysis of para-di-xylylene. The process is presented by the following scheme:

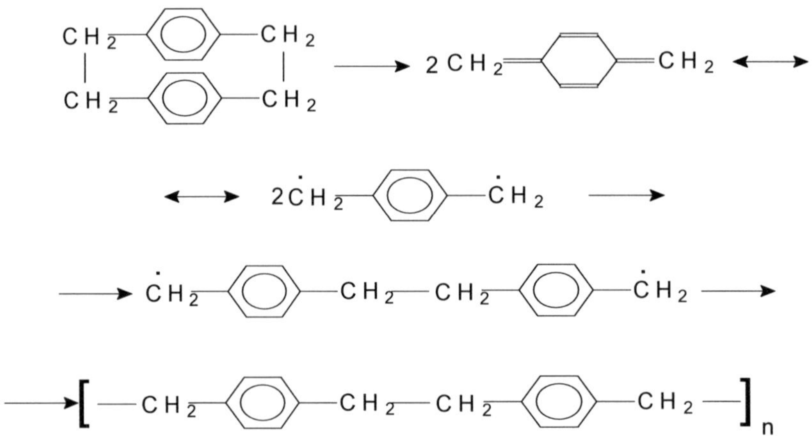

Polymeric films containing of aggregates of metal atoms were obtained by joint and layer-by-layer condensation on cooled surface (Sergeev et.al, 1995, Sergeev 2006). Co-condensate samples were polymerized by heating up to 110-130 K or by light irradiation at 80 K. The polymer poly-para-xylylene films with incorporated metal particles could be withdrawn from the reaction vessel and studied by different physical and chemical methods. The particle size histogram demonstrates rather narrow size distribution over the range 2-8 nm. The average diameter of the particles was estimated as 5,5 nm. The nanosize particles of Zn, Cd, Ag, Mg and Mn were also incorporated in poly-para-xylylene films.

We have developed also cryochemical and chemical synthetic methods for incorporating of metal particles into polyarylamide gels (Sergeev et.al, 1998; Sergeev et.al, 1999; Sergeev, 2006). It is interesting to compare properties of Ag-MA and Pb-MA polymer stabilized sol systems. Lead particles in contrast to silver do not initiate m methylacrylate (MA) polymerization. The behavior of bimetallic Ag-Pb-MA system with the respect to MA polymerization resembled the properties of Pb-MA system rather than for Ag-MA system, bimetallic system does not initiate polymerization of MA monomer. Thus, bimetallic system possesses nonadditive changes of individual nanoparticle properties. Probably, the presence of lead in the system inhibits silver-induced polymerization of MA. It is important that lead particle size didn't exceed 5 nm for both Pb/MA and Ag/Pb/MA systems. The diameter of silver nanoparticles formed under the same conditions is 10-15 nm.

Polymer films including metal nanoparticles open new possibilities for synthesis of the materials with promising properties. High sensitivity for ammonia and water vapors was shown for films containing lead nanoparticles (Sergeev et.al, 1997). Such films were proposed as new ammonia sensors having a response exhausted 3-4 orders of magnitude in their electrical conductivity (Bochenkov, 2002; Bochenkov 2005). As an example of the importance of cryosynthesis conditions, the microstructures of two samples containing nearly the same amounts of lead, but deposited at different rates are presented in Fig.4. The microstructure influences the electrical properties, thus, only sample (b) was found to be sensitive to humidity. After deposition and controlled annealing, the particles can be oxidized totally or partially to form a highly porous sensitive layer. A careful control over the deposition parameters, such as the condensation rate, evaporation rate and substrate temperatures allows us to obtain condensates with the required structure and also opens up the possibility of chemical modification of the surface and grain boundaries.

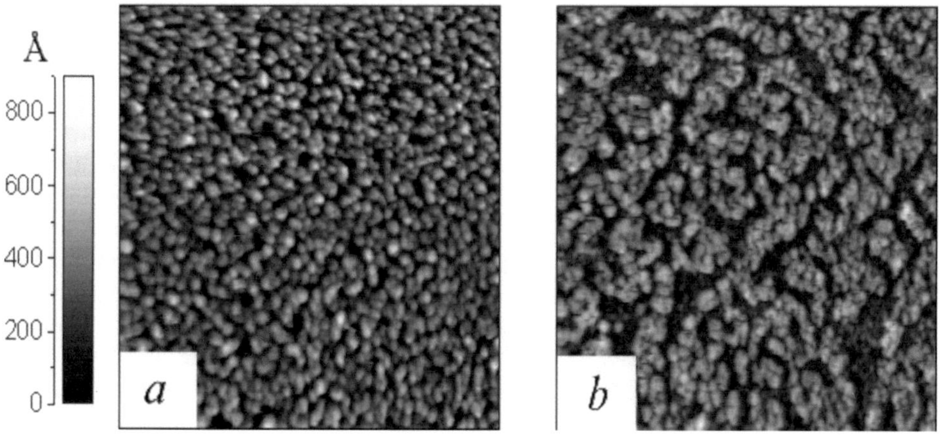

Fig. 4. AFM images of Pb condensates, deposited at 80 K after annealing to room temperature and exposure to the air. Scan area is 5mm 5 m m. (a) 34.3 ML deposited at 0.05 ML/s, no conductance onset during the deposition was observed. (b) 28.1 ML deposited at 0.20 ML/s, the conductance onset was observed. (Bochenkov, 2005)

5. Metal-mesogenic nanosystems formed by cryocondensation technique

New metal-mesogenic hybrid nanosystems were obtained based on silver, copper and mesogenic alkylcyanobiphenyls derivatives, which formed diversely organized solid and liquid crystalline phases with different type of molecular organization. In temperature range 80-300 K the controlling of metal particles size in range 2-30 nm and their morphology and aggregation processes were made (Shabatina, 2002; Shabatina, 2003a). Metal containing cyanobiphenyl film samples (l=20-50 μm) were obtained by reagent's vapor co-condensation on the cooled surfaces of quartz, KBr or CaF_2 or polished copper under molecular beam conditions. IR, UV-Vis and ESR spectroscopic studies of the samples were realized in situ, in vacuum, using special spectral cryostats. The systematic IR-, UV- and ESR-study of specific interactions were carried out for low temperature co-condensates of mesogenic cyanobiphenyls and cyanophenylpyridines with transition metals of the first group (Ag, Cu)

at 80-350K. The spectroscopic data combing with the results of quantum chemistry calculations showed for alkylcyanobiphenyl system the formation of sandwich-like silver complex at low temperatures and the structure with head-to-tail arrangement of two ligand molecules was proposed. It was established that annealing of the samples up to 150-200 K led to complex thermal degradation and metal nanocluster formation. The kinetics of complex thermal degradation and metal cluster's growth were studied by ESR technique. It was shown the kinetics retarded character revealed the existence of activation energy distribution for complex thermal degradation. The main value of process activation energy was estimated as 30 kJ/mole, the silver cluster formation was shown also for UV-irradiation of the samples at low temperatures. UV-Vis spectroscopic study of silver/CB and copper/CB systems at 80 K showed the appearance of new absorbance band at 360 and 420 nm due to complexes formation. These bands disappeared at 200 K and the new wide bands with maxima at 440 and 560-600 nm were detected at 200-300 K for silver and copper co-condensates, correspondingly. The aggregation of metal atoms and/or small clusters via complex decomposition and formation of nanosize metal particles could cause it.
Low temperature layer-by -layer co-condensation of silver, mesogenic compound 4-pentyl-4-cyanobiphenyl (5CB) and 4-octyl-4-cyanobiphenyl (8CB) and para-xylylene monomer followed by heating of the obtained film sample resulted in encapsulation of such hybrid metal-mesogenic system into polymer film. Transmission electron and probe microscopic study (TEM, AFM) of the samples show the existence of two kinds of metal particles stabilized in nematic (orientationally ordered) alkylcyanobiphenyl matrix at elevate temperatures. These were globular silver particles with diameter of 15-30 nm and highly anisometric rod-like metal particle with length more than 200 nm. The preferential growth of rod-like metal particles regulated by the increase in metal to ligand ratio. The use of smectic (layered organized) phases of higher homology 4-octyl-4-cyanobiphenyl (8 CB) led to the formation of flat quasi-fractal aggregates.

6. Conclusions. Prospects of thin films contatining nanosized metal particles in nanochemistry, catalysis and electronic materials

This survey of literature data and of the results obtained by the authors of the review shows that the methods of cryochemistry and cryonanochemistry make it possible to carry out and control self-assembling processes of metal atoms to form subnano- and nanosized aggregates, perform competitive chemical interactions of atoms, small clusters and nanosized metal particles with different organic and inorganic substances. At present time the effective approaches of cryochemistry are being developed aimed on stabilization of metal atoms, dimmers, trimmers and higher clusters and metal particles in inert gas matrices and polymer films and by certain organic substances layers at different substrates. The use of low temperatures and controlled condensation of reagent vapors allowed us to obtain and stabilized metal particles of 1 nm and less in size. Such hybrid organic-inorganic film materials can find their applications in catalysis and as highly sensitive chemical sensors.
The accumulation of data on the reactions of the variety of different metals for a wide range of temperatures in condensed systems, including solid co-condensate films, will allow one to reveal the fundamental peculiarities of nanochemistry, and at first the manifestation of size effects and clearing its nature for the competitive reactions of metal species and the

periodicity in the variation of reactivity of these objects. This information together with the comparative analysis of characteristics of ligand-free and ligand-stabilised metal nanoparticles makes it possible to separate the effects of metal core and stabilizing organic shell effects on the chemical and catalytic activity of the system at different temperatures. This approach should be coupled with the quantum chemical modeling of the systems under investigation.

The development of cryonanochemistry is aimed on the elaboration of new methods for synthesis of nanostructured film materials with the unique electronic, optical, magnetic, electrophysical and mechanical characteristics. It is necessary to carry out highly selective catalytic, photo- and thermo -induced chemical transformations as the basis in the production of nanodevices and surface modification processes. The quest of promising applications is associated with the development of methods for the formation of ordered assembles of metal particles of definite sizes and shapes incorporated in thin polymeric films, liquid crystals, carbonaceneous and biological systems. The use of low temperature vapor condensation technique offers a possibility of the direct introduction of metal atoms and clusters into organic and polymer films without using additional chemical reagents and solvents. This can result in the development of environmentally clean methods for the synthesis of nanostructured film materials with special properties. Of great importance is the use of such methods for production of multifunctional hybrid films based on chemically modified metal particles and biologically active substances, which includes the development of new systems for medical diagnostics and target delivery of drug substances.

The works have been financially supported by Russian Foundation of Basic Research, grants 04-03-32748, 05-03-32293, 08-03-00798 and 08-03-00712 and INTAS programme.

7. References

Bochenkov V.E. &G.B.Sergeev, (2005). Preparation and chemiresistive properties of nanostructured materials. *Adv.Coll.Int.Sci.*, 116, N1-3, 245-254.

Vlasov A.V., Shabatina T.I., Ivanov A.Yu., Sheina G.G., Nemukhin A.V. & Sergeev G.B. (2005). Interaction of lantanide atoms with 4-pentyl-4-cyanobiphenyl in low temperature co-condensates. *Mendeleev Commun.* N1, pp. 10-11.

Klabunde K.J. (1994). *Free atoms, Clusters and nanosized particles*, Academic Press, San-Diego.

Mikhalev S.P., Soloviev V.N. & G.B.Sergeev. (2004). Cryoreactions of magnesium atoms, clusters and nanoparticles with polyhalomethanes. *Mendeleev Commun.*, N2, pp.48-50.

Ozin G.A. & Arsenault A.C.(2005) *Nanochemistry: A chemical Approach to Nanomaterials.* RCS Publ., Toronto.

Poole Ch.P. & Owens F.J.(2003) *Introduction to nanotechnology.* Wiley-Interscience, ISBN 5-94836-021-0.

Rogov Mikhalev S.P., Granovskii A.A., Soloviev V.N., Nemukhin A.V., Sergeev G.B., (2004),. *Moscow University Chemistry Bulletin*, 45, N 4, pp. 214-224.

Sergeev B.M., Sergeev G.B. & Prusov A.N. (1998). Cryochemical synthesis of bimetallic nanoparticles in the silver-lead-metthylacrylate system.*Mendeleev Commun.*, p. 1.

Sergeev G.B., Zagorsky V., Petrukhina M., Zavyalov S., Grigori'ev E.& Trahtenberg L. (1997). Preliminary study of the interaction of metal nanoparticle-containing poly-para-xylylene films with ammonia. *Analitycal Commun.* 34, pp. 113-114.

Sergeev G.B., Sergeev B.M., Shabatina T.I., Nemukhin A.V. (1999). Cryosynthesis and properties of metal-organic nanomaterials. *Nanostructured Materials*, 12, pp. 1113-1115.

Sergeev G.B.& Shabatina T.I. (2002). Low temperature surface chemistry and nanostructures. *Surface Science*. 500, pp. 628-655

Sergeev G.B. (2003). Cryochemistry of metal nanoparticles. *J.Nanoparticle Research*, 5, 529-537.

Sergeev G.B. (2006) *Nanochemistry*. Elsevier Publ., Amsterdam, ISBN-13:978-0-444-51956-6 250 pp.

Sergeev G.B. & Shabatina T.I. (2008). Cryochemistry of nanometals. Colloids and Surfaces A: Physicochemical and Engineering Aspects, 2008, 313-314, pp. 18-22.

Shabatina T.I., Timoshenko V.A., Morosov Yu.N. & Sergeev G.B., (2002).Thermal andlight-induced nanocluster formation in silver-mesogenic cyanobiphenyl films. *Material Science and Engineering C*, 22, N2, pp. 193-196.

Shabatina T.I. (2003). Low temperature reactions in mesogenic cyanobiphenyls. *Moscow University Chemistry Bulletin*. 57, N5, pp. 20-36.

Shabatina T.I. & Sergeev G.B. (2003). Reactions at low temperatures in chemistry of nanosystems. *Rus. Chem. Rev*. 72, pp. 643-663.

Shabatina T.I., Timoshenko V.A., Morosov Yu.N. & Sergeev G.B., (2003). The ESR-study of complexation and nanoclusters growth in silver-liquid crystal system. *Mol.Cryst. and Liq. Cryst.*, 390, 43-47.

Shabatina T.I. (2007) Metastable complexes of d- and f-block metals with cyanobiphenyls. *Struct.Chem.*, 18, pp. 511

Shabatina T.I.; Mascetti J.; Ogden J.S. & G.B.Sergeev. (2007). Competitive cryochemical reactions of transition metal atoms, clusters and nanosized particles. *Russian Chemical Reviews*, 76 (12), pp. 809-825.

Timoshenko V.A., Shabatina T. I., Morosov Yu.N. &. Sergeev G.B., (2005) The ESR-study of chemical interactions in triple solid'Ag-CCl_4-5CB' cocondensate mesogenic system. *Applied Surface Science*, 246, pp. 420-424.

12

High Performance Organic Thin-Film Transistors and Nonvolatile Memory Devices Using High-κ Dielectric Layers

Albert Chin
National Chiao-Tung University
Taiwan, ROC

1. Introduction

Pentacene-based organic thin-film transistors (OTFTs) are attractive because of their inherit merits of low cost, small weight and visible-light transparency, for potential use in applications such as organic displays, flexible displays and low-cost integrated circuit (IC) (Klauk et al, 1999; Zhou et al, 2005). The low thermal budget and rapid processing have strong advantages of energy saving and environment friendly (Chang et al, 2008), which are in sharp contrast to the prolonged 600°C annealing in conventional solid-phase crystallized (SPC) poly-crystalline silicon (poly-Si) TFTs (Hung et al, 2005). Although low thermal budget poly-Si TFTs can also be formed on plastic substrate using excimer laser annealing (Lemmi et al, 2004), the uniformity of threshold voltage (V_t) and mobility are the major concerns for TFTs on different poly-Si grains. Alternatively, although even single crystal sub-μm MOSFETs can be realized on plastic substrate (Kao et al, 2005) by fabrication first, thinning down the substrate, transferring and bonding, this method still requires high thermal budget for device fabrication.
However, conventional OTFTs require a high operating voltage and show a poor sub-threshold swing, which are opposite to the low power technology trend and detract from their suitability in IC operation. Besides, the relative low transistor current is difficult to drive the need high operation current of organic light-emitting diode (OLED). To address these issues, high dielectric constant (κ) material is required for OTFTs to improve the device performance. Using high-κ HfLaO as the gate dielectric, the pentacene OTFTs fabricated on SiO_2 showed even comparable device performance with SPC poly-Si TFTs, with extra merit of much better sub-threshold swing for low voltage and low power application (Chang et al, 2008). Similar good device performance was also achieved using high-κ HfLaO on pentacene OTFTs, fabricated on low-cost flexible polyimide substrates and useful for portable low power electronics (Chang et al, 2009).
Besides the logic TFTs, non-volatile memory function is also necessary for system-on-panel (SOP) application (Yin et al, 2008). Previously, OTFT memory devices have been reported using polymer insulators as the charge trapping layer (Baeg et al, 2006). Nevertheless, the OTFT memory devices require low program/erase (P/E) voltages, low reading voltages and

long data retention. Using the high-κ dielectrics for pentacene non-volatile memory fabricated on flexible polyimide substrate, low P/E voltage and reasonable long data retention were reached (Chang et al, 2008). The low P/E voltage results from the high gate capacitance using high-κ layers, while the small band-gap high-κ HfON trapping layer with deep trapping energy yields acceptable data retention. Further performance improvement of OTFT-based non-volatile memory is expected using advanced device design such as charge-trapping-engineered flash (CTEF) (Lin et al, 2008). The good device performance of logic OTFT and OTFT-based non-volatile memory should be useful to realize the flexible displays and low-cost IC in the near future.

2. High-κ Dielectric OTFT

2.1 Device Requirements

The technology goals for OTFT ICs are to achieve high speeds, large drive current and low power consumption. The high speed is necessary for SOP without using external Si logic ICs, while the low power is required for portable electronics. The high circuit speed arises from the high drive current of the OTFT, since the circuit delay (τ) is determined by

$$\tau = C_{load}V_{max}/I_{drive} \quad (1)$$

Here the C_{load}, V_{max} and I_{drive} are the load capacitance, maximum voltage and drive current respectively. The current of OTFT is expressed as the following equation of a Metal-Oxide-Semiconductor Field-Effect Transistor (MOSFET)

$$I_d = \mu C_{ox}(W/L)[(V_g-V_t)V_d-V_d^2/2] \quad (2)$$

The μ and C_{ox} are the mobility and gate capacitance density, respectively. The L, W, V_g and V_d are the transistor's gate length, gate width, gate voltage and drain voltage, respectively. The source voltage is connected to ground (0 V). The maximum transistor drive current for the OTFT operated at saturation region ($V_g-V_t<V_d$) is simplified as

$$I_d = \mu C_{ox}(W/L)(V_g-V_t)^2/2 \quad (3)$$

A low V_t is beneficial for high drive current and low voltage operation, which is limited to a minimum 0.11 V of $4kT/q$ based on the thermal noise considerations. The increasing V_g can give the required high drive current but unfortunately increases the power consumption. Although the desired high drive current is reachable by decreasing L, the minimum L is limited to μm range from the practical consideration of low cost lithography. The increasing W can also increase the drive current, but this is opposite to the needed high resolution of display pixel. This is a fundamental challenge to use OTFT to drive high current OLED (Nomura et al, 2004), because of the low drive current from relatively low mobility.
One useful method to improve the drive current is to increase C_{ox}:

$$C_{ox} = \varepsilon_0 \kappa / t_{ox} \quad (4)$$

Although higher C_{ox} can be reached by decreasing the dielectric thickness (t_{ox}), this is restricted by the exponentially increasing leakage current. Therefore, the most effective method is to use high-κ material as the gate dielectric.

Currently, the high-κ gate dielectric has been used for Intel's 45 nm node MOSFET, with a small equivalent oxide thickness (EOT) of 1 nm. Here the EOT is defined as

$$\mathrm{EOT} = t_{ox}(\kappa_{sio2}/\kappa) = t_{ox}(3.9/\kappa) \quad (5)$$

The κ_{sio2} is the dielectric constant for SiO_2, which has a value of 3.9.

It is important to notice that the combination of inorganic high C_{ox} and good mobility organic channel is the few practical method to realize the integrated OTFT with OLED, because of the needed high operation current for OLED. The high-κ dielectric OTFT can be implemented to the existing process line of Si TFT for low cost and light weight displays shown in following sections.

2.2 OTFT Device Design

One important advantage of OTFT is the significantly lower thermal budget and fast device process than SPC poly-Si TFT. However, the low hole mobility and poor sub-threshold swing are the drawbacks that give unwanted low drive current and slow turn on when operated at low voltage.

To overcome these problems, we used high-κ HfLaO as the gate dielectric for OTFTs. To mimic poly-Si TFTs fabricated on glass panel, we first fabricated the metal-gate and high-κ OTFTs on a thick SiO_2 layer grown on Si wafers. This is because of the different substrate size and processing equipments used for commercial glass panel and high-κ dielectric process for Si MOSFET. Figure 1 shows the schematic device diagram of the high-κ dielectric and organic channel OTFT, where a bottom gate electrode is used.

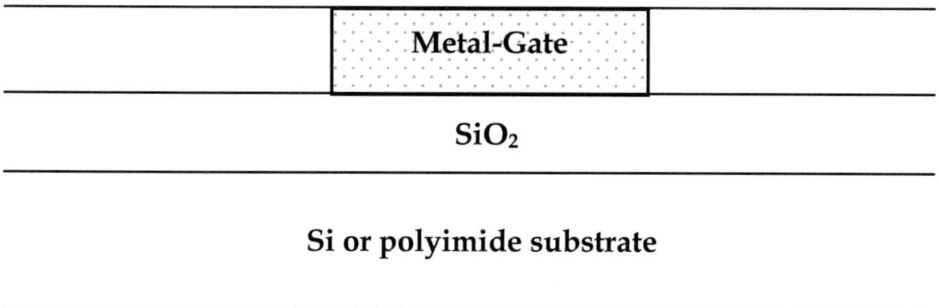

Fig. 1. Schematic device diagram of the OTFT using high-κ gate dielectric and metal gate.

Currently, the pentacene is the commonly used organic material for OTFT. The first question is what kind of gate electrode should be used. The metal gate offers a simple and low temperature process compared with doped poly-Si gate used in sub-μm MOSFET. However, the choice of proper metal-gate with adequate work-function is important to reach low leakage current of pentacene/high-κ/metal-gate OTFT.
Figure 2 shows the schematic diagram of energy-levels of molecular pentacene. For comparison, those of the single crystalline Si are also plotted. The energy levels for small-molecular materials like pentacene are defined as the lowest unoccupied molecular orbital (LUMO) and the highest occupied molecular orbital (HOMO), which are different from the electronic bands in crystalline lattice such as Si. The appreciable energy bands in polymer material may be obtained for highly crystalline defect-free material but this is difficult to reach. The important current conduction for OTFT is mainly along the direction of the polymer chain. For pentacene, the majority carriers for conduction are the holes that are occupied at HOMO energy level with work-function very close to the valence band of Si crystal with only 0.1 eV difference.

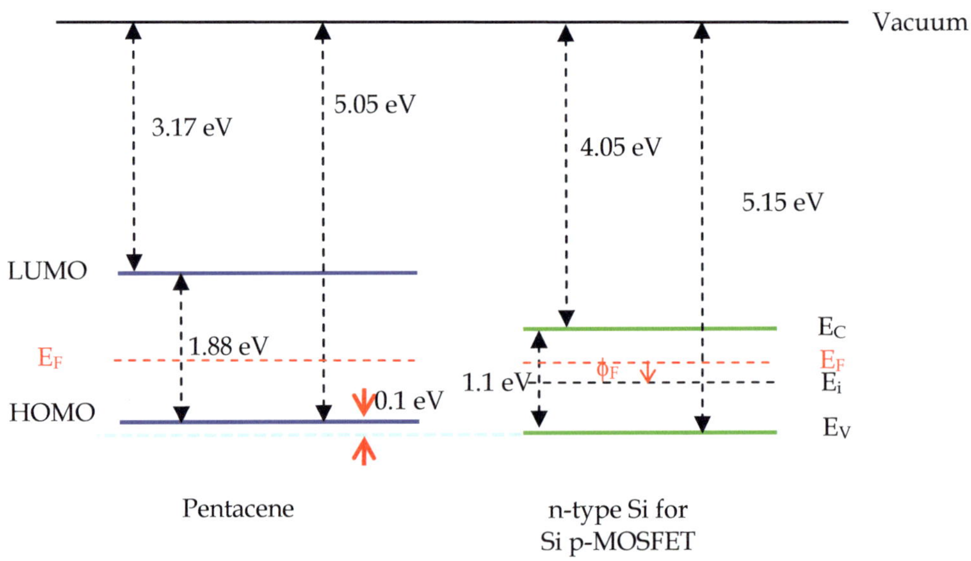

Fig. 2. Schematic energy-level diagrams of molecular pentacene and single crystal Si. An n-type doping in Si body is needed to form the p-MOSFET, but pentacene is difficult to dope into n-type.

For low V_t pentacene OTFT, the gate electrode should have work-function close to HOMO energy similar to a hole conductive p-type Si MOSFET (p-MOSFET). The V_t of the Si p-MOSFET is expressed as

$$V_t = \phi_{MS} - Q_{ox}/C_{ox} - 2\phi_F - Q_{dpl}/C_{ox} \qquad (6)$$

Here ϕ_{MS} (ϕ_M-ϕ_S), Q_{ox}, Q_{dpl} and $2\phi_F$ are the work-function difference of metal-gate and pentacene, oxide charge, depletion region charge and surface bending voltage to turn on

OTFT, respectively. The ϕ_F is energy difference of Fermi level (E_F) to intrinsic level (E_i), which is related to doping concentration (N_D) and intrinsic carrier density (n_i) in Si by

$$\phi_F = (kT/q)\ln(N_D/n_i) \quad (7)$$

Note that the Si body of p-MOSFET is doped with opposite n-type doping. This lowers the leakage current in a p-MOSFET at off-state, when transistor is biased at V_g=0 V and V_{ds}<0 V. The n-type doping moves E_F toward E_C and forms energy barriers of p^+-n junctions in source and drain that lowers the off-state leakage current.

The V_t equation can also give important guideline to design the pentacene OTFT, although some of the important parameters are unknown in pentacene. Nevertheless, the pentacene is generally p-type and difficult to dope into n-type. Therefore, no leakage current blocking p^+-n junctions can be easily formed. One alternative method to lower the off-state leakage current is to form the partially depleted OTFT structure shown in Fig. 3(a), which has an opposite surface bend bending to accumulation mode shown in Fig. 3(b). This can be reached using relative low work-function metal-gate compared with the high work-function one used for low V_t Si p-MOSFET. However, the needed low leakage current is traded off the slightly high V_t, since higher V_g is required to form the strong hole accumulation mode and turn on the OTFT shown in Fig. 3(c). Such partially or even fully depleted channel has previously led to the invention of low leakage current Ge-on-Insulator (GOI) MOSFET (Huang et al, 2003), where the very high leakage current from small energy bandgap Ge (0.66 eV) is the major concern of the Ge MOSFET (Chin et al, 2005).

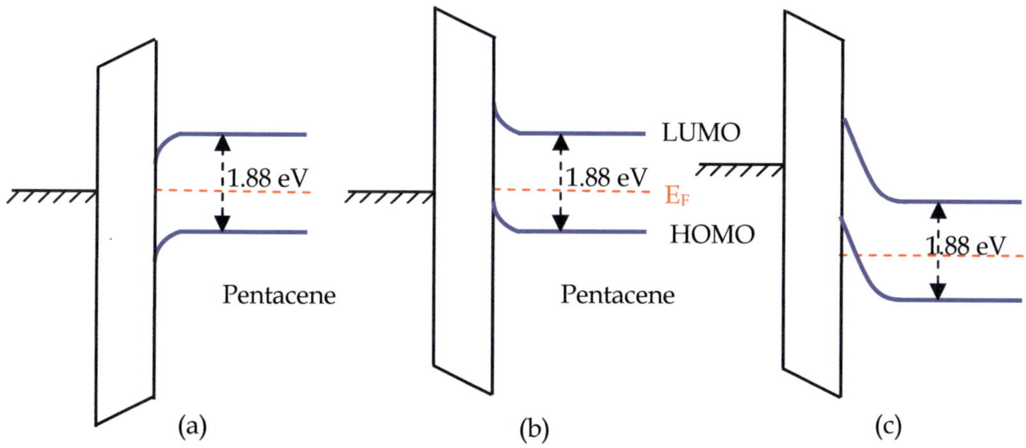

Fig. 3. Schematic energy-level diagrams of pentacene/oxide/metal-gate OTFT with (a) low and (b) high work-function metal-gate under equilibrium and V_g=0; (c) under a negative V_g bias to reach strong hole accumulation.

Although many elements in the periodic table have low work-function to meet the requirement of low off-state leakage, we choose the TaN as the gate electrode. This TaN electrode has been developed previously for metal-insulator-metal (MIM) capacitors used for Analog/RF and Dynamic Random Access Memory (DRAM) applications (Chiang et al, 2006), where similar bottom electrode device structure and process sequence were used. The reason to choose TaN is because it has larger bond enthalpy than pure metals, which also has one of the strongest bond enthalpies among various metal-nitrides in the periodic table (Yu et al, 2005). Such robust metal-gate is required to lower the oxidation rate of gate electrode during subsequent high-κ gate dielectric deposition and post-deposition anneal (PDA), under an O_2 ambient. The oxidized metal-surface will degrade the EOT and leakage current by trap-assisted tunneling.

The second question is the choice of source-drain contact metals. Figure 4 shows the work-function of atoms in the periodic table. There are only 3 stable elements of Ir, Pt and Au that have work-function larger than the HOMO energy. The higher work-function than HOMO allows direct hole conduction via Schottky contact, without using p^+ doping in pentacene. Such direct source-drain metal contact provides unique advantages for low temperature and fast process compared with the high thermal budget of n^+-doped SPC poly-Si TFT (Hung et al, 2005), which is the enable technology on low temperature flexible substrate. Note that the Ni metal may be a low cost choice since it has reported 5.0~5.1 eV work-function (Chiang et al, 2007) close to that of HOMO. However, the vital OTFT drive current may decrease exponentially with increasing energy barrier of metal-pentacene energy difference, if the metal work function is slighter lower than HOMO level. Here the Au is the most widely used source-drain contact metal for pentacene, due to the significantly lower melting temperature than Pt and Ir for easy deposition using a simple thermal evaporation.

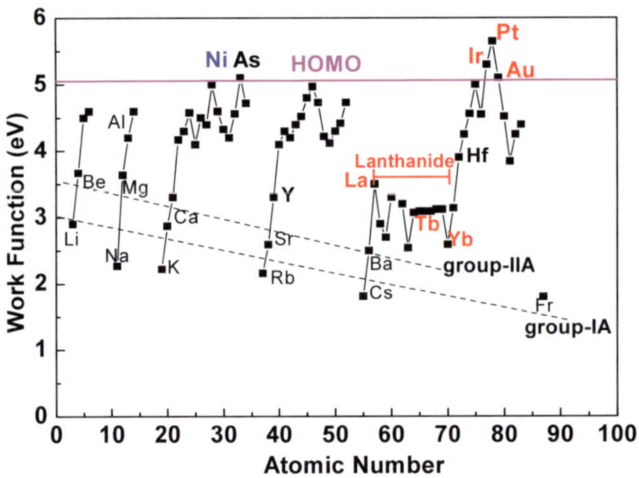

Fig. 4. Metal work-function as a function of atomic number in the periodic table.

The next question is the choice of proper high-κ gate dielectric for OTFT. Figure 5 shows the potential binary gate oxides for pentacene OTFT. Generally, the energy bandgap of high-κ dielectric and valence or conduction band offset (ΔE_V or ΔE_C) to pentacene decrease monotonically with increasing κ value. The higher κ value provides the wanted higher drive for OTFT. Nevertheless, a large enough ΔE_V or ΔE_C energy barrier of ~1 eV is required to reach the low gate leakage current for practical application. Fortunately, the ΔE_V is larger than ΔE_C for most high-κ dielectrics that is useful to reach low gate leakage current in p-type OTFT. Therefore, the HfO_2, ZrO_2, La_2O_3 (Chin et al, 2000) and TiO_2 (Cheng et al, 2008) are the few good candidates for pentacene OTFT. However, the κ value of TiO_2 or $SrTiO_3$

depends strongly on the growth temperature and the degree of dielectric crystallization. The high process temperature of 450~550°C for higher κ $SrTiO_3$ (Chiang et al, 2006) is not suitable to integrate the OTFT onto organic substrate. Here the HfO_2 is the most used high-κ gate dielectric for Intel's 45 nm node MOSFET using gate-last technology. The La_2O_3 capping (Wu et al, 2000) on HfSiON has been widely used for low V_t 32 nm node gate-first high-κ n-MOSFET, such as IBM, Toshiba, IMEC, SEMATECH etc.

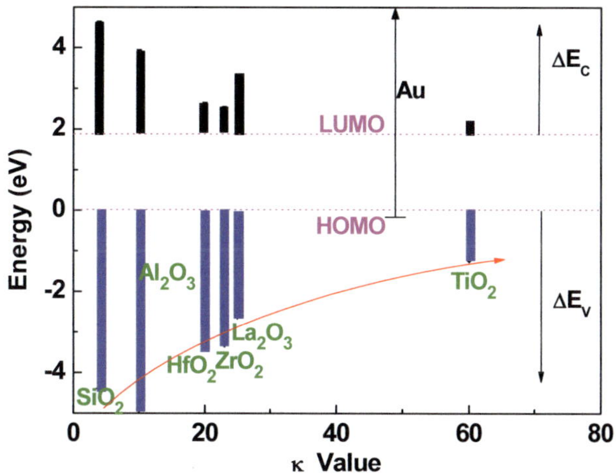

Fig. 5. Energy bandgap of oxides and band alignment to pentacene. Generally, both the bandgap and band offset become smaller with increasing κ value.

2.3 High Performance OTFT on SiO_2

The OTFT on SiO_2 has potential to replace the amorphous Si TFT on glass panel. The process of pentacene/high-κ/metal-gate OTFT starts with a 50 nm thick TaN gate electrode that was formed by sputtering deposition on the SiO_2/Si substrate through a shadow metal mask. An NH_3 plasma was further applied to the TaN surface (Chiang et al, 2005). Such NH_3 plasma treatment is the key factor to achieve low leakage current and small EOT in previous DRAM capacitors (Chiang et al, 2005; Cheng et al, 2008; Lin et al, 2009). Then the HfLaO gate dielectric with 20 nm thickness was deposited by electron beam evaporation at room temperature and followed by a 350°C O_2 PDA for 10 min. The adding La_2O_3 into HfO_2 is especially important to decrease the leakage current at low process temperature (Chang et al, 2008). After that the pentacene active layer with 70 nm thickness was deposited by electron beam evaporation through a shadow mask. The OTFT was finished by depositing 50 nm thick Au onto the pentacene and forming the source/drain contact electrodes.

Figure 6 shows the output characteristics of the fabricated pentacene/high-κ/metal-gate OTFT. Well-behaved I_d-V_d characteristics were measured under dark and air ambient, even at a low operation voltage of 2 V. This is the lowest reported operation voltage of OTFT for low power ($I_d \times V_d$) circuit application. Figure 7 shows the transfer characteristics of the pentacene/HfLaO/TaN OTFT. Here the V_t and mobility can be extracted from the linear $I_d^{1/2}$-V_g plot using *eq.* (3). This pentacene/high-κ/metal-gate OTFT shows excellent device integrity of a low V_t of -1.3 V, a record small sub-threshold swing of 78 mV/decade, good μ_{FE} of 0.7 cm²/Vs and a large on- and off-state current ratio (I_{on}/I_{off}) of 1.0×10^5 (Chang et al,

2008). The low V_t and small sub-threshold swing ensure the OTFT to operate at low voltage. Such small sub-threshold swing is the best reported data among Si TFT (Hung et al, 2005) and OTFT (Klauk et al, 1999; Zhou et al, 2005; Kawasaki et al, 2006) to date, to the best knowledge. This small sub-threshold swing is close to the theoretical ideal value of 60 mV/decade for MOSFET at room temperature and even comparable with that of a 0.18-μm single crystalline Si MOSFET fabricated from an IC foundry (Kao et al, 2006).
To further understand the record low sub-threshold swing $[d(\log_{10}I_d)/dV_g]^{-1}$ of OTFT, we have examined the relation

$$\text{Sub-threshold swing} = (kT/q)ln10[1+(C_{dpl}+C_{it})/C_{ox}] \quad (8)$$

Here C_{dpl} is the gate depletion capacitance density of pentacene. The C_{it} is the capacitance density from the pentacene/high-κ interface trapped charges that equals to interface trap density D_{it} by C_{it}/q. To measure the gate capacitance density C_{ox}, an Au/HfLaO/TaN capacitor was fabricated side-by-side along with the pentacene/HfLaO/TaN OTFT. A C_{ox} of 950 nF/cm^2 and a high-κ value of 24 were measured that gave a small EOT of 3.6 nm.

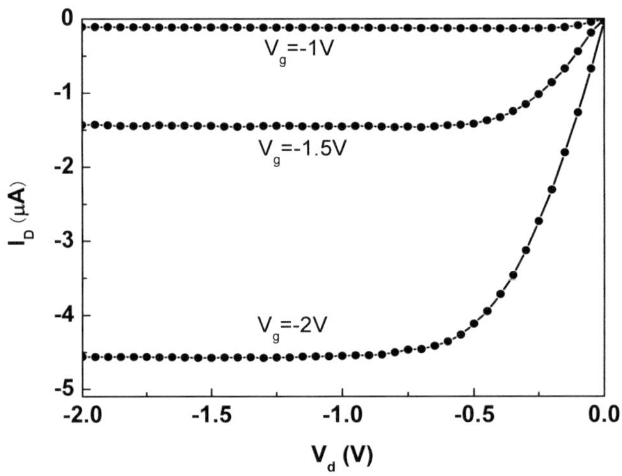

Fig. 6. Output characteristics of the pentacene/HfLaO/TaN OTFT. The device channel length and width are 80 μm and 2000 μm, respectively.

The C_{dpl} is smaller than high-κ C_{ox} and negligible, which leads to a D_{it} of 2×10^{12} eV^{-1}/cm^2 even processed at low temperature. This value is typical for high-κ gate dielectric MOSFET on single crystalline Si (Cheng et al, 2007). This good electrical interface property is the merit of using HfLaO as the gate dielectric for OTFT.

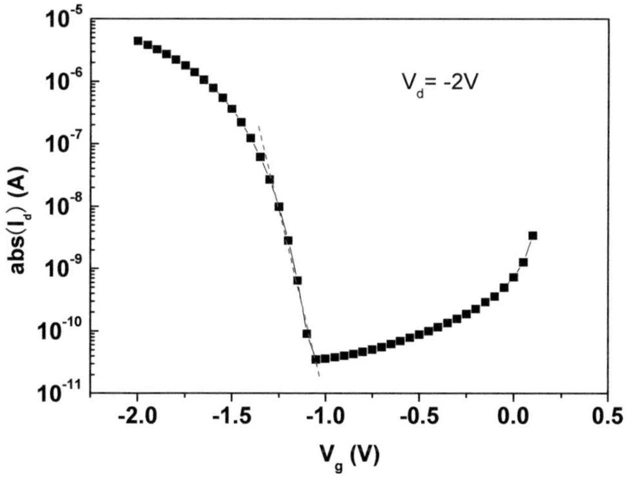

Fig. 7. Transfer characteristics of the pentacene/HfLaO/TaN OTFT with channel length and width of 80 μm and 2000 μm respectively. The sub-threshold swing is defined as the slope of I_d-V_g curve in voltage per current increase for 10X (a decade).

Table 1 summarizes the important device data of this p-channel OTFT and conventional n-channel poly-Si TFTs (Choi et al, 1999; Lin et al, 1999; Chang et al, 2003). The pentacene/high-κ/metal-gate OTFT has better device performance in terms of lower V_t and much smaller sub-threshold swing. This OTFT device also has comparable normalized drive current (μC_{ox}) with the poly-Si TFTs. Here the μC_{ox} is normalized to the channel length L, width W and over-drive V_G-V_T:

$$\mu C_{ox} = 2I_d\,(L/W)\,/\,(V_G-V_T)^2 \tag{9}$$

Thus, the high C_{ox} is as important as the high mobility for high drive current. Besides, higher mobility above 5 cm²/Vs is reachable by optimizing the chemistry (Lee et al, 2006). The excellent device characteristics ensues the good potential for OTFT used for display.

Active channel	Gate dielectric	V_t, V	Sub-threshold swing, V/decade	μ, cm²/Vs	μC_{ox}, nF/cm²	I_{on}/I_{off}
Pentacene (This work)	HfLaO	-1.3	0.078	0.7	674	1.0×10^5
poly-Si (Choi et al, 1999)	LPCVD SiO_2	5.6	1.4	20	863	3.5×10^5
poly-Si (Lin et al, 1999)	PECVD TEOS SiO_2	8.1	1.97	12	716	3.0×10^5
poly-Si (Chang et al, 2003)	PECVD TEOS SiO_2	-	2.67	3	259	-

Table 1. Device data of p-channel pentacene/HfLaO/TaN OTFTs and compared with n-channel poly-Si TFTs.

2.4 OTFT on Flexible Substrate

The important advantages of OTFT are its flexibility and low process temperature that can be used for flexible displays and low-cost flexible electronics. Based on the excellent device performance of pentacene/high-κ/metal-gate OTFTs on SiO_2 shown in previous section, the process temperature was further decreased to 200°C to fabricate the OTFTs on low-cost flexible polyimide substrates (Kapton HPP-ST, Dupont). This substrate is much more economical than other polyimide (Kapton E-type, DuPont) and polyethylene naphthalate (PEN) (Teonex Q65 PEN, Dupont) substrates, although the cost is traded off the degraded performance of poor surface roughness and impurity out-diffusion.

To improve the impurity diffusion, the polyimide substrates (Kapton HPP-ST, DuPont) with 125 μm thickness were annealed at 200°C in a vacuum environment. To lower the internal stress, a thin 100 nm SiO_2 was deposited on the polyimide substrate at room temperature by electron beam evaporation. Then the pentacene/HfLaO/TaN OTFTs were fabricated on SiO_2/polyimide by following the similar process steps in previous section on SiO_2/Si, with lower 200°C PDA temperature and thicker 30 nm HfLaO gate dielectric (Chang et al, 2009). The thicker HfLaO is to compensate the degraded gate leakage current when processed at lower temperature. The schematic diagram of fabricated devices is shown in Fig. 1, where the OTFTs were measured in the dark and air ambient.

Figure 8 shows the output characteristics of the fabricated OTFT on SiO_2/polyimide. Well-behaved I_d-V_d characteristics were obtained, under a low voltage operation of 2.5 V for low power electronics application. Figure 9 shows the I_d-V_g characteristics of the pentacene/high-κ/metal-gate OTFT on SiO_2/polyimide. The V_t and mobility were extracted from the linear $I_d^{1/2}$-V_g plot.

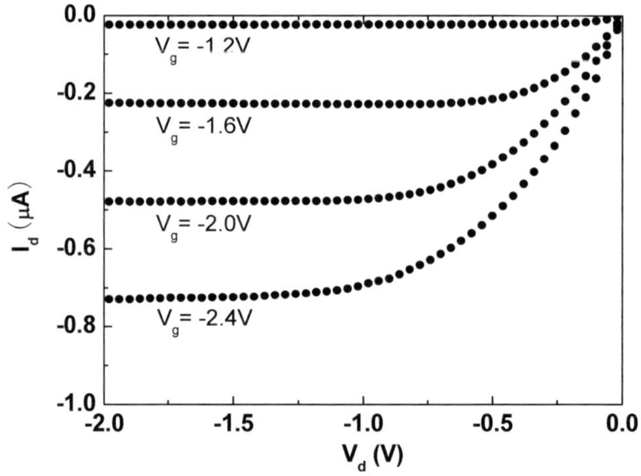

Fig. 8. Output characteristics of the pentacene/HfLaO/TaN OTFT on SiO_2/polyimide. The device channel length and width are 100 μm and 2000 μm, respectively.

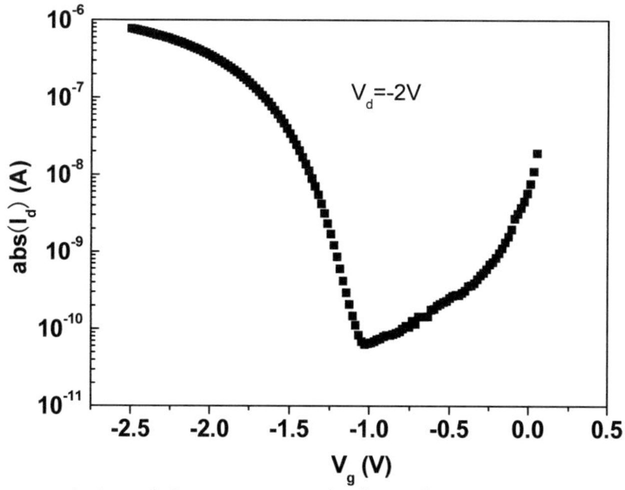

Fig. 9. Transfer characteristics of the pentacene/HfLaO/TaN OTFT on SiO_2/polyimide, with device channel length and width of 100 μm and 2000 μm respectively. The increasing current at V_g>-1 V suggests the electron conduction, but the small current is due to the high barrier source-drain electrodes.

The pentacene/HfLaO/TaN OTFT on SiO_2/polyimide shows good device performance of a low V_t of -1.25 V, a small sub-threshold swing of 130 mV/decade, a μ_{FE} of 0.13 cm^2/Vs and a I_{on}/I_{off} of 1.2×10^4 (Chang et al, 2009). The low operation voltage of -2.4 V is due to the low V_t and small sub-threshold swing. Such small sub-threshold swing is better than other reported values for flexible OTFTs (Choi et al, 2005, Kang et al, 2005; Klauk et al, 2005; Mizukami et al, 2006; Na et al, 2007), which is due to the using high-κ HfLaO dielectric with a high capacitance density of 450 nF/cm^2 and a small EOT of 7.7 nm. The smaller mobility and I_{on}/I_{off} than similar OTFT fabricated on SiO_2/Si were due to the rougher HfLaO surface of 4.2 nm measured by Atomic Force Microscopy (AFM), which was originated from the poor 9.0 nm surface roughness of low cost polyimide substrate (Kapton HPP-ST, Dupont). The drain current at V_g>-1 V may be attributed to the ambipolar behavior of electron conduction (Schön & Kloc 2001), although the current is limited by inadequate high work-function source-drain electrodes. However, this ambipolar conduction may be useful for pentacene-based non-volatile memory discussed in next session.

Table 2 compares the important device data of our work with other flexible pentacene OTFTs using various gate dielectrics and fabricated on high quality polyimide (Kapton E-type) and PEN (Teonex Q65 PEN, DuPont) substrates. The pentacene/HfLaO/TaN flexible OTFT shows comparable device performance of normalized drive current and I_{on}/I_{off} with other OTFTs fabricated on high cost substrates, with the added merit of the best sub-threshold swing for fast turn on and low voltage operation. The good device performance of low voltage operated OTFT, on low cost polyimide substrate, should lead to future more economic low-power flexible electronics.

Flexible substrate	Gate dielectric	V_t, V	Sub-threshold swing, V/decade	μ, cm^2/Vs	μC_{ox}, nF/cm^2	I_{on}/I_{off}
Polyimide Kapton HPP-ST (This work)	HfLaO	-1.25	0.13	0.13	58	1.2×10^4
Polyimide Kapton E-type (Choi et al, 2005)	BZN	0.1	0.3	0.5	110	3.5×10^5
Polyimide Kapton E-type (Kang et al, 2005)	Mn-doped $BaSrTiO_3$	-1	-	0.32	35	$<1\times10^3$
PEN (Klauk et al, 2005)	PVP	-	0.6	0.1	6	1×10^4
PEN (Mizukami et al, 2006)	Ta_2O_5	0.8	-	0.25	35	-
PEN (Na et al, 2007)	$TiSiO_2$	-0.88	0.315	0.67	95	1×10^4

Table 2. Device data of pentacene active channel OTFTs using various gate dielectrics and fabricated on different flexible substrates.

3. Organic Non-volatile Memory Devices Using High-κ Dielectric

3.1 Device requirements

Based on the good device performance of flexible OTFT, we can further apply to the non-volatile memory. Such non-volatile memory function is indispensible for SOP (Baeg et al, 2006). The first task for OTFT-based non-volatile memory is to choice the proper device structure that can meet the device requirement of low write power, long retention, good endurance and fast write speed.

Among various new types non-volatile memory devices such as Phase Change Random Access Memory (PRAM) and resistive RAM (RRAM), the SiO_2/poly-Si/SiO_2 floating gate flash memory device still has irreplaceable advantage of the lowest write current and being one-transistor (1T) in size compared with other one-transistor-one-resistor (1T1R) designs. Such low current is mandatory for page writing NAND logic and low power portable electronics. This floating gate flash memory shown in Fig. 10 (a) employs the well-known device physics and manufacturing experience of the Si industry. Digital data can be programmed into the device by injecting charges over tunnel oxide, or erased by removing the stored charges or injecting different polarity charges into the trapping layer. The program or erase function can be obtained by applying a large gate electric field across the tunnel oxide. This charge transfer is readable by measuring the V_t of the transistor.

The fundamental challenge of a conventional floating-gate flash device is that all the storage charges can leak out via a single oxide defect by trap-assisted tunneling shown in Fig. 10(a). This is due to the drawback of electrically conductive poly-Si charge storage layer. Unfortunately, such oxide defect is inevitable that can be generated by continuous device operation and measurable from the stress-induced leakge current (SILC). One method to address this issue is to use the [metal-gate]-oxide-nitride-oxide-semiconductor (MONOS) device structure. As shown in Fig. 10(b), the programmed charges are stored in discrete and electrically-isolated zero-dimensional quantum traps within the Si_3N_4, while only the trapped charges, close to oxide defects generated by write cycles, can leak out. Besides, the physically separated traps allow the storage of 2 bits/cell in both source and drain sides to increase the memory density.

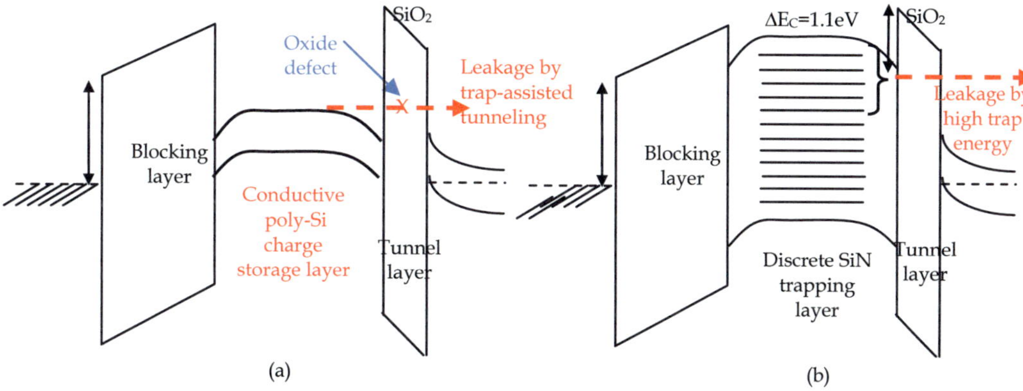

Fig. 10. Schematic energy band diagram of the electron trapping (a) poly-Si floating gate and (b) MONOS non-volatile memory devices.

In addition to the low write current and good endurance of multiple write cycles, the non-volatile memory devices also need good data retention and fast write speed. However, the charge-storage energy in Si_3N_4 traps is much higher than conventional poly-Si floating gate flash as shown in Fig. 10 that causes the stored charges to leak out. One method to maintain the good data retention, without scarifying the write speed, is to engineer the trap energy level. Here the deeper trapping energy level, the better data retention.

This can be reached by replacing the conventional Si_3N_4 with a high-κ metal-nitride dielectric shown in Fig. 11, where the later should have deeper ΔE_C or ΔE_V to barrier oxide layers for better charge retention. However, there are few metal-nitrides in the periodic table that can meet this requirement. Most metal-nitrides are metallic and similar to conductive poly-Si floating gate, which failed for discrete trapping MONOS application. The few candidates left are BN, AlN, GaN, InN etc, but the BN is difficult to process because of the strong bonding energy. Note that the AlGaN and GaInN have been widely used for blue-light LED. The using deep trapping energy AlN and AlGaN were initiated by Chin's group (Lai et al, 2005; Chin et al, 2005). Further lowering the P/E voltage with good retention can be reached by using higher κ synthesized HfON trapping layer (Lai et al, 2006). The significantly better retention using deep ΔE_C AlGaN trapping layer was also affirmed by Samsung's work (Joo et al, 2006). The alternative high-κ metal-nitride trapping layer was also listed in the *International Technology Roadmap for Semiconductors* (*ITRS*, 2007) for future generation charge-trapping flash (CTF) non-volatile memory.

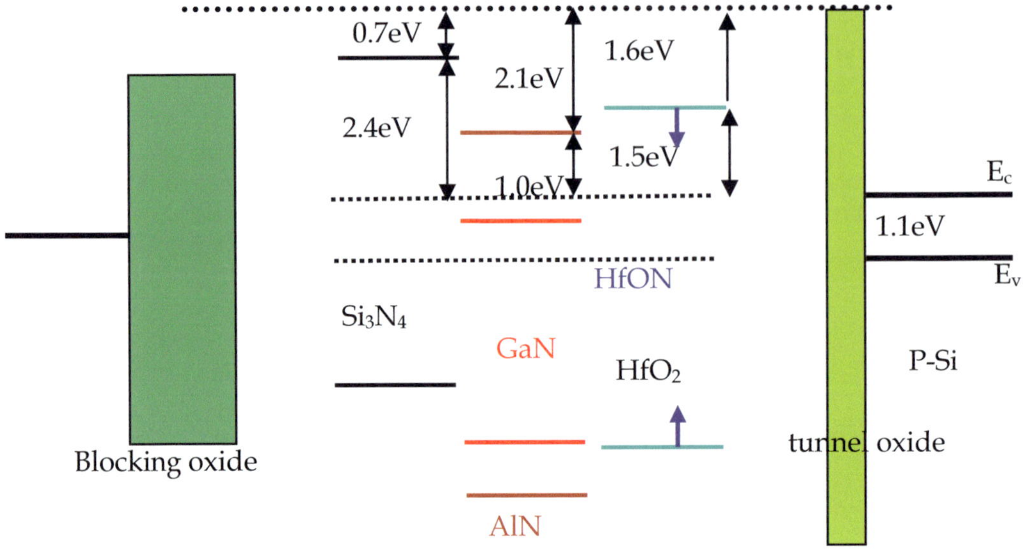

Fig. 11. Schematic energy band alignments for different trapping layers of Si_3N_4, AlN, GaN and HfON within the MONOS structure.

3.2 OTFT MONOS Non-volatile Memory on Flexible Polyimide

Similar to the OTFT on flexible polyimide, the bottom gate TaN-HfLaO-HfON-HfO_2-Pentacene (MONOP) memory was fabricated on a low-cost flexible polyimide substrate (Kapton HPP-ST). The device fabrication starts the substrates annealing at 200°C. Then 100 nm SiO_2 was deposited and 50 nm TaN gate electrode was formed. After applying the NH_3^+ plasma to TaN surface, the 20 nm HfLaO blocking oxide, 20 nm HfON trapping layer and 6 nm HfO_2 tunnel layer were deposited. Then a 200°C O_2 PDA was applied to improve the dielectric quality. After that 70 nm pentacene active layer was deposited at 70°C. The device was finished by making Au source-drain electrodes and forming ohmic-like contact for the hole injection. Here a thicker HfLaO blocking layer is important to reduce gate leakage current during erase.

The transfer hysteresis curves of OTFT MONOP memory device under P/E are shown in Fig. 12. From the initial transfer characteristics at a saturation V_d of -3 V, the V_t, sub-threshold swing, mobility, and I_{on}/I_{off} were -1.4 V, 160 mV/decade, 0.1 cm²/Vs and 1×10^4, respectively. These data are very close to the pentracene/HfLaO/TaN OTFT on the same flexible polyimide substrate shown in previous section. When a program V_g of -12 V and 1 ms was applied, the holes accumulated at HfO_2/pentacene interface can tunnel through the HfO_2 by the large gate electric field and trapped in the HfON layer. This increases the $|V_t|$ and shifts the I_d-V_g curve in negative direction. When an erase V_g of 12 V for 100 ms was applied, the trapped holes in the HfON may tunnel out over the HfO_2 by the electric field and lowered the $|V_t|$. Alternatively, the minority carriers of electrons may also be generated in the depletion region of pentacene, tunnel into the HfON and annihilate the trapped holes (Chin et al, 2006). Therefore, the V_t value can be shifted, reversibly, by applying an appropriate program or erase V_g. Nevertheless, the erase is much more difficult than program since the density of minority carriers is less in large bandgap pentacene.

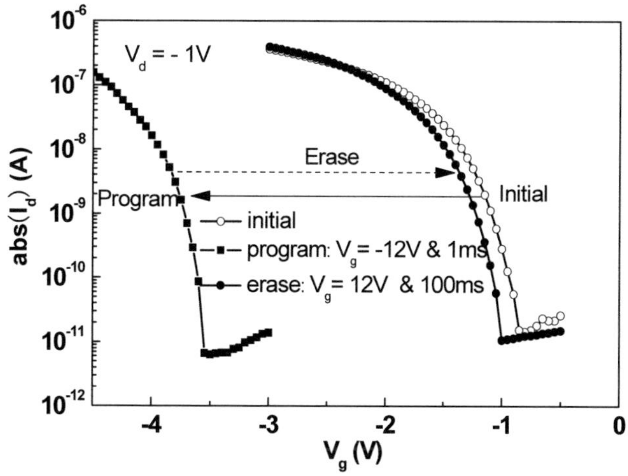

Fig. 12. Transfer hysteresis curves for a pentacene OTFT memory device under V_g = -12 V, 1 ms program and V_g = 12 V, 100 ms erase at V_d = -1 V. The channel length and width were 150 and 1500 μm, respectively.

Good retention characteristics are necessary for non-volatile memory. Figure 13 shows the retention data under a program and erase condition of -12 V for 1 ms and 12 V for 100 ms. Here the V_t was extracted from the linear I_d-V_g characteristics. The initial memory window of 2.4 V was obtained that decreased to 0.78 V after 48 hrs retention. The fast charge loss for the first 10^3 s is possibly due to the defects in the low temperature formed HfO_2 and the rough tunnel oxide surface by low cost polyimide substrate. The charge loss becomes stable at time longer than 10^3 s. If no other charge loss mechanism, an extrapolated 10-year retention window of 0.55 V may be reachable that is large enough for sense amplifier.

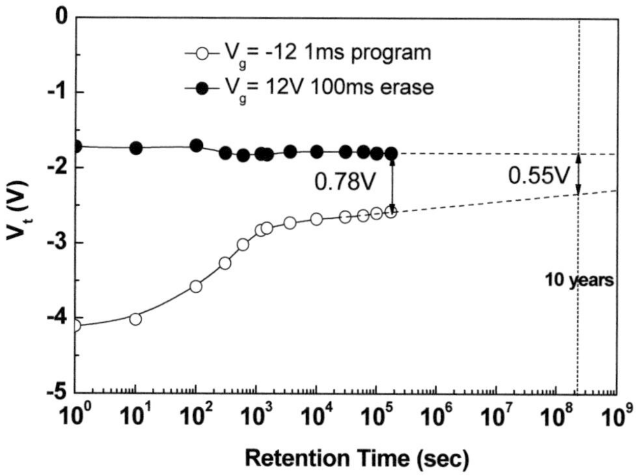

Fig. 13. Retention characteristics for pentacene MONOP OTFT device.

4. Conclusions and Future Directions

The high transistor current of OTFT is the key factor to drive the high operation current OLED and high resolution display. This can be achieved by increasing the mobility of pentacene and gate dielectric C_{ox} using high-κ dielectric. Although the highest reported mobility for pentacene is better than amorphous Si, it is still inferior to poly-Si or high mobility InGaZnO (Nomura, et al, 2004). This limit makes the using high-κ gate dielectric more important and inevitable. The using metal-gate is also necessary to decrease the OTFT leakage current and improve the I_{on}/I_{off} by forming the partially depleted OTFT. Good device integrity of pentacene/HfLaO/TaN OTFT on SiO_2 is reached with a low V_t of -1.3 V, the record best sub-threshold swing of 78 mV/decade, a good μ_{FE} of 0.7 cm^2/Vs and a large I_{on}/I_{off} of 1.0×10^5. The pentacene/HfLaO/TaN OTFT on low-cost flexible polyimide shows still reasonable good device performance of a low V_t of -1.25 V, a small sub-threshold swing of 130 mV/decade, a μ_{FE} of 0.13 cm^2/Vs and a I_{on}/I_{off} of 1.2×10^4. Based on the good OTFT device performance, the MONOP non-volatile memory shows an initial memory window of 2.4 V and a 48 hrs retention window of 0.78 V, under a low P/E condition of -12 V for 1 ms and 12 V for 100 ms. Further improving the retention may be achievable using advanced CTEF non-volatile memory structure and using smoother flexible substrate.

One of the reasons to give pentacene the higher mobility than amorphous Si is due to the short range ordering. Further improving mobility might be achievable by forming other ordering in organic martial such as Si Cage Clusters (Hiura, Miyazaki, & Kanayama, 2001).

The author would like to thank Mr. M. F. Chang's device fabrication and the support from National Nano Program, National Science Council of Taiwan, R.O.C.

5. References

Baeg, K. J.; Noh, Y. Y.; Ghim, J.; Kang, S. J.; Lee, H. and Kim, D. Y. (2006). Organic non-volatile memory based on pentacene field-effect transistors using a polymeric gate electret. *Adv. Mater.*, Vol. 18, No. 10, pp. 3179-3183.

Chiang, K. C.; Chin, Albert; Lai, C. H.; Chen, W. J.; Cheng, C. F.; Hung, B. F.; & Liao, C. C. (2005). Very high κ and high density TiTaO MIM capacitors for analog and RF applications, *Proceedings of Symp. on VLSI Technology*, pp. 62-63, Kyoto, Japan.

Chiang, K. C.; Huang, C. C.; Chen, G. L.; Chen, W. J.; Kao, H. L.; Wu, Y. H.; Chin, Albert & McAlister, S. P. (2006). High performance $SrTiO_3$ metal-insulator-metal capacitors for analog applications. *IEEE Trans. Electron Device*, Vol. 53, No. 9, pp. 2312-2319.

Chiang, K. C.; Cheng, C. H.; Jhou, K. Y.; Pan, H. C.; Hsiao, C. N.; Chou, C. P.; McAlister, S. P.; Chin, Albert & Hwang, H. L. (2007). Use of a high work-function Ni electrode to improved the stress reliability of analog $SrTiO_3$ metal-insulator-metal capacitors. *IEEE Electron Device Lett.*, Vol. 28, No. 8, pp. 694-696.

Chang, K. M.; Yang, W. C. & Tsai, C. P. (2003). Electrical characteristics of low temperature polysilicon TFT with a novel TEOS/oxynitride stack gate dielectric. *IEEE Electron Device Lett.*, Vol. 24, No. 8, pp. 512-514.

Chang, M. F.; Lee, P. T.; McAlister, S. P. & Chin, Albert (2008). Low subthreshold swing HfLaO/pentacene organic thin-film transistors. *IEEE Electron Device Lett.*, Vol. 29, No. 3, pp. 215-217.

Chang, M. F.; Lee, P. T.; McAlister, S. P. & Chin, Albert (2008). A flexible organic pentacene non-volatile memory incorporating high-κ dielectric layer. *Appl. Phys. Lett.*, Vol. 93, No. 23, p. 23302.

Chang, M. F.; Lee, P. T.; McAlister, S. P. & Chin, Albert (2009). Small-subthreshold-swing and low-voltage flexible organic thin-film transistors which use HfLaO as the gate dielectric. *IEEE Electron Device Lett.*, Vol. 30, No. 2, pp. 133-135.

Cheng, C. F;. Wu, C. H.; Su, N. C; Wang, S. J.; McAlister S. P. & Chin, Albert. (2007). Very low V_t [Ir-Hf]/HfLaO CMOS using novel self-aligned low temperature shallow junctions. *Proceedings of International Electron Devices Meeting (IEDM)*, pp. 333-336, Washington DC, USA.

Cheng, C. H.; Lin, S. H.; Jhou, K. Y.; Chen, W. J.; Chou, C. P.; Yeh, F. S.; Hu, J.; Hwang, M.; Arikado, T.; McAlister, S. P. & Chin, Albert (2008). The 300°C-processed high density TiO_2 MIM capacitors with low leakage current," *IEEE Electron Device Lett.*, Vol. 29, No. 8, pp. 845-847.

Chin, Albert; Wu, Y. H., Chen, S. B., Liao, C. C. & Chen, W. J. (2000). High quality La_2O_3 and Al_2O_3 gate dielectrics with equivalent oxide thickness 5-10Å, *Proceedings of Symp. on VLSI Technology*, pp. 16-17, Hawaii, US.

Chin, Albert; Laio, C. C.; Chiang, K. C.; Yu, D. S.; Yoo, W. J.; Samudra, G. S.; McAlister, S. P. & Chi, C. C. (2005). Low voltage high speed SiO_2/AlGaN/$AlLaO_3$/TaN memory with good retention. *Proceedings of International Electron Devices Meeting (IEDM)*, pp. 165-168, Washington DC, USA.

Chin, Albert; Kao, H. L.; Tseng, Y. Y.; Yu, D. S.; Chen, C. C.; McAlister, S. P. & Chi, C. C. (2005). Physics and Modeling of Ge-on-Insulator MOSFETs, *Proceedings of European Solid State Device Research Conf. (ESSDERC)*, pp. 285-288, Grenoble, France.

Choi, Y. W.; Lee, J. N.; Jang, T. W. & Ahn, B. T. (1999). Thin-film transistors fabricated with poly-Si films crystallized at low temperature by microwave annealing. *IEEE Electron Device Lett.*, Vol. 20, No. 1, pp. 2-4.

Choi, Y. W.; Kim, I. D.; Tuller, H. L.; & Akinwande, A. I. (2005). Low-voltage organic transistors and depletion-load Inverters with high-K pyrochlore BZN gate dielectric on polymer substrate. *IEEE Trans. Electron Devices*, Vol. 52, No. 12, pp. 2819–2824.

Hiura, H.; Miyazaki, T.; Kanayama, T. (2001). *Phys. Rev. Lett.*, Vol. 86, No. 9, pp. 1733-1736.

Huang, C. H.; Yang, M. Y.; Chin, Albert; Chen, W. J.; Zhu, C. X.; Cho, B. J.; Li, M.-F. & Kwong, D. L. (2003). Very Low Defects and High Performance Ge-On-Insulator p-MOSFETs with Al_2O_3 Gate Dielectrics, *Proceedings of Symp. on VLSI Technology*, pp. 119-120, Kyoto, Japan.

Hung, B. F.; Chiang, K. C.; Huang, C. C.; Chin, Albert & McAlister, S. P. (2005). High-performance poly-silicon TFTs incorporating $LaAlO_3$ as the gate dielectric. *IEEE Electron Device Lett.*, Vol. 26, No. 6, pp. 384-386.

International Technology Roadmap for Semiconductors (*ITRS*), 2007, Process Integration, Devices, & Structure Chapter, pp. 40-44. www.itrs.net

Joo, K. H.; Moon, C. R.; Lee, S. N.; Wang, X.; Yang, J. K.; Yeo, I. S.; Lee, D.; Nam, O.; Chung, U. I.; Moon, J. T. & Ryu, B. I. (2006). Novel charge trap devices with NCBO trap layers for NVM or image sensor. *Proceedings of International Electron Devices Meeting (IEDM)*, pp. 979–982, San Francisco, CA, USA.

Kang, K. T.; Lim, M. H.; Kim, H. G.; Choi, Y. W.; Tuller, H. L.; Kima, I. D. & Hong, J. M. (2005). Mn-doped $Ba_{0.6}Sr_{0.4}TiO_3$ high-κ gate dielectrics for low voltage organic transistor on polymer substrate. *Appl. Phys. Lett.*, Vol. 87, No. 24, p. 242908.

Kao, H. L.; Chin, Albert; Hung, B. F.; Lee, C. F.; Lai, J. M.; McAlister, S. P.; Samudra, G. S.; Yoo, W. J. & Chi, C. C. (2005). Low noise RF MOSFETs on flexible plastic substrates. *IEEE Electron Device Lett.*, Vol. 26, No. 7, pp. 489-491.

Kao, H. L.; Chin, Albert; Liao, C. C.; Chen, C. C.; McAlister, & Chi, C. C. Electrical stress effects and device modeling of 0.18μm RF MOSFETs. (2006). *IEEE Trans. Electron Device*, Vol. 53, No. 4, pp. 636-642.

Klauk, H.; Gundlach, D. J.; Nichols, J. A. & Jackson, T. N. (1999). Pentacene organic thin-film transistors for circuit and display applications. *IEEE Trans. Electron Devices*, Vol. 46, No. 6, pp. 1258-1263.

Klauk, H.; Halik, M.; Zschieschang, U.; Eder, F.; Rohde, D.; Schmid, G. & Dehm, C. (2005). Flexible organic complementary circuits. *IEEE Trans. Electron Devices*, Vol. 52, No. 4, pp. 618–622.

Kawasaki, M.; Imazeki, S.; Ando, M.; Sekiguchi, Y.; Hirota, S.; Uemura, S. & Kamata, T. (2006). High-resolution full-color LCD driven by OTFTs using novel passivation film. *IEEE Trans. Electron Devices*, Vol. 53, No. 3, pp. 435–441.

Lai, C. H.; Chin, Albert; Chiang, K. C.; Yoo, W. J.; Cheng, C. F.; McAlister, S. P.; Chi, C. C. & Wu, P. (2005). Novel SiO_2/AlN/HfAlO/IrO_2 memory with fast erase, large ΔV_{th} and good retention. *Proceedings of Symp. on VLSI Technology*, pp. 210-211, Kyoto, Japan.

Lai, C. H.; Chin, Albert; Kao, H. L.; Chen, K. M.; Hong, M.; Kwo, J.; & Chi, C. C. (2006). Very low voltage SiO_2/HfON/HfAlO/TaN memory with fast speed and good retention," *Proceedings of Symp. on VLSI Technology*, pp. 54-55, Hawaii, USA.

Lemmi, F.; Chung, W.; Lin, S.; Smith, P.M.; Sasagawa, T.; Drews, B.C.; Hua, A.; Stern, J.R. & Chen, J.Y. (2004). High-performance TFTs fabricated on plastic substrates. *IEEE Electron Device Lett.*, Vol. 25, No. 7, pp. 486-488.

Lee, S.; Koo, B.; Shin, J.; Lee, E.; Park, H. & Kim, H. (2006). Effects of hydroxyl groups in polymeric dielectrics on organic transistor performance. *Appl. Phys. Lett.*, Vol. 88, No. 16, p. 162109.

Lin, C. W.; Yang, M. Z.; Yeh, C. C.; Cheng, L. J.; Huang, T. Y.; Cheng, H. C.; Lin, H. C.; Chao, T. S. & Chang, C. Y. (1999). Effects of plasma treatments, substrate types, and crystallization methods on performance and reliability of low temperature polysilicon TFTs, *Proceedings of International Electron Devices Meeting (IEDM)*, pp. 305-308, Washington DC, USA.

Lin, S. H.; Chin, Albert; Yeh, F. S. & McAlister S. P. (2008). Good 150°C Retention and Fast Erase Charge-Trapping-Engineered Memory with Scaled Si_3N_4, *Proceedings of International Electron Devices Meeting (IEDM)*, pp. 843-846, San Francisco, CA, USA.

Lin, S. H.; Chiang, K. C.; Chin, Albert; & Yeh, F. S. (2009). High density and low leakage current MIM capacitor using stacked TiO_2/ZrO_2 insulators. *IEEE Electron Device Lett.*, Vol. 30, No. 7, pp. 715-717.

Mizukami, M.; Hirohata, N.; Iseki, T.; Ohtawara, K.; Tada, T.; Yagyu, S.; Abe, T.; Suzuki, T.; Fujisaki, Y.; Inoue, Y.; Tokito, S. & Kurita, T. (2006). Flexible AM OLED panel driven by bottom-contact OTFTs, *IEEE Electron Device Lett.*, Vol. 27, No. 4, pp. 249-251.

Na, J. H.; Kitamura, M.; Lee, D. & Arakawa, Y. (2007). High performance flexible pentacene thin-film transistors fabricated on titanium silicon oxide gate dielectrics. *Appl. Phys. Lett.*, Vol. 90, No. 16, p. 163514.

Nomura, K.; Ohta, H.; Takagi, A.; Kamiya. T.; Hirano M. & Hosono H. (2004). Room-temperature fabrication of transparent flexible thin-film transistors using amorphous oxide semiconductors. *Nature*, Vol. 432, pp. 488-392.

Schön, J. H. & Kloc, H. (2001). Fast organic electronic circuits based on ambipolar pentacene field-effect transistors. *Appl. Phys. Lett.*, Vol. 79, No. 24, p. 4043.

Wu, Y. H.; Yang, M. Y.; Chin, Albert & Chen, W. J. Electrical characteristics of high quality La_2O_3 dielectric with equivalent oxide thickness of 5Å. (2000) *IEEE Electron Device Lett.*, Vol. 21, No. 7, pp. 341-343.

Yin, H; Kim, S.; Lim, H.; Min, Y.; Kim, C. J.; Song, I.; Park, J.; Kim, S. J.; Tikhonovsky, A.; Hyun, J. & Park Y. (2008). Program/erase characteristics of amorphous gallium indium zinc oxide nonvolatile memory. *IEEE Trans. Electron Devices*, Vol. 55, No. 8, pp. 2071-2077.

Yu, D. S.; Chin, Albert; Wu, C. H.; Li, M.-F.; Zhu, C.; Wang, S. J.; Yoo, W. J.; Hung B. F. & McAlister, S. P. (2005). Lanthanide and Ir-based dual metal-gate/HfAlON CMOS with large work-function difference, *Proceedings of International Electron Devices Meeting (IEDM)*, pp. 649-652, Washington DC, USA.

Zhou L.; Park, S.; Bai, B.; Sun, J.; Wu, S. C.; Jackson, T. N.; Nelson, S.; Freeman, D. & Hong, Y. (2005). Pentacene TFT driven AM OLED displays. *IEEE Electron Device Lett.*, Vol. 26, No. 9, pp. 640-642.

13

High performance color conversion polymer films and their application to OLED devices

Masaru Nagai
Fuji Electric Advanced technology Co., Ltd.
Japan

1. Introduction

When dye molecules absorb light, they emit light at a different, generally red-shifted, wavelength. This is a principal characteristic of dye molecules and is applied in many fields, such as laser systems, lighting, and displays. In this article, we report highly efficient color-conversion polymer films and their application to full-color, active-matrix (AM), organic light emitting diode (OLED) displays.
Since they were first reported by Tang and van-Slyke [1], OLEDs have attracted much attention for application to next-generation displays. During the past decade, several commercial products have been developed using small OLEDs fabricated by vacuum evaporation. However, OLED manufacturing is still problematic, especially for full-color displays. Figure 1 outlines three typical methods for fabricating full-color, OLED displays. Side-by-side patterning for red-green-blue (RGB) using shadow masks has been the most commonly used method [2], but using shadow masks is costly. Moreover, the substrate size and display resolution are limited.
Using OLEDs with color filters (CFs) eliminates the need for shadow masks and is suitable for large-scale high-resolution displays [3, 4]. The CFs are used with white OLEDs to attain full-color functionality. However, the transparent spectral band of the CFs must be narrowed to expand color gamut, leading to a reduction in the amount of light emitted by the white OLEDs. Therefore, balancing high efficiency with a wide color gamut is difficult. A good way to overcome this trade-off without losing the advantages of the color filter method is to use a color conversion material (CCM) that converts the blue spectrum of the OLED emission into red and green spectra [5-7]. If a polymer is used as the CCM, patterning can be achieved using a low-cost printing technology, such as ink jetting.
Use of a CCM is thus a promising alternative for manufacturing large-scale, full-color OLED displays. In this article, we report high-efficiency CCM films made using various polymers and their application to AM-OLED displays.

2. Proposed CCM systems

Several types of CCM films have been proposed for OLED use. The simplest type is a thin solid film composed of single-dye molecules. This is a single-step conversion system of single-dye molecules. With such films; there should be sufficient overlap between the absorbance spectrum of the material and the emission spectrum of the backlight OLED, the emission spectrum of the material should have a wide color gamut, and the films should have high photoluminescence quantum efficiency (PL_{eff}). Finding a suitable CCM is thus difficult, particularly with respect to the requirement for high-photoluminescence quantum efficiency. With most fluorescent dyes, the PL_{eff} in film is considerably smaller than that in solution due to the concentration quenching effect.

A photo-resist-dispersion-type CCM was proposed to overcome this problem. In this system, fluorescent dyes are dispersed in photoresist polymers with a dispersion density sufficiently low to avoid concentration quenching. Photolithography is used for the material patterning. However, when the dispersion density is reduced, the film thickness increases to more than 10 μm in order to obtain sufficient backlight absorption. This increased thickness creates various types of defects in the OLED device, such as current leakage [8]. The photosensitivity of the photoresist polymer is another problem. The photo-initiators react with the dyes and degrade the PL_{eff} [9].

To overcome these drawbacks, we have developed an advanced CCM system that uses host materials and guest dyes. The host material absorbs OLED-emitted light, and the excited energy is transferred from the host to the guest dye in a Förster fashion, resulting in radiation from the guest dyes. Using a low concentration (~1%) of guest dyes prevents concentration quenching. Since a photoresist polymer is not used, durability is improved compared to the systems that do use one. In short, this advanced CCM system results in both high PL_{eff} and improved durability without increasing the film thickness significantly.

3. Experiments

In the experiments we used to evaluate the use of various polymers as a CCM, we dissolved the polymer in tetrahydrofuran (THF) to 1.0 wt% and coated it on a glass substrate by spinning (500 rpm for 5 s; then 800 rpm for 20 s). The coating was done in dry N2 atmosphere (due point < –40□). The oxygen concentration was less than 50 ppm. The film samples were encapsulated with glass covers in dry N2 atmosphere. The photoluminescence (PL) and absorbance spectra of the samples were measured. The PL_{eff} of the samples and solutions was measured using an integrating sphere with a diffraction grating and a Xe lamp as an excitation source [10]. The excitation wavelength for the measurement was 470 nm, corresponding to the backlight peaking. To evaluate advanced CCM systems, a mixture of small dye molecules and host polymers in THF was used. The host/guest ratio was tuned so as to obtain the maximum PL_{eff}. American Dye Source (ADS) Co., Ltd supplied the polymers used in this research.

4. Materials

4.1 CCM-green conversion

There are severe restrictions with CCM green conversion (CCM-G). Sufficient overlapping between the absorbance spectrum of the material and the emission spectrum of the backlight OLED is required to achieve sufficient photon transfer. If blue OLEDs are used for the backlight, the CCM-G should have a very short Stokes' shift (~ 50 nm). Since achievement of an advanced CCM host/guest energy transfer system is impossible when blue OLEDs are used for the backlight, single-step color conversion using a single-dye system may be the only way to implement CCM-G. Reducing concentration quenching in

dye films is thus very important for using CCM-G. We examined the use of various types of polymers for CCM-G. We focused on polyfluorene derivatives, which are well known for their high PL_{eff} [11, 12].
Figure 2 diagrams the chemical structures of potential candidates for CCM-G polymers, and Table 1 summarizes their performances. For most fluorescent polymers, the PL_{eff} in film is much less than that in solution. Few polymers result in a PL_{eff} in film higher than 60%. Moreover, the maximum PL peak in film is red-shifted 20-30 nm compared to that in solution.

A comprehensive evaluation of the performances showed that fluorene/MEH-PPV copolymer (CCM(G)-1) is the best polymer for our purpose. This polymer has a very short Stokes' shift and a very high PL_{eff} (70%) in film, the same as that in solution. This strongly suggests that concentration quenching would not occur in film.
Figure 3 plots the PL and absorbance spectra of a film fabricated using CCM(G)-1. There was an absorbance peak around 480 nm, which is nearly identical to the emission peak of the backlight blue OLEDs. Moreover, the film showed bright green emission peaking at 510 nm.
The absorbance and PL peaks for polyfluorene, 370 and 420 nm, respectively, are much shorter than the excitation wavelength. Thus, the absorption and emission of CCM(G)-1 arose mainly due to the MEH-PPV units in the chains. In other words, fluorene units in the chains acted as "spacers" between MEH-PPV units and reduced their mutual interactions. This apparently caused higher PL_{eff} in film made of CCM(G)-1.
CCM(G)-2 exhibited an even higher PL_{eff} in film—90%. This excellent PL_{eff} is also explained in the same manner. Unlike the CCM(G)-1 structure, a phenyl unit is connected to each side of the fluorene units. These phenyl units further weakened the MEH-PPV units—MEH-PPV units interaction. Thus, use of this polymer resulted in even higher PL_{eff} than using CCM(G)-1. Unfortunately, its absorbance and PL peaks do not match our target values.

4.2 Drying conditions for CCM(G) films

In the experiments using CCM(G), THF was used as a solvent since it is easily removed by spinning due to its low boiling point (66°C). A low-boiling-point solvent, however, is hard to use for ink-jet printing because it can clog the nozzles. Generally speaking, solvents used for ink-jet printing should have a boiling point higher than 150°C. If a high-boiling-point solvent were used for patterning the CCM, a drying process would be needed after spinning to remove the solvent residue. Our experiments showed that the PL_{eff} of CCM films is significantly affected by the drying conditions. Using a tetralin solution, which is typically used as an ink-jetting solvent, we investigated the relationship between the PL_{eff} of CCM(G)-1 films and the drying conditions.

CCM(G)-1 was dissolved in tetralin to 1.0 wt% and coated on a glass substrate by spinning in the same way as described above. Next, it was heat dried using a hot plate in dry N2 atmosphere. The temperature was varied between 25°C (room temperature) and 200°C, and the drying time was 30 minutes.
As shown in Fig. 4, the PL_{eff} of the CCM(G)-1 films decreased as the drying temperature was increased. The films dried at room temperature had the same order of PL_{eff} as that of THF films. However, they had some fluidity due to insufficient solvent removal. The films dried at 100°C had the same PL_{eff} as those dried at room temperature and were stiff. The PL_{eff} of

films dried above this temperature decreased as the temperature was increased. Films made of a polyfluorene derivative aggregate when heated, and the decrease in PL_{eff} was apparently caused by polymer aggregation and the resultant strengthened interactions between MEH-PPV units [13-15]. The stronger interactions promoted the transfer of exited energy to other units without radiation.

We also investigated the effect of the processing atmosphere on the PL_{eff} of CCM-G films. From the manufacturing point of view, fabrication of CCM films in air is preferable. We thus first performed all processes and measurements in air for CCM(G)-1 films. Unfortunately, this significantly reduced the PL_{eff} (to 30%).

We further investigated the effect of air exposure on the PL_{eff}. As shown in Table 2, air exposure during spin coating and drying had little effect on the PL_{eff}. However, once the films had been dried, air exposure significantly degraded the PL_{eff}.

Moisture and oxygen adversely affect the characteristics of MEH-PPV derivatives [16-18]. Moisture and oxygen on the non-dried films had trouble diffusing and penetrating the films due to the presence of residual solvent. In contrast, they easily penetrated the dried films and thus degraded them.

4.3 Monomer diffusion method

To enhance the PL_{eff} of the CCM(G) films, we developed a "monomer diffusion method". This method can be used for any type of CCM polymer film. Thermosetting acrylate monomers are used to reduce concentration quenching in the film.

Figure 5 schematically illustrates the four steps in this method. 1) A CCM polymer film is formed on a substrate. 2) Liquid-state acrylate monomers are coated on the substrate without using a solvent. The substrate is then heated in two steps: 3) at a temperature lower than the thermosetting point and 4) at a higher temperature. In the first heating step, the monomers diffuse into the CCM polymer film and act as spacers in the polymer chains. In the second heating step, cross-links are formed between the remaining monomers. The acrylate monomers used must have sufficient transparency and an adequate refractive index, both of which depend on the adjacent CCM film.

The PL_{eff}'s of the CCM(G)-1 films with and without monomer diffusion are compared in Fig. 6 for three drying temperatures. The weight of the acrylate monomers was 1200, and their refractive index was 1.65. The thicknesses of the films and acrylate-monomer layers were 0.80 and 1.0 μm, respectively. Using this method substantially improved the PL_{eff} for each drying temperature. The highest PL_{eff} was 80% for the film dried at 100°C. Interestingly, this method restored the PL_{eff} of degraded films. Acrylate monomers apparently diffused into the aggregated portion of the films and separated the adjacent units, thereby reducing the strength of their interactions.

Cross-sectional images obtained by focused ion beam (FIB) analysis of CCM(G)-1 films without and with an acrylate layer and dried at 100°C are shown in Fig. 7. The white layer on the films is a carbon protection layer that was formed to minimize ion-beam damage. Comparison of the two images reveals that diffusion of the acrylate monomers doubled the thickness of the CCM(G)-1 film.

In another experiment, we investigated the effect of preliminary mixing of the acrylate monomers in the CCM polymer solution on the PL_{eff} in film. As shown in Table 3, increasing the fraction of acrylate monomers to 30:1 had little effect on the PL_{eff} for CCM(G)-1 film. Only when the fraction was increased to 40:1 there was a substantial effect. However, such a thick film would adversely affect device performance, and it would be difficult to form by ink-jet printing. The acrylate monomers and CCM polymers were apparently separated in the preliminary mixed solution and also in the films, so the monomers did not act as spacers and thus did not reduce the interaction between the CCM polymer units. In contrast, when the monomers were added to the CCM solid films, the individual monomers diffused into the CCM polymers and acted as spacers.

We also examined the use of epoxy monomers, another typical thermosetting material. However, they significantly degraded the CCM(G) films. These monomers apparently have an adverse chemical effect on polyfluorene-MEH-PPV derivatives.

4.4 Durability of CCM(G) films

We examined the durability of CCM(G) polymer films by encapsulating the CCM substrate with a glass cover and placing a sharp-cut filter (Hoya Co., Ltd.; L42) on the cover. The light from a Xe lamp was passed through the filter, and the light irradiated the target film with a luminance of 10,000 cd/m^2. The filter had a transmission wavelength of 420 nm, and most of the light shorter than this wavelength was removed from the irradiation. During the irradiation, the substrate was kept at 60°C to accelerate film degradation. These test conditions accelerated CCM film degradation ten-fold compared to OLED device driving with 1,000-cd/m^2 backlight luminescence.

As shown in Fig. 8, the cyano-substituted polymers, CCM(G)-4 and -5, showed a rapid decrease in PL_{eff} over time. Cyano groups seem to be sensitive to light irradiation. The fluorene/MEH-PPV derivatives with no cyano units showed better light durability. For CCM(G)-1, the decrease in PL_{eff} was only 20%.

To investigate the effect of impurities on durability, we further purified the CCM(G)-1 polymer using the reprecipitation method, with THF as the good solvent and acetonitrile as the poor solvent. The reprecipitation was continued until all traces of impurities had been removed, as measured using liquid chromatography. As shown in Fig. 8, the durability of CCM(G)-1 was greatly improved following this further purification—the PL_{eff} remained unchanged for the duration of the test.

These fluorene/MEH-PPV derivatives, however, had absorbance durability problems. Their absorbance was significantly degraded by light irradiation, and material purification did nothing to alleviate this. As shown in Fig. 9, the absorbance of CCM(G)-1 film decreased as irradiation proceeded, and the peak was blue shifted.

As shown in Fig. 10, the absorption constant of the films depended on the film thickness. The thinner the film, the larger the decrease. Therefore, this problem can be alleviated by increasing the film thickness.

4.5 CCM-red conversion

If blue to red conversions are to done with a single-step process, the material must have a very long Stokes' shift (>150 nm). Few fluorescent polymers have such a long Stokes' shift.

Figure 11 shows the chemical structures for candidate single-step CCM-R polymers, and Table 4 summarizes their performances. All of them had unacceptable performance. The PL_{eff} was 20% at best in solution and in film.

The long Stokes' shift reflects the higher potential of radiation-less deactivation of excited polymers. Therefore, the poor PL_{eff} is not surprising. In short, the single-step CCM-R system is not particularly suitable.

An advanced CCM system should be more suitable for red conversion. In this system, the two functions of a CCM, absorption and emission are achieved using two different materials, a host polymer and a guest dye, respectively. This leads to a significant advantage: the restrictions mentioned in section 2 are greatly eased, so there are more candidate materials that can be used. There are two steps in selecting the two materials to use. First, candidates for the guest dye are selected on the basis of their emission spectra without considering their absorbance peaks. Next, candidates for the host polymer are identified on the basis of their absorbance spectra. The transfer of excited energy from the host polymer to the guest dye is another factor in the selection process. This transfer is strongly related to the energy alignment between the host polymer and guest dye.
Pyrromethene derivatives are potentially suitable as a guest dye. They have a narrow emission spectrum, so their use results in a sufficiently wide color gamut without excessive emission losses [19]. The chemical structure and emission spectrum of a possibly useful pyrromethene derivative (650) are shown in Fig. 12.
We examined the use of five CCM-G polymers as the host material for this guest dye. Table 5 summarizes the PL_{eff}'s of the host/guest mixture films, and Fig. 13 shows the energy diagrams. The HOMO level of the material was determined by photoelectron spectroscopy measurement (AC2, Riken Keiki Co., Ltd.). The LUMO level was determined by using the HOMO value and the band gap energy. The CCM(G)-4/guest system had an excellent PL_{eff} (as high as 80%). In this system, the guest-HOMO-LUMO level was completely confined within the host-HOMO-LUMO level. In this mixed film, emissions came from only the guest dye. The CCM(G)-1, and -2 had particularly low PL_{eff} values. In these systems, the host-HOMO level was higher than the guest-HOMO level. Therefore, to achieve sufficiently high performance with the advanced CCM system, the guest-HOMO-LUMO level must be completely confined within the host-HOMO-LUMO level.

5. CCM-OLED displays

5.1 Display preparation

We prepared top-emission, full-color, 2.8" QVGA (240×320), active-matrix (AM), OLED displays with CCM-R, and -G polymers. Figure 14 shows the process flow. The a-Si thin-film transistors (TFTs) and color filter substrates were fabricated independently. A blue OLED was fabricated on the TFT substrate, and CCM film was deposited on the color filter substrate by ink-jet printing. In the final stage, the two substrates were joined together using thermosetting acrylate polymer. This polymer was used both for the transparent adhesive filler and the diffusion layer.

5.2 Ink-jet printed CCM

Ink-jet printing is generally used for polymer OLEDs and color filters due to its cost advantages. Obtaining uniformly thick printed films is critical because their performance is strongly affected by their thickness. In contrast, it is not critical for CCM films. The photon conversion efficiency of CCM film is expressed as

$$Ec = Ea \times PL_{eff} = \{1-\exp(-\alpha T)\} \times PL_{eff}, \quad (1)$$

where Ea is the absorptance, α is the absorption constant (per cm), and T is the film thickness. As T increases, Ec rapidly converges to a specific value. In other words, Ec is not affected by fluctuations in film thickness in thicker regions. As shown in Fig. 15, beyond a thickness of 400 nm, Ea was basically constant.
Images taken with a fluorescent microscope (NIKON INSTECH Co., Ltd.; TE2000-S) of ink-jetted CCM(G)-1 films illuminated by 470-nm light are shown in Fig. 16. The film thickness was significantly nonuniform due to the interaction between the CCM solution and the substrate/bank surfaces. The thickness tended to increase near the banks. The luminance of the insufficiently thick film (a) was not uniform, while that of the sufficiently thick film (b) was uniform. For CCM(G)-4 films, we obtained the same results.
In accordance with these results, we used a thickness of 400 nm for both the red (CCM(G)-4/pyrromethene-650) and green (CCM(G)-1) CCM films in the OLED device we prepared.

5.3 Backlight blue OLED

The schematic structure of the OLED is shown in Fig. 17. Common anode was connected to the a-Si TFT. A transparent oxide electrode made of indium-zinc-oxide (IZO) was used for the common anode. It was fabricated by plasma sputtering. To avoid damage to the OLED during sputtering, a buffer layer was fabricated under the electrode. Doping was used to reduce the driving voltage and to improve the injection performance of both the hole-injection layer and the electron-injection layer.
Figure 18 shows the EL spectrum of the OLED, and Figure 19 shows the a) current density-voltage and b) luminescence efficiency-current density curve of the pure blue emitting OLED, respectively.

5.4 Display performance

Figure 20 shows a photograph of the AM-OLED display. Uneven areas were not evident to the naked eye in any of the ink-jet-printed pixels.
Backlight conversion efficiency Eb was used an index of the color-conversion ability of the CCM films:

$$Eb = \text{CCM efficiency (cd/A)/OLED efficiency (cd/A)}. \tag{2}$$

For CCM-G, the Eb value was nearly constant at 1.9. This value was greater than 1.0 due to the luminosity effect. For CCM-R, the value was 0.53.
Figure 21 shows the color gamut of the OLED display. The pure blue emitting OLED and the better emission spectra of the CCM polymers enabled the achievement of an NTSC gamut ratio of 100%.
The display had a significantly wider viewing angle than that of the conventional side-by-side patterned displays, as shown in Fig. 22. This was mainly because the fluorescent light emitted from the CCM layer has Lambertian distribution.

6. Conclusion

High-performance, color-conversion polymer systems were identified: single-step CCM for green and advanced CCM for red. They have high PL quantum efficiencies (70–80% in film), and their PL and absorbance spectra are well suited for CCM-OLED devices. A monomer diffusion method was developed for enhancing the PL_{eff} in film. AM CCM-OLED displays were fabricated using ink-jet printing that had an NTSC gamut ratio of 100%. These polymers are promising color-conversion materials that can be used to fabricate efficient CCM-OLED devices.

7. References

[1] C. W. Tang and S. A. Vanslyke, Appl. Phys. Lett., 51, 913 (1987)
[2] S. Miyaguchi, S. Ishizuka, T. Wakimoto, J. Funaki, Y. Fukuda, H. Kubota, K. Yoshida, T. Watabe, H. Ochi, T. Sakamoto, M. Tsuchida, I. Ohshita and T. Tohma, J. Soc. Inf. Disp., 7, 221 (1999)
[3] M. Arai, K. Nakaya, O. Onitsuka, T. Inoue, M. Kodama, M. Tanaka and H. Tanabe, Synth. Met., 91, 21 (1997)
[4] M. Nagai, J. Electrochem. Soc., 154, J116 (2007)
[5] C. Hosokawa, M. Eida, M. Matsuura, K. Fukuoka, H. Tokailin, M. Funahashi and T. Kusumoto, J. SID, 6, 257 (1998)
[6] H. Kimura, K. Kawaguchi, T. Saito, M. Nagai, T. Asakawa, C. Li, H. Hashida and Y. Taniguchi, SID Symposium Digest of technical papers, 39, 299 (2008)
[7] M. Nagai, C. Li, N. Kanai, T. Asakawa, H. Hashida, Y. Kawamura, K. Kawaguchi and H. Kimura, IDW '08 Proc., 1013 (2008).
[8] M. Nagai, J. Electrochem. Soc., 154, J387 (2007)
[9] K. Sakurai, H. Kimura, K. Kawaguchi, Y. Taniguchi, M. Kobayashi, T. Suzuki, Y. Kawamura, H. Sato and M. Nakatani, IDW '04 Proc., 1269 (2004)
[10] N. C. Greenham, I. D. W. Samuel, G. R. Hayes, R. T. Phillips, Y. A. R. R. Kessener, S. C. Moratti, A. B. Holmes and R. H. Friend, Chem. Phys. Lett., 241, 89 (1995)
[11] T. Ahn, S. Y. Song and H. K. Shim, Macromol., 33, 6764 (2000)
[12] Q. Peng, Z. Y. Lu, Y. Huang, M. G. Xie, D. Xiao, S. H. Han, J. B. Peng and Y. Cao, J. Mater. Chem., 14, 396 (2004)
[13] T. Q. Nguyen, I. B. Martini, J. Liu and B. J. Schwartz, J. Phys. Chem.B, 104, 237 (2000)
[14] W. Zheng, M. Angelopoulos, A. J. Epstein and A. G. MacDiarmid, Macromol, 30, 7634 (1997)
[15] B. A. Weir, E. A. Marseglia, S. M. Chang and A. B. Holmes, Synth. Met., 101, 154 (1999)
[16] R. K. Khillan, Y. Su and K. Varahramyan, Mat. Res. Soc. Symp. Proc., 814, 297 (2004)
[17] D. Vacar, A. Dogariu and A. J. Heeger, Chem. Phys. Lett., 290, 58 (1998)
[18] V. I. Klimov, D. W. McBranch, N. N. Barashkov and J. P. Ferraris, Chem. Phys. Lett., 277, 109 (1997)
[19] T. H. Allik, S. Chandra, T. R. Robinson, J. A. Hutchinson, G. Sathyamoorthi and J. H. Boyer, Mat. Res. Soc. Symp. Proc., 329, 291 (1994)

Technology	Side-by-side patterned RGB	White OLED with CFs	Blue OLED with CCM s
Schematics	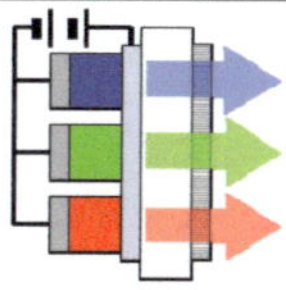	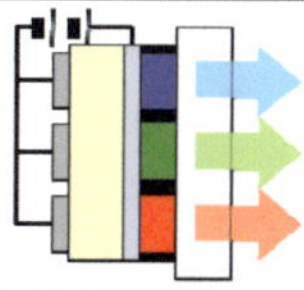	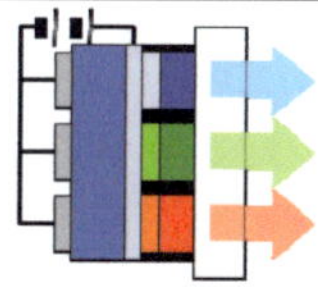
Efficiency	high	low	medium
Color Gamut	wide	narraw	wide
Resolution	<150ppi	>180ppi	>180ppi
Characteristic Manufacturing Process	OLED patterning by metal masks	OLED evaporation without patterning	CCM layers to be added
Mother Glass Size	G3	G4 available	G4 available
Production Yield	low	high	high

Fig. 1. Typical methods for fabricating full-color OLED displays.

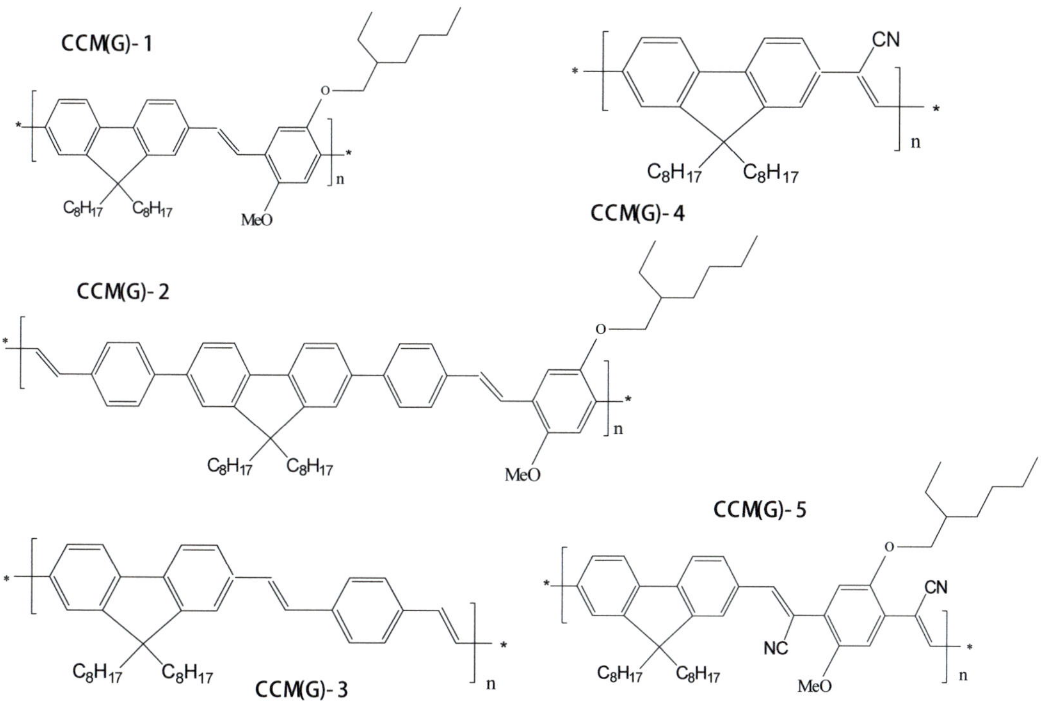

Fig. 2. Chemical structures of candidate CCM-G polymers.

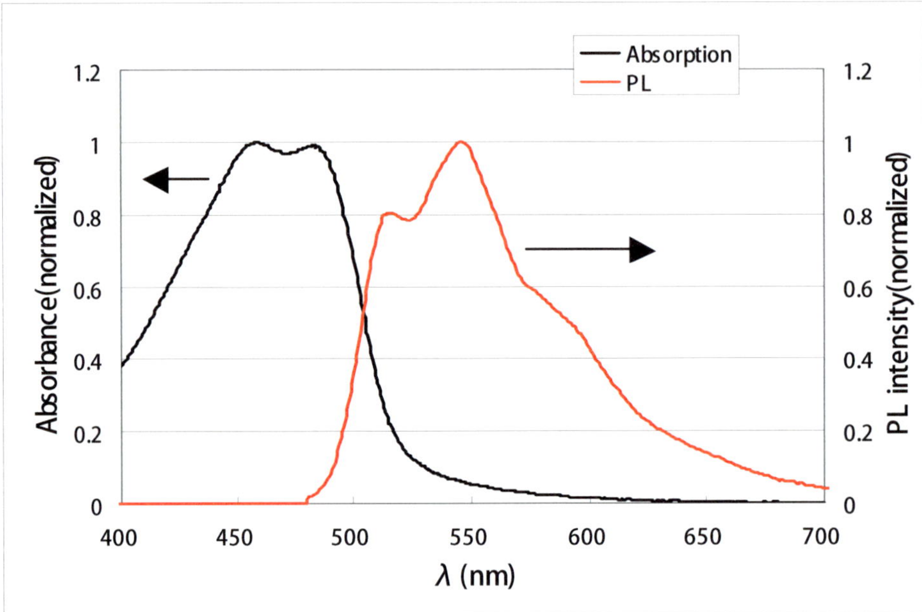

Fig. 3. Absorbance and PL spectra of CCM(G)-1 film.

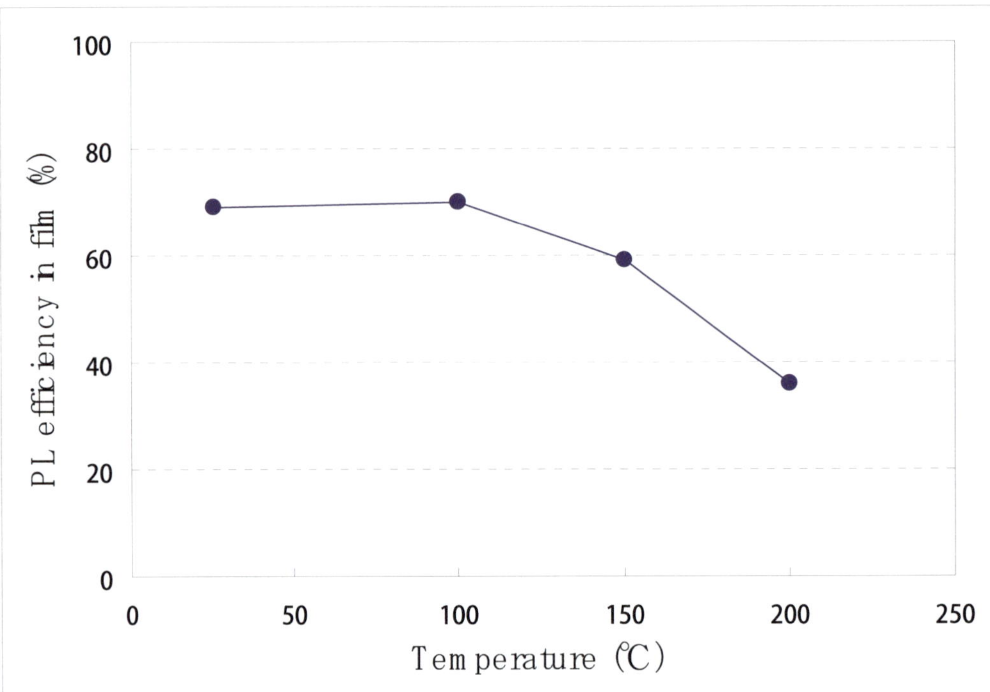

Fig. 4. PL efficiency of CCM(G)-1 film vs. drying temperature.

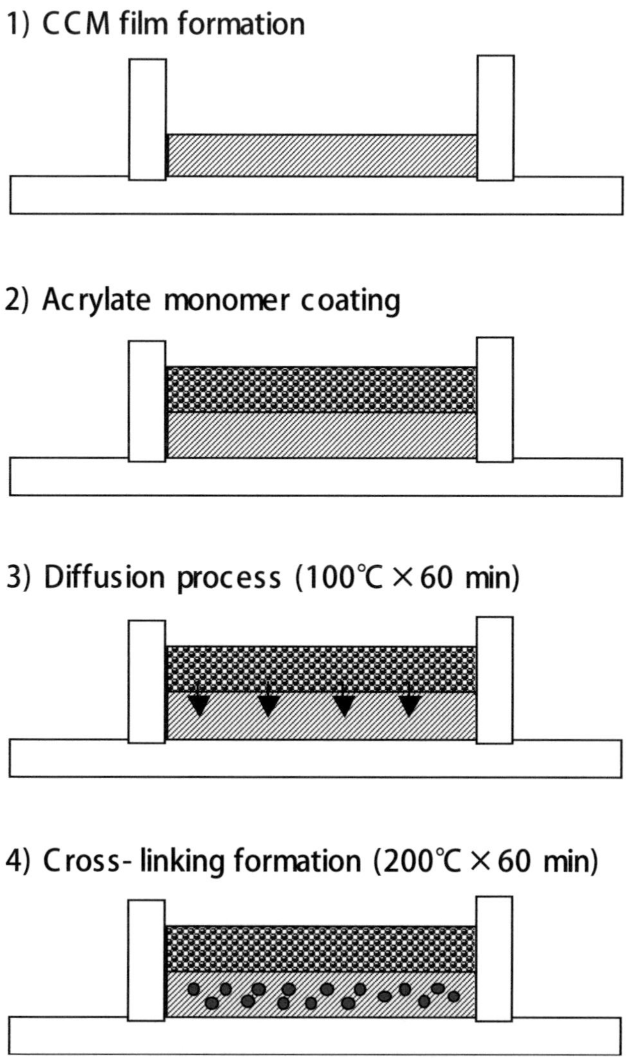

Fig. 5. Schematic showing steps in monomer diffusion method.

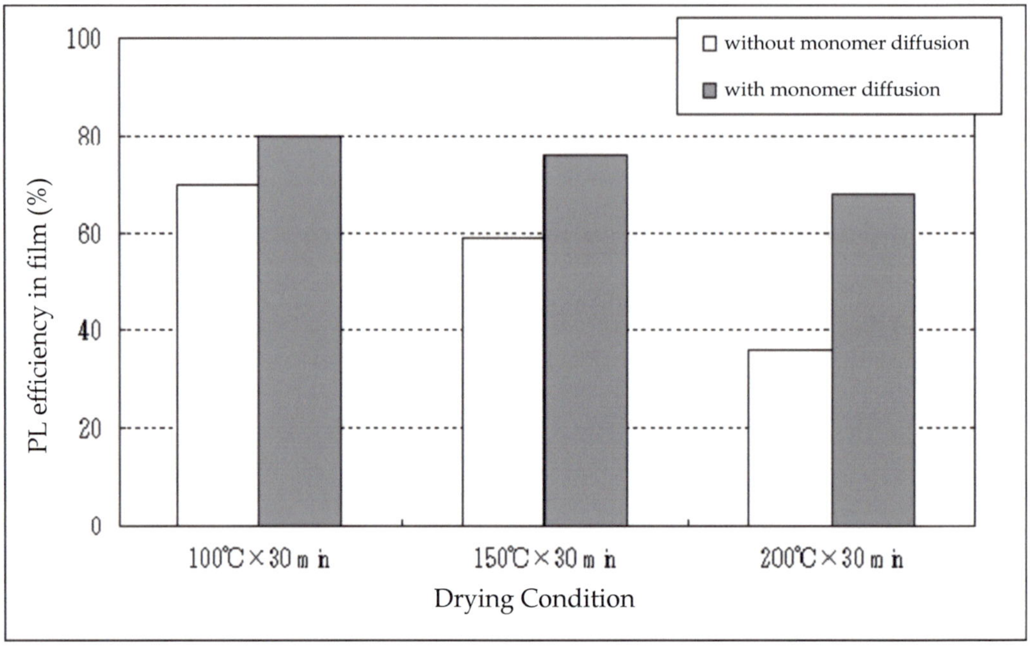

Fig. 6. Effect of monomer diffusion on PL efficiency of CCM(G)-1 film dried with various temperatures.

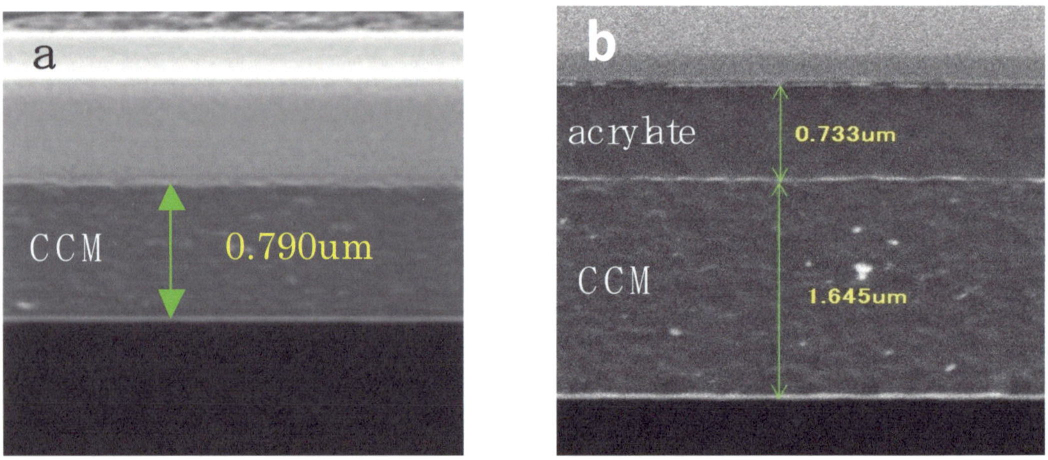

Fig. 7. Cross-sectional images of CCM(G)-1 film (a) without and (b) with acrylate layer.

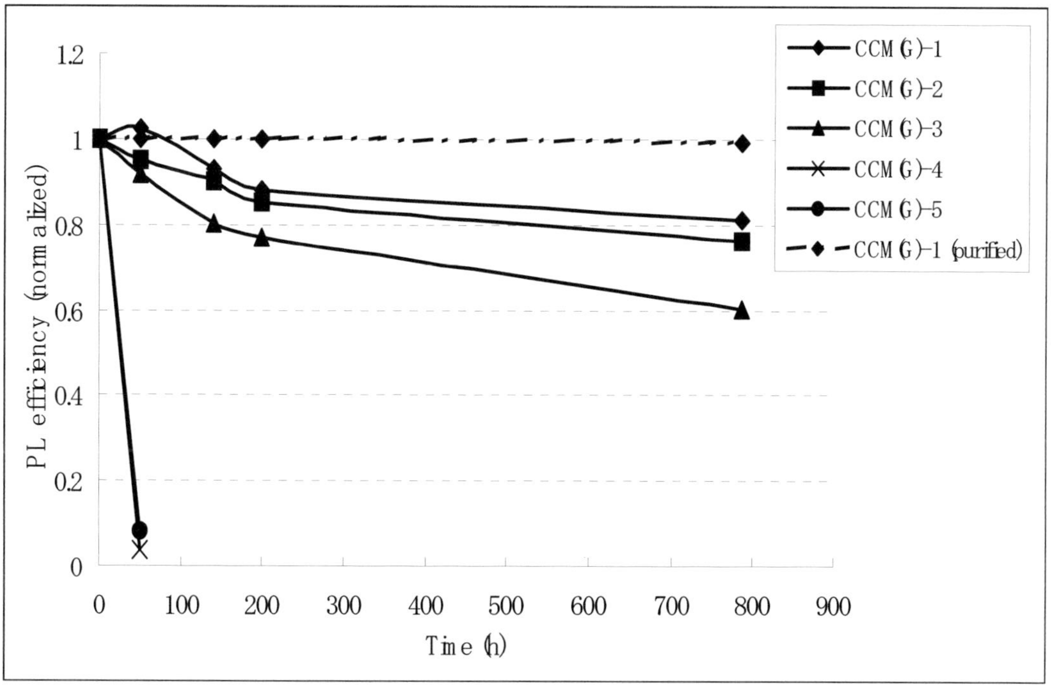

Fig. 8. Decrease in PL efficiency for CCM polymers.

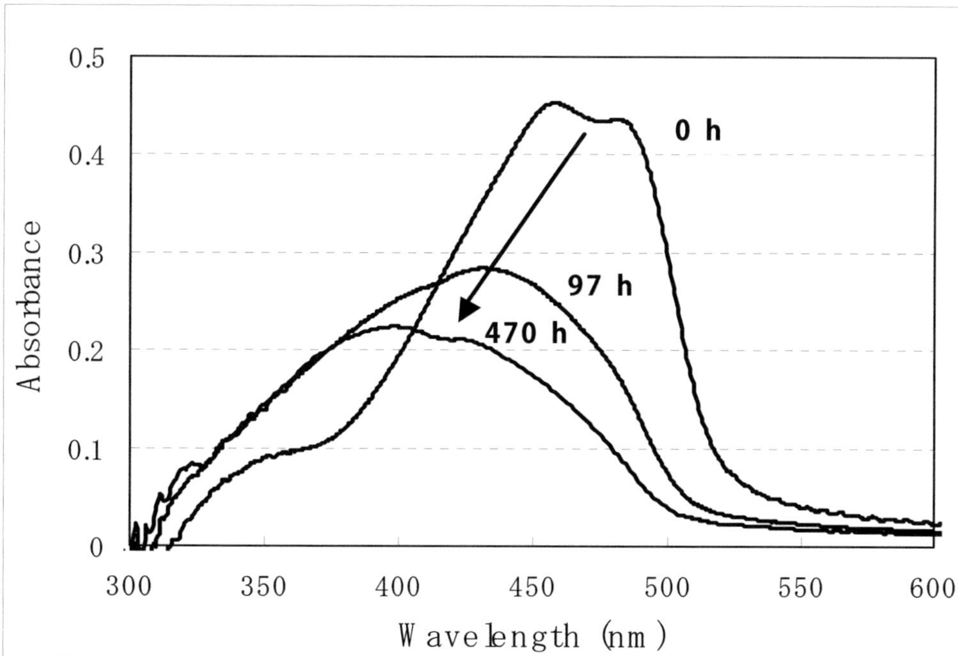

Fig. 9. Degradation in absorbance spectra for CCM(G)-1 film at 97 and 470 hours.

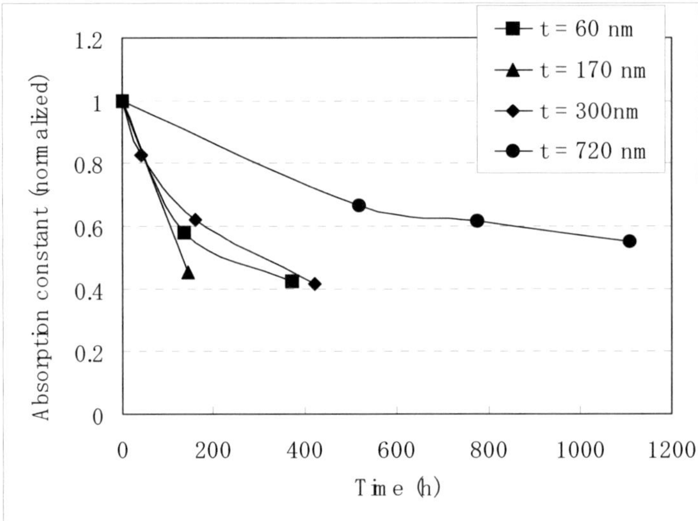

Fig. 10. Degradation in absorption constant over time for four CCM(G)-1 film thicknesses.

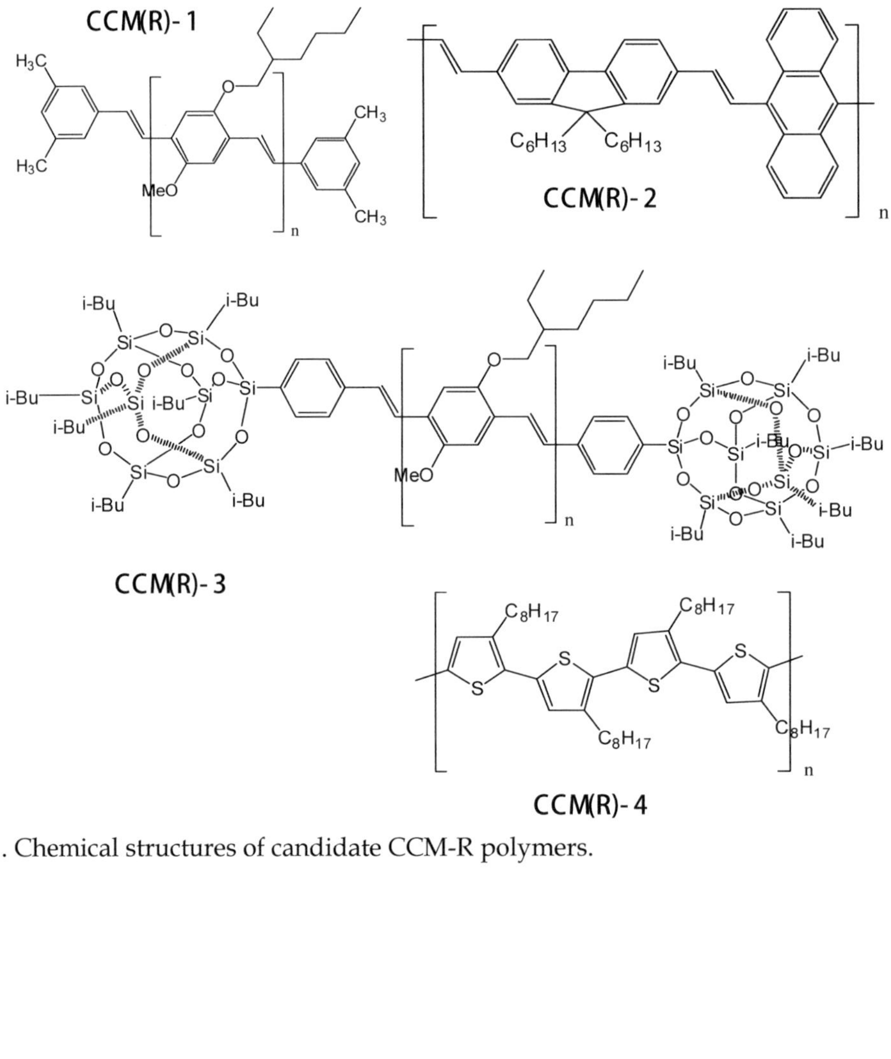

Fig. 11. Chemical structures of candidate CCM-R polymers.

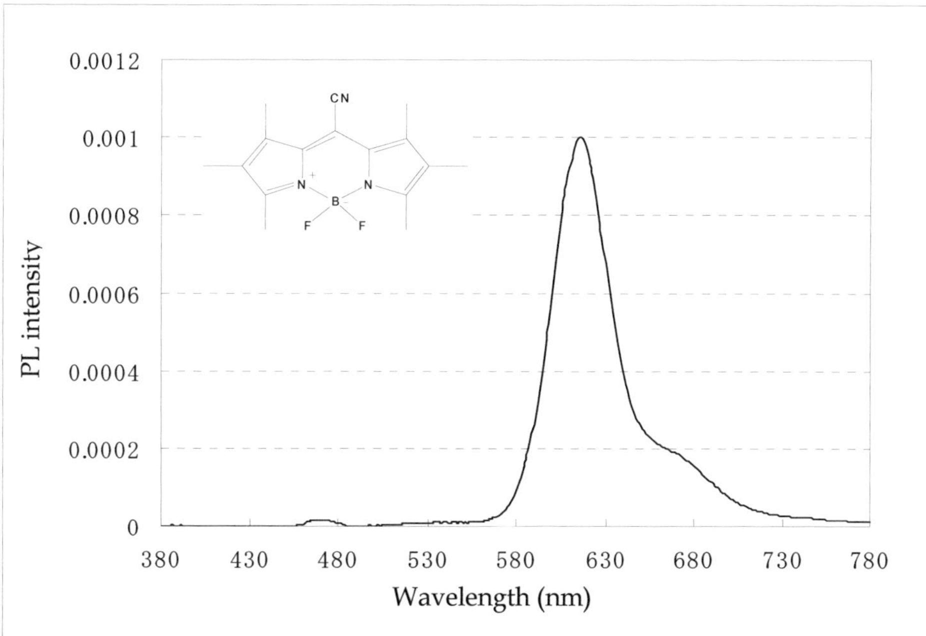

Fig. 12. PL spectrum and chemical structure of pyrromethene 650.

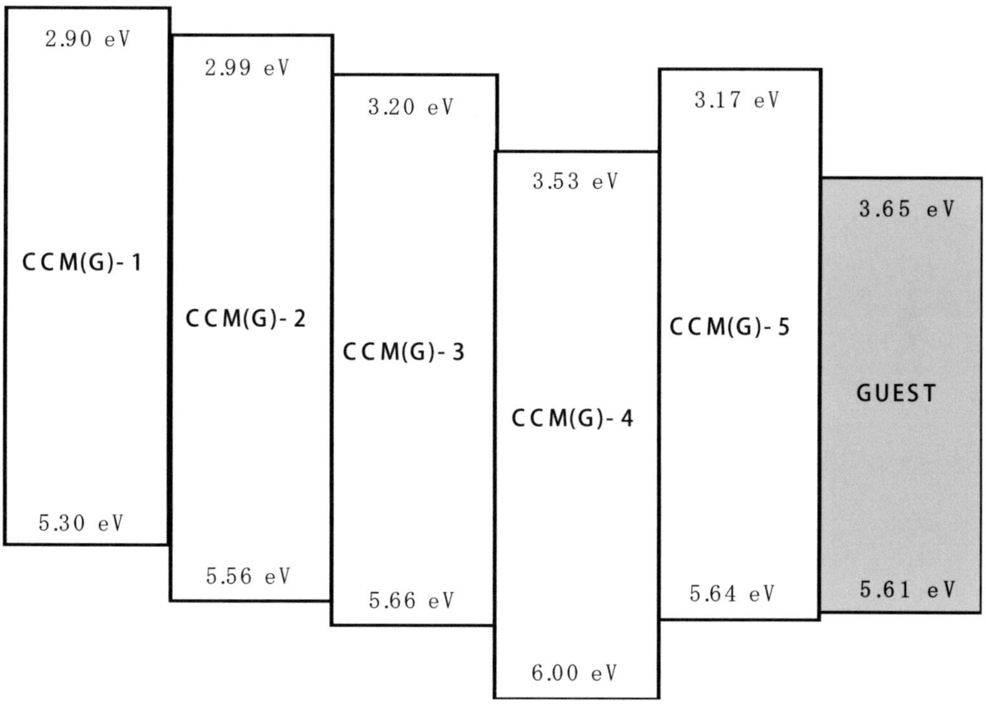

Fig. 13. Energy diagrams for advanced CCM(R) systems.

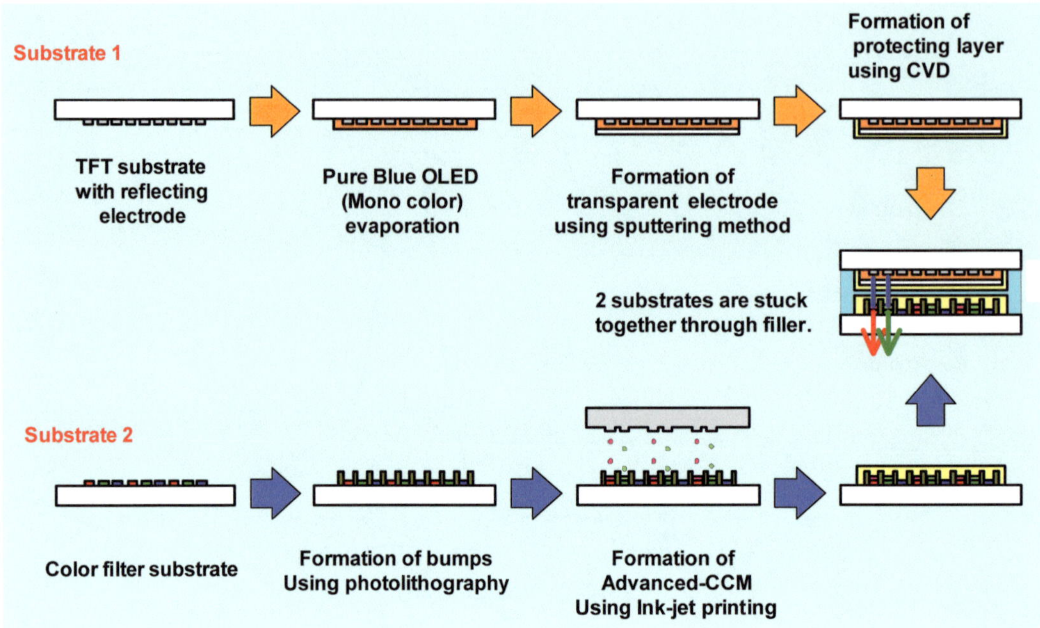

Fig. 14. Process flow for preparing CCM-OLED displays.

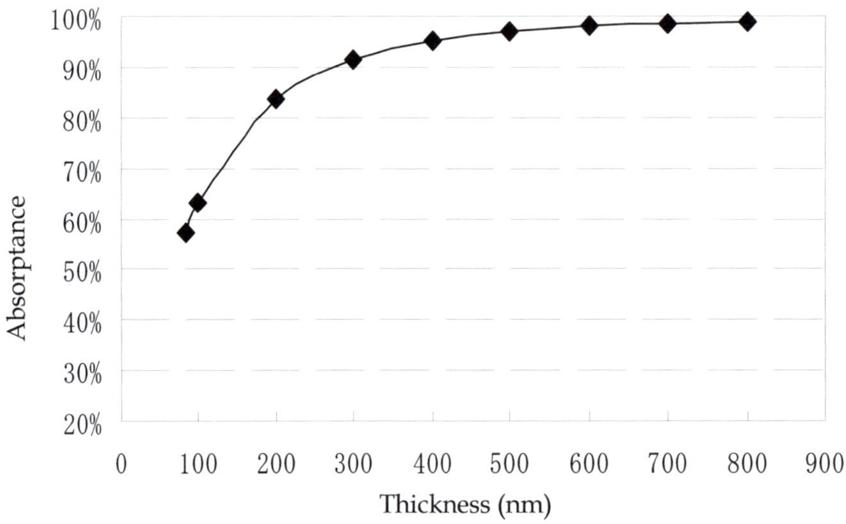

Fig. 15. Calculated Ea of CCM(G)-1 as a function of film thickness.

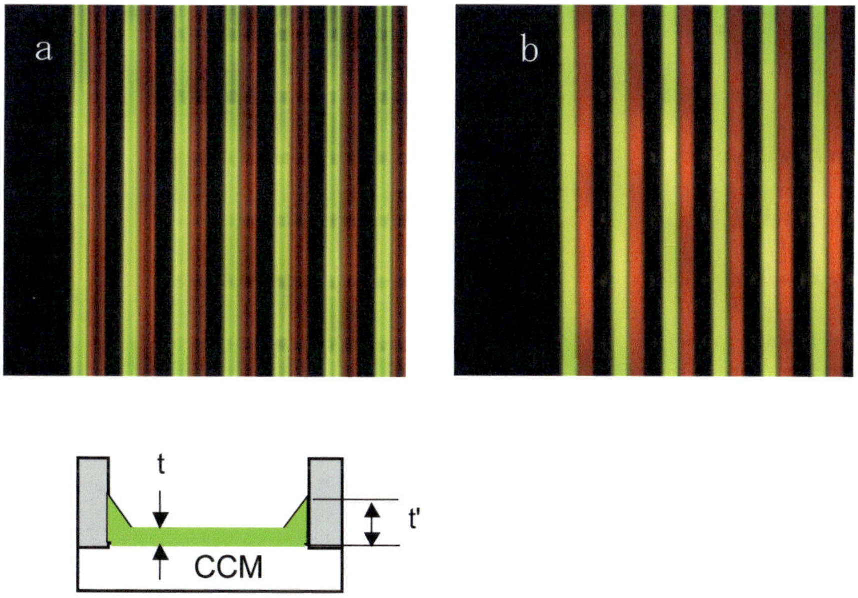

Fig. 16. Fluorescent-microscopy images of ink-jetted CCM films: (a) t, 500 nm; t′, 100 nm and (b) t, 1500 nm; t′, 400 nm.

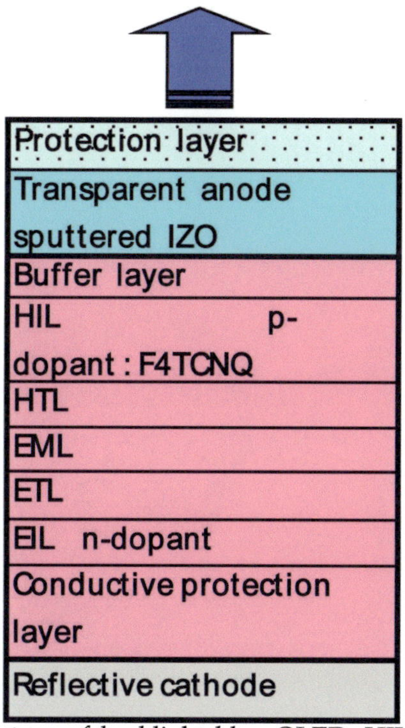

Fig. 17. Cross-sectional structure of backlight blue OLED. HIL: hole injection layer, HTL: hole transport layer, EML: emitting layer, ETL: electron transport layer, EIL: electron injection layer.

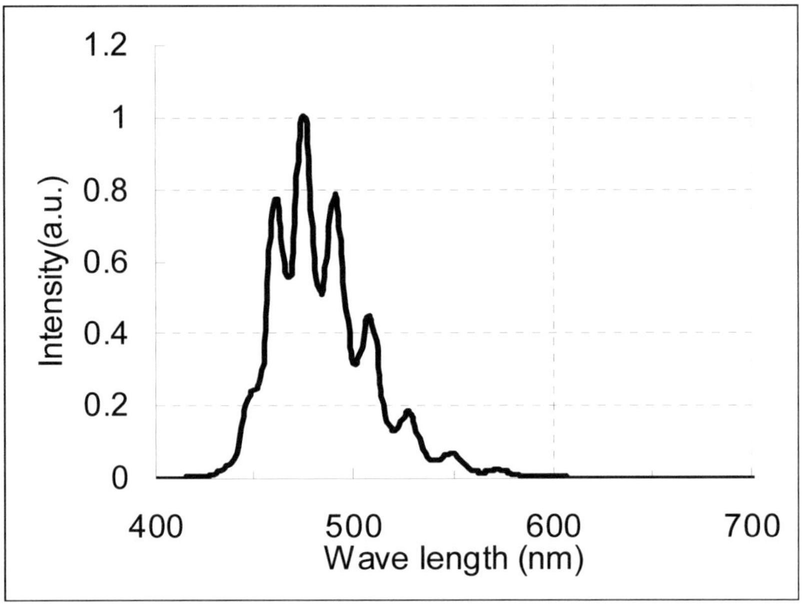

Fig. 18. EL spectrum of backlight blue OLED.

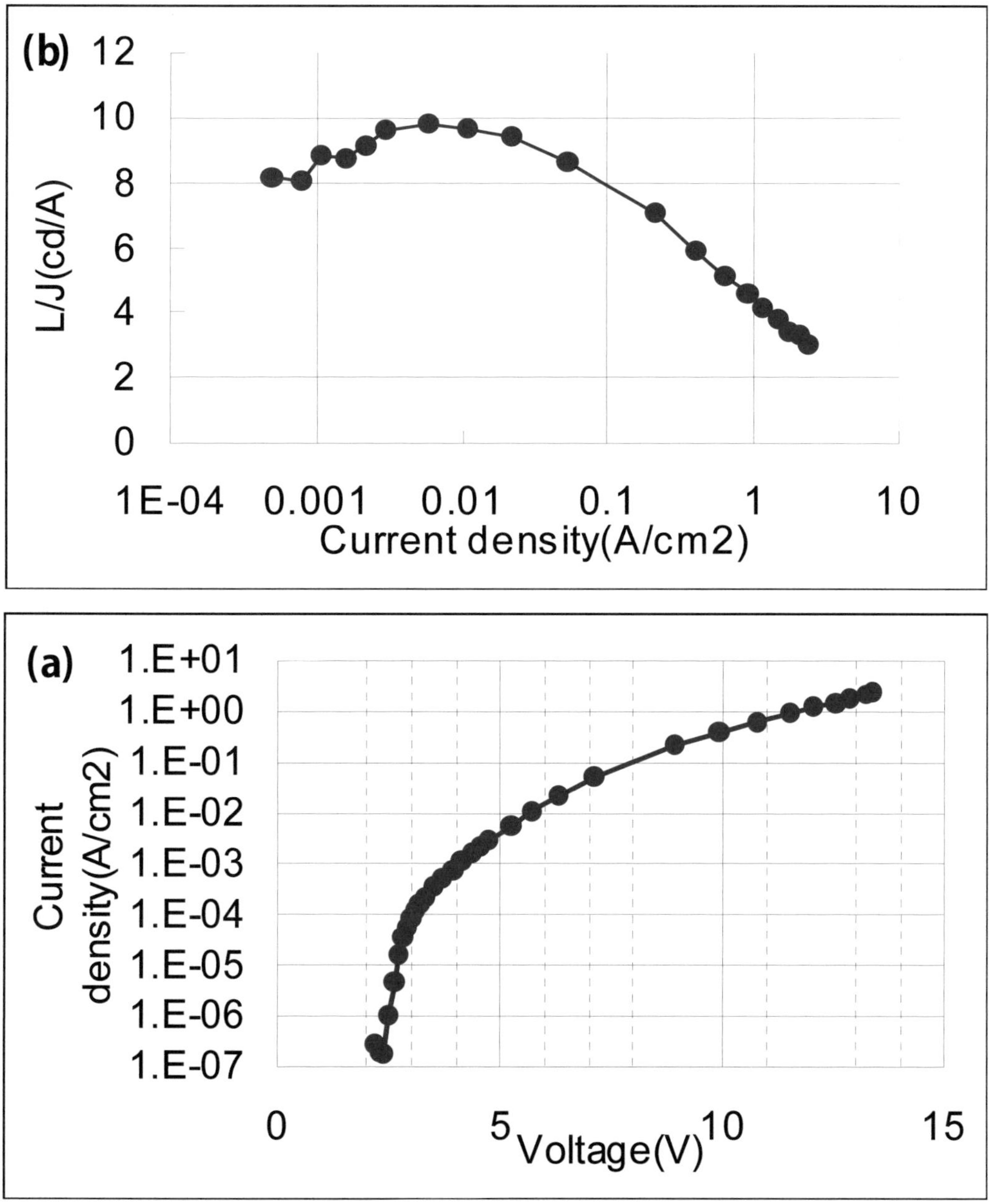

Fig. 19. (a) Current density (I) and (b) current efficiency (L)-voltage (V) characteristics of backlight blue OLED.

Fig. 20. 2.8″ a-Si TFT AM-OLED display.

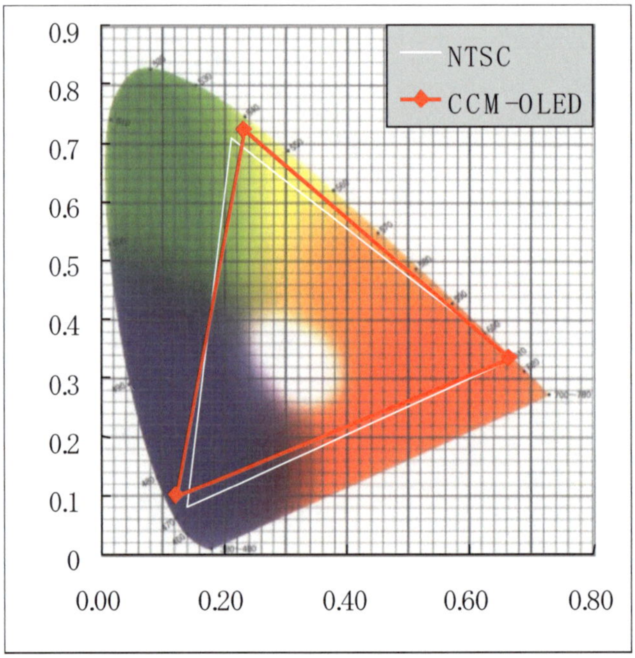

Fig. 21. Color gamut of AM-OLED display.

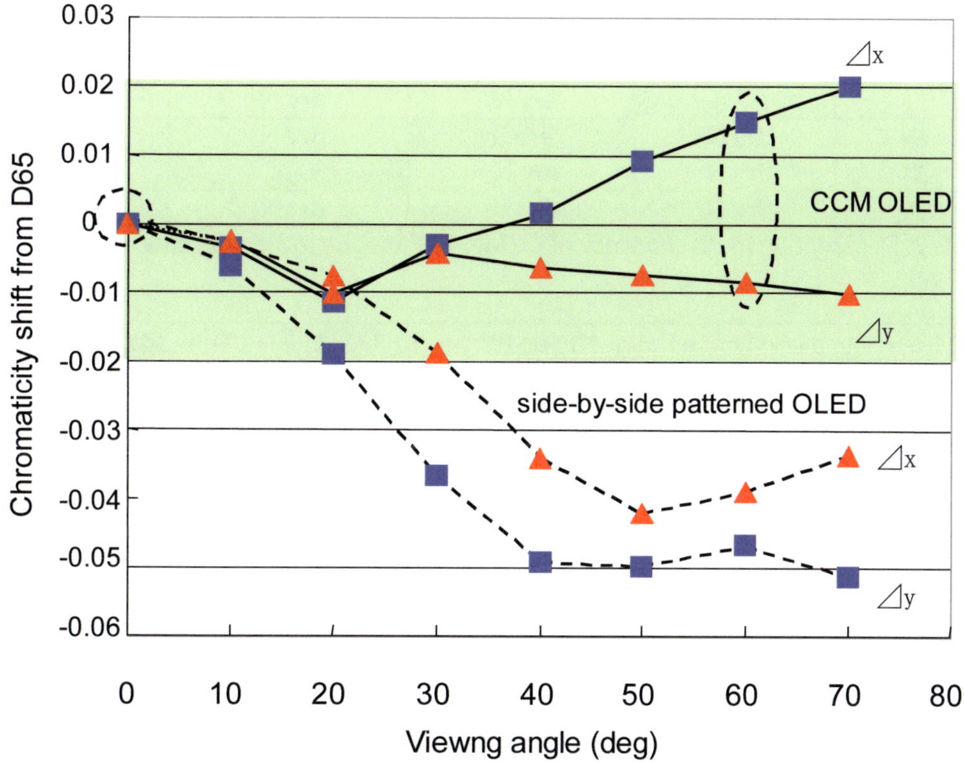

Fig. 22. Shift in chromaticity with viewing angle of CCM and side-by-side patterned OLED displays.

***Tables**

CCM-G	Molecular Weight (Mw)	Film			Solution		
		PL_{eff} (%)	Abs. Peak (nm)	PL Peak (nm)	PL_{eff} (%)	Abs. Peak (nm)	PL Peak (nm)
1	68,000	72	480	510	73	480	520
2	57,000	90	425	560	88	430	550
3	110,000	50	457	520	72	460	490
4	60,000	54	439	536	76	440	500
5	65,000	39	410	550	70	410	510

Table 1. Performance of CCM-G polymers.

PL_{eff} (%)	Spin Coating	Drying (100°C × 30 min)	Substrate Handling (less than 30 min)	Measurement (less than 10 min)
72	dry N_2	dry N_2	dry N_2	dry N_2
71	in Air	dry N_2	dry N_2	dry N_2
68	dry N_2	in Air	dry N_2	dry N_2
33	dry N_2	dry N_2	in Air	dry N_2
30	dry N_2	dry N_2	dry N_2	in Air

Table 2. Degradation in PL efficiency of CCM(G)-1 film due to air exposure.

CCM: acrylate monomer : tetralin	1 : 0 : 100	1 : 10 :100	1 : 20 : 100	1 : 30 : 100	1 : 40 : 100
PL_{eff} in film (%)	70	70	71	71	79

Table 3. PL efficiencies of CCM(G)-1 film and of CCM(G)-1 film made using acrylate monomer preliminary mixed solution.

CCM-R	Molecular Weight (Mw)	Film			Solution		
		PL_{eff} (%)	Abs. Peak (nm)	PL Peak (nm)	PL_{eff} (%)	Abs. Peak (nm)	PL Peak (nm)
1	42,000	21	485	638	22	485	597
2	33,000	5	462	616	4	460	574
3	50,000	19	492	634	19	460	601
4	49,000	6	563	731	16	624	500

Table 4. Performance of single-step CCM(R) polymers.

Guest	Pyromethene 650				
Host	CCM(G)-1	CCM(G)-2	CCM(G)-3	CCM(G)-4	CCM(G)-5
PL_{eff} (%)	12	35	54	80	48

Table 5. Performance of advanced CCM(R) films.

14

Micro/nano scale phase front inscription on polymer thin layer for flexible beam shaping

Jun Ki Kim[1] and Kyunghwan Oh[2]
[1]*Harvard medical school, Massachusetts General Hospital U.S.A*
[2]*Yonsei University, Department of Physics, Republic of Korea*

1. Introduction

As the demands for non-spherical lenses or diffractive optical devices have been increased, much investigation efforts over Beam transforming and the wavefront control in optical technologies have been attempted. Thus, novel and various technologies have been proposed in order to transform the shape and power distribution of a given light beam. Through waveguide-branching or phase front matching methods which were previously reported, the controlling of either phase front curvature or power distributions had been achieved. However, beam reshaping technique using these methods relied mainly on the use of bulk optical system, which required complicated design and fabrication processes at the risk of system size as well as cost.

Along with fast development in fiber optic communications and sensory systems, various attempts in order to cope with these weaknesses have been made to incorporate the bulk-optic technique into optical fibers. Direct mechanical deformation of fiber ends into spherical or wedge-shaped surfaces have been attempted as one of first attempt in fiber optics for applications in laser-diode to optical fiber coupling. By utilizing conventional micro-lithography and etching techniques, reflowing technique of photoresist on the fiber ends have been also attempted. However, laser direct writing process over optical fiber endfaces suffers from not only sophisticated optical alignments and expensive femto-second laser systems but also surface damages after fabrication process.

Recently, the polymeric phase-front modification technique using optical fiber composition was investigated by the authors in order to overcome the disadvantages of described methods. As the proposed methods are suitable for beam forming and beam pattern control in the fiber optic system, it was confirmed that the device showed strong potentials for flexible and economic optical phase-front control without resorting to conventional lithography and etching techniques.

Thus, in this chapter, micro/nano scale phase front inscription techniques were introduced and investigated for flexible beam shaping on polymer thin layer. The numerical simulation of the diffraction patters out of azo-polymer layer on the fiber was analyzed. In parallel, a new method to inscribe linear and concentric circular surface relief gratings (SRGs) to manipulate the propagation properties of a beam was described. The principles, fabrication procedure, and characterization of beam propagation and beam patterns from linear and circular azo-polymer SRGs are discussed both experimentally and theoretically.

2. Formation of surface relief grating (SRG) on Azo polymer thin layer

2.1 Azobenzene-functionalized polymers

Azo-polymer complexes having unique mass shift property induced by photo-reaction have been widely used for inscribing of the periodic optical structures. Thus, there have been many reports that use them to generate spontaneous surface modulation by exposing different light intensity on an Azo-polymer thin film. In essence, the polymer material will reversibly deform so as to minimize the amount of material exposed to the light. This phenomenon is not a kind of laser ablation, since it readily occurs at low power as well as the transformation is reversible. Although this is clearly related to the azobenzene isomerization, the exact mechanism of this phenomenon is still unresolved.
Azoxy has a double bond group of atom. Especially, it has a double bond of Nitrogen in the both terminal of the molecular formula as shown in Figure 1 as chemical structures.
The epoxy-based azo polymer PDO3 was synthesized from diglycidyl ether of bisphenol A and 4-(4′-nitrophenylazo) phenyl amine for the investigations. The *Tg*s of the azo polymers are about 106 °C for PDO3. Figure 2 shows the UV–visible absorption spectra of the azo films. As the absorption band of the azo-polymer is in the range of blue-green band, Ar-ion laser source is generally utilized as an inscribing laser beam.

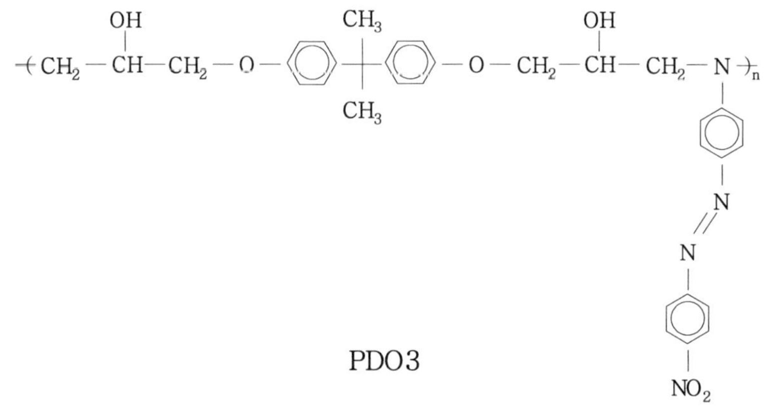

Fig. 1. The chemical structures of PDO3

Typical surface deformation images of Azo-polymer under various engraving conditions are shown in Figure 3. Figure 3(a) depicts modulated surface induced by one-dimensional Gaussian beam, and Figure 3(b) and 3(c) show the typical surface deformation induced by the linearly and circularly polarized Gaussian beams, respectively. The engraved diameter and the modulation depth are dependent upon engraving conditions such as polarization condition, launched laser power, laser beam diameter and exposure time.

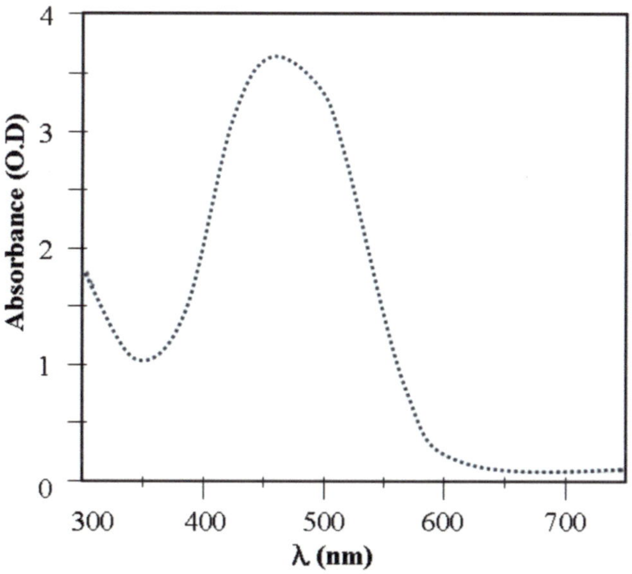

Fig. 2. Absorbance of azopolymer

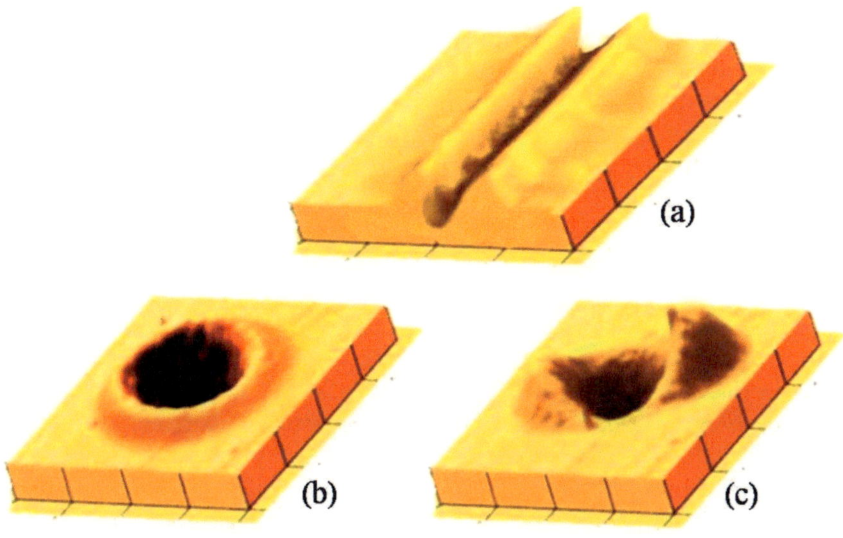

Fig. 3. Typical surface deformation images of Azo polymer induced by Gaussian beam

Table 1 describes the diffraction efficiencies and surface modulation of the gratings which are recorded under different recording conditions. The different polarizations defined by an angle α, with respect to s polarization.

In comparison with photoresist films, Azo-polymer layers produce surface relief grating (SRG) pattern by absorption of blue-green photons, where the actual mass of layer is modulated rather than refractive index. In a single-step writing process, topographic structures on the azo-polymer layers can be formed upon exposure to the appropriate optical patterns. This process, therefore, has a significant advantage over other techniques which typically require complicate process, such as baking, exposure and developing, etc.

Recording conditions	Difraction efficiency (%)	Surface modulation (Å)
α = 0°	<0.01	<100
α =16°	5.5	1470
α =45°	27	3600
α =90°	15.2	2540
Unpolarized	16.5	2560
Circularly polarized	30	3500

Table 1. The diffraction efficiencies and surface modulations under different recording conditions.

2.2 Formation of linear and circular pattern

Most significant merit for developing interferometric lithography on optical fibers is that both linear and concentric phase-fronts over the azo-polymer layers could be inscribed in a single exposure without using any photo-mask, or further post-processes. For the linear SRGs, linear fringe patterns were generated by using conventional two-beam interference set-up based on bulk-optics. In contrast to these linear SRGs, compact fiber-optic pattern generation method was utilized in the circular SRGs. The concentric interference pattern was generated within the cross section of the fiber, using a coreless fiber segment and adjusting its length.

In both linear and concentric SRGs, three main steps for preparing a sample are required; Azo-polymer thin film coating on fiber end facet, and optical interference pattern generation and exposure to the inscribing laser beam. It is noteworthy that these one-step and direct exposure methods of various phase-fronts over optical fiber could provide strong economical mass production capability to the both fiber array and planar waveguides.

An optical fiber was, firstly, cleaved to form an optically flat endface, which makes a right angle with respect to the fiber axial direction, using an ultrasonic high precision fiber cleaver. The cleaved endface served as a substrate for azo polymer thin film coating. PDO3 polymer containing azo-benzene group was solvated in cyclohexanone and the filtered 10wt% PDO3 solution was dropped on the fiber-end-surface and then spin-coated. The film was, then, dried in a vacuum oven at 80°C for 1-hour. Thin film layer of azo-polymer with flat and smooth surface could be possible due to relatively low viscosity of the solution. The thickness of the thin film layer was 900nm having a flatness of ±20nm over the entire circular optical fiber endface.

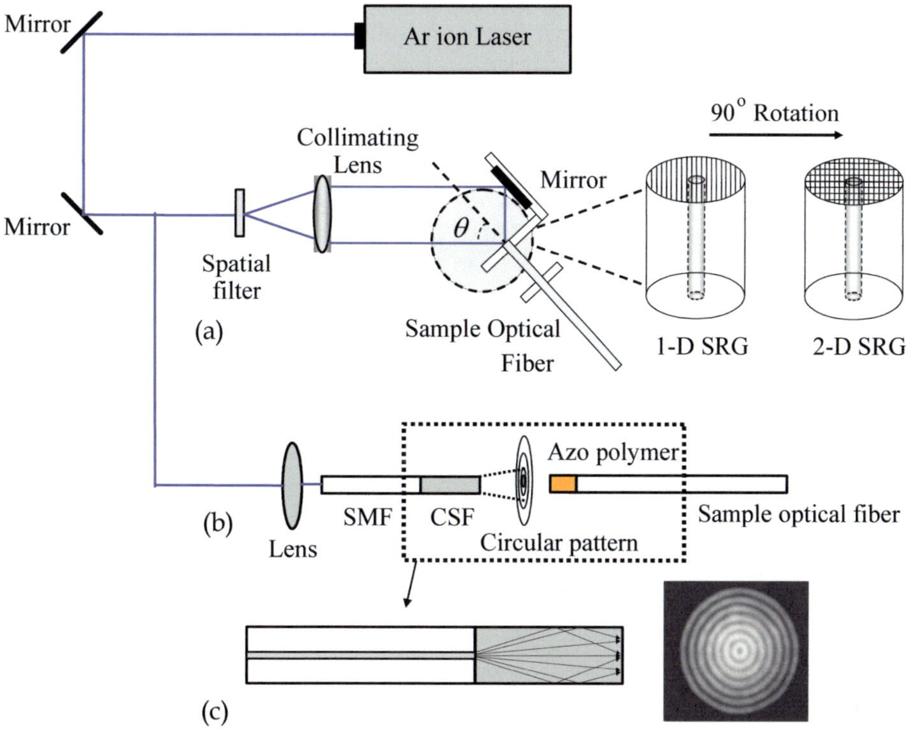

Fig. 4. Experimental setup

Generation for optical interference pattern and exposure process is shown in Figure 4. It is notable that the proposed process is direct one-step exposure to generate SRGs obviating photo masks and etching processes for both linear, (a), and concentric, (b), (c). Schematic diagram for fabrication of linear SRGs is presented in Figure 4-(a). An Ar-ion laser was used as a writing beam. The laser beam intensity was 100 mW/cm^2 and operating wavelength was 488nm which is in the range of absorption band of the azo-polymer. The laser beam was expanded by a spatial filter and collimated by a collimator. Linear interference patterns were formed due to optical path difference between the direct beam and reflected beam at the mirror. The pitch of interference was adjusted by changing the incident angle θ, which determined the period (Λ) of SRG pattern on azo-polymer film. The incident angle was adjusted to 7° in order to inscribe a uniform linear pattern with a pitch of 2μm. Once the linear interference pattern was exposed to azo-polymer, 1-dimensional SRG was formed by the mass-shift. After adjusting the alignment of the fiber at an angle of 90°, another pattern can be superimposed to form a 2-dimensional SRG. After this double exposure process, a well-defined 2D grating was fabricated.

The schematic diagrams for concentric SRGs pattern generations are shown in Fig 4-(b), and (c). The pattern generation method for concentric interference is based on a compact all-optical fiber device, which is contrast to the linear SRGs based on conventional bulk optics in Fig 4-(a). The device was composed of conventional single mode fiber (SMF) and coreless silica fiber (CSF) without GeO_2 doped core. The LP01 mode exits from the SMF core and it passes through CSF with expanding the beam diameter, which can be approximated by Gaussian beam propagation.

As the beam further propagates along CSF, part of the beam hits the air-glass interface, and then reflects into CSF, generating circular interference patterns. The schematic diagrams for concentric SRGs pattern generations are shown in Fig 4-(b), and (c). The pattern generation method for concentric interference is based on a compact all-optical fiber device, which is contrast to the linear SRGs based on conventional bulk optics in Fig 4-(a). The device was composed of conventional single mode fiber (SMF) and coreless silica fiber (CSF) without GeO2 doped core. The LP01 mode exits from the SMF core and it passes through CSF with expanding the beam diameter, which can be approximated by Gaussian beam propagation.

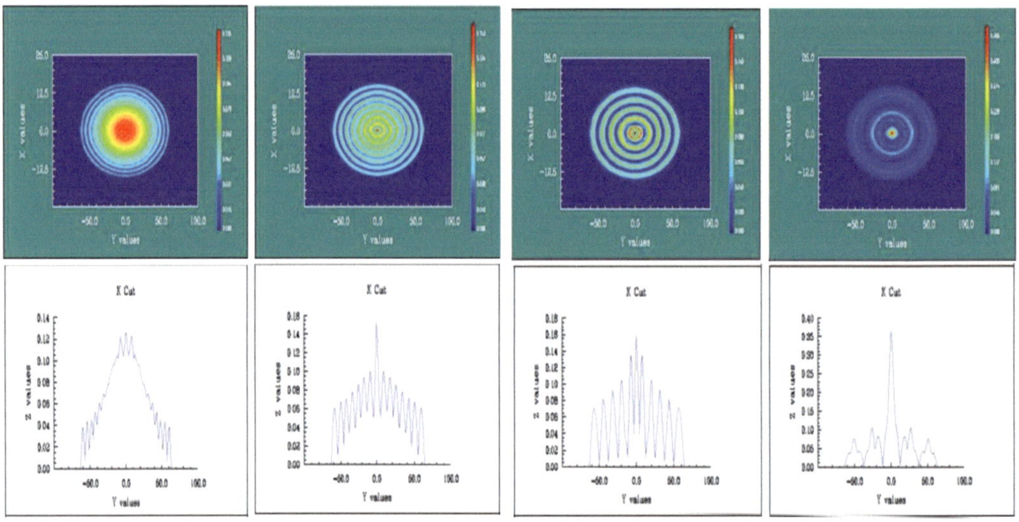

(a) L=700μm **(b) L=1000μm** **(c) L=1500μm** **(d) L=3000μm**

Fig. 5. The numerical estimation of the intensity distributions

In order to further investigate the concentric interference pattern generation in terms of variation of CSF lengths, a commercial beam propagation method (BPM) package, BeamPROP™, was utilized for numerical analysis. The results are summarized in Figure 5. Here we assumed the fundamental mode of SMF can be approximated as a Gaussian beam as it propagates along CSF, which has been widely accepted for calculation of light propagation in the free space out of SMF. The outer diameter of both SMF and CSF were 125μm. The intensity profile and fringe spacing at the surface of CSF were directly dependent on the CSF length for the given diameter as shown in Figure 5. Gaussian distribution profile without concentric fringes is maintained under the condition of CSF length less than 700μm. As the CSF length approaches near 700μm, fringe patterns appear overlaid on the Gaussian beam. For CSF length longer than 900μm, the entire cross-section of CSF is filled with concentric interference fringes. As the CSF length increases furthermore, less numbers of fringes and longer pitch were predicted. Furthermore, by varying the distance between the CSF end face and azo-polymer end face, the fringe spaces could be adjustable. Therefore, the proposed composite fiber-optic device can provide very versatile and flexible control over the concentric interference pattern generation, by varying the geometrical dimension of CSF, diameter and segment length.

2.3 Surface relief grating (SRG) on optical fiber surface

The scanning electron microscope (SEM) images and blow up images of the fabricated SRGs are shown in Figure 6. One and Two-dimensional linear SRG are shown in Figure 6-(a) and (b), respectively. The bright center regions of the figures denote the location of the fiber core. Firstly, one-dimensional (1-D) linear SRG was fabricated on the fiber-end-surface. The pitch was 2μm and the modulation depth was in the range of 450~500nm. For this 1-D SRG, another linear interference pattern was exposed after rotating the fiber at an angle of 90 degrees from the initial position, to form two-dimensional (2-D) linear SRGs as in Figure 4-(a). As a result, periodic 2-D SRG patterns with 2μm -by-2μm were engraved on the target surface shown in Figure 6(b). Through the SRG pattern, the fundamental mode guided by the SMF spreads out to the entire cross section of the engraved facet and derives effective modification of the phase front of incident beam on it.

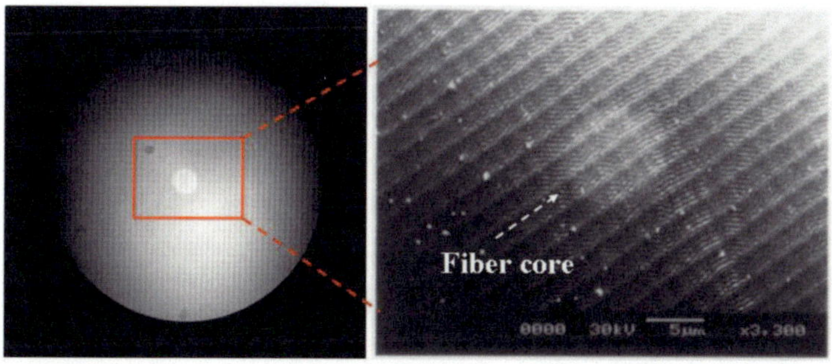

(a)

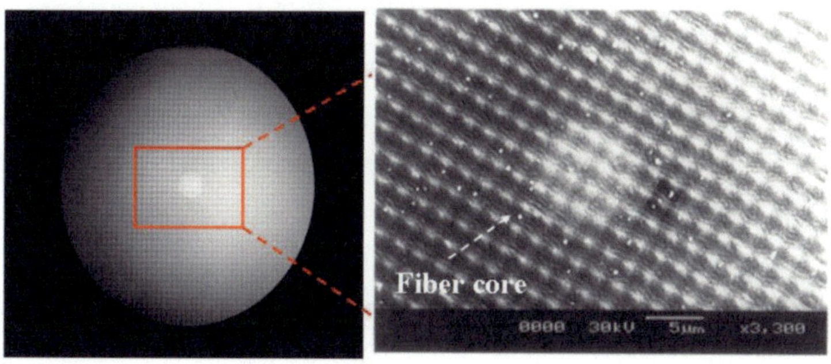

(b)

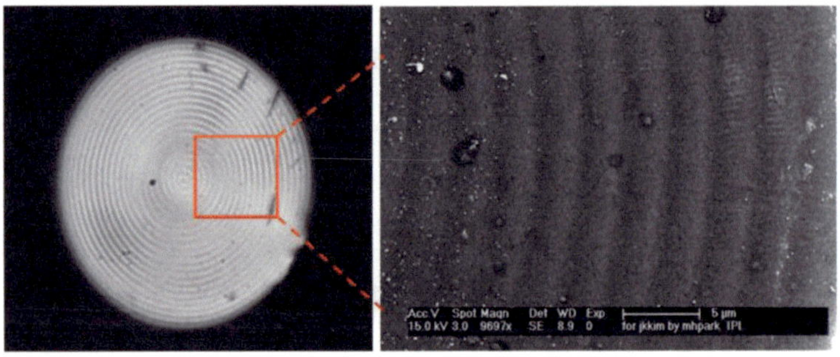

(c)

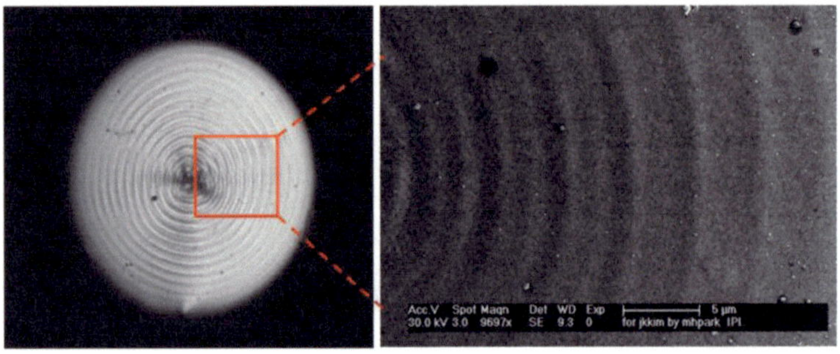

(d)

Fig. 6. SEM images on the fiber-end-surface with 1-D/2-D linear and concentric SRG

In concentric SRGs, on the while, the composite SMF-CSF pattern generators with the CSF lengths of 1000μm and 1500μm was used, and the SEM images of fabricated SRGs are shown in Figure 6(c) and (d), respectively. The pitches of the engraved concentric SRG patterns were about 2.7 and 4.3μm for Figure 6(c) and (d), respectively. The concentric pattern generation using all-optical fiber device shows same concentric SRG properties, as numerically predicted in Figure 5. Thus, it was experimentally confirmed that direct exposure of both linear and concentric interference pattern at 488nm Ar-ion laser can successfully form the corresponding SRGs by using azo-polymer thin films over SMFs.

3. Measurements and simulations

3.1 Measurements of diffraction patterns out of SRGs

The optical field propagating through the SMF will experiences the spatially periodic modulation provided by the SRGs at the prepared endface of the fiber, resulting in unique diffraction patterns. Fabricated SRGs on SMF were further examined in terms of their diffraction pattern both experimentally and theoretically. In order to investigate impact of the SRGs over beam patterns, the far-field diffraction patterns from the SRG on SMFs were measured by a CCD camera using a laser source at 635 nm.

In Figure 7(a), schematic diagram of the experimental setup for measurement of diffraction patterns is shown, where L is the distance from the SRG to the measured diffraction patterns, and D is the distance between the 0th and the 1st order diffraction beam pattern, and θ_m is the diffraction angle. From the diffraction patterns, we can calculate the diffraction angle

(θ_m) by measuring the distance data (*D* and *L*), and finally obtain the SRG pitch (Λ) from the following diffraction grating equation

$$\Lambda = m\lambda \,/\, \sin\theta_m \tag{1}$$

where *m* denotes the order of principle maxima.

For the case of linear SRGs, measured diffraction patterns and the line profile is shown in Figure 7-(b),(c). In Figure 7-(b),(c), clear diffraction beam pattern of the zero'th and the first order from the 1D/2D linear SRG was observed. The circular pattern at the center regions is identified as the zero'th order and two side lobes correspond to the first order diffraction pattern in Figure 7(b). Similarly, the four first order side lobes perpendicular to one another, was measured as shown in Figure 7(c).

The far field diffractive patterns and the line profiles of the concentric SRGs are shown in the Figure 7 (d),(e). In circular SRGs, the orders of diffraction patterns are dependent upon the CSF lengths of 1000 and 1500μm making different concentric patterns. For the case of CSF length of 1000μm, the diffraction pattern has the central zero'th order and one thin ring, which corresponds to 1'st order. In the case of 1500μm CSF, two rings were clearly dissolved. They correspond to 1'st and 2'nd order diffraction, overlaid with the zero'th order circular pattern. In comparison to diffraction patterns from linear SRGs, as in Figure 7-(b),(c), those from the concentric SRGs do not show clear and definitive images, which is attributed to relatively low contrast in the concentric interference pattern generators. See Figure 5-(b) and (c). Minima of the fringes do have finite intensity and subsequently SRG would have shallow contrast to make the diffraction pattern less definitive.

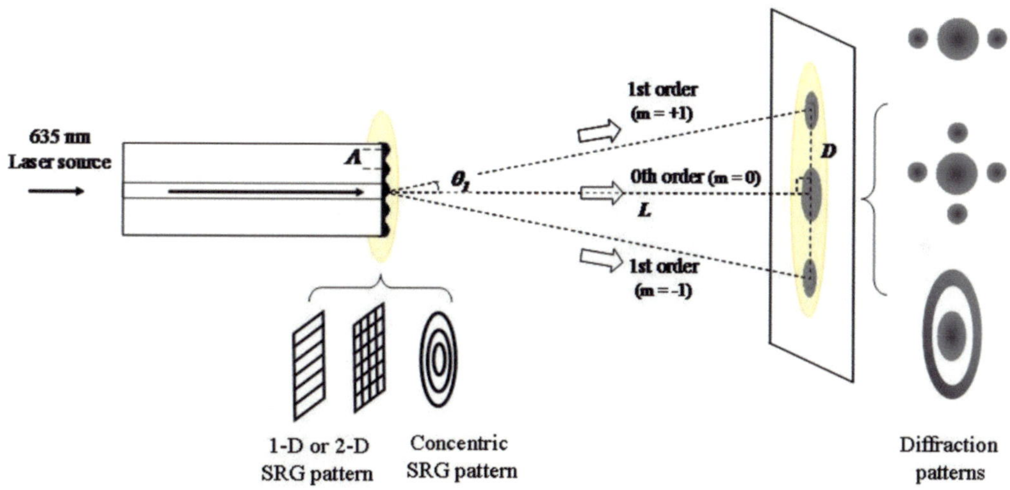

(a)

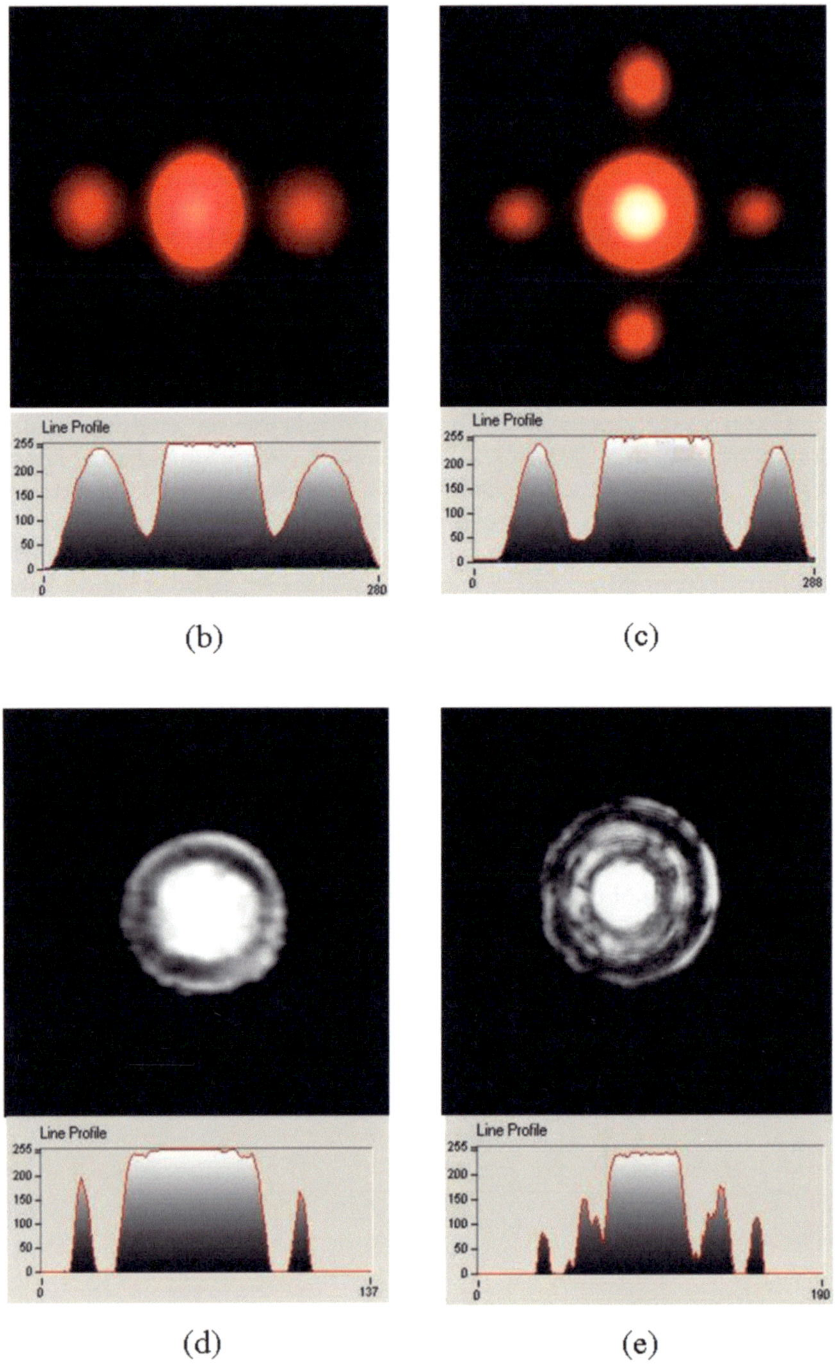

Fig. 7. Schematic diagram for measuring diffraction beam patterns and line profiles

	L [mm]	D [mm]	θ [°]	Λ [μm] (calculated)	Λ [μm] (measured)
1-D SRG	10	3.56	19.60	1.89	2.0
2-D SRG	10	3.39	18.73	1.98	2.0
CSRG with 1000μm	11.68	3.0	14.41	2.55	2.7
CSRG with 1500μm	19.11	3.0	8.92	4.10	4.3

Table 2. Measured data from diffraction pattern measurements

The measured diffraction parameters are shown in Table 2, from which we calculated the pitches (Λ) of SRGs. For linear 1-D/2-D SRGs, pitches were calculated as 1.89μm and 1.98μm, respectively. For concentric SRGs made by the SMF-CSF interference pattern generators of CSF length of 1000 and 1500μm, the pitches were calculated as 2.55 and 4.10μm, respectively. Compared with pitches measured by SEM images in Figure 6 in previous section, these numbers are in a reasonably good agreement as summarized in the last two columns of Table 2.

3.2 Theoretical analysis using simulation tools

Diffraction patterns from the fabricated SRGs on optical fiber endfaces were also theoretically investigated using a commercial simulation package, LightTools™. In the simulation tool, the diffraction of the guided mode from the given SRG was investigated in terms of the irradiance beam patterns using the illumination analysis.
Actual dimension of the SRGs were used in the simulation with an approximation that the incident light is Gaussian, instead of LP_{01} mode of the SMF, which is very common in the free space optic analysis. The illumination analysis in the LightTools is based on a Monte Carlo ray trace. From randomly selected points on the surface or Volume into randomly selected angles in space, it traces the desired number of rays. Diffraction patterns from the linear 2-D SRGs, Figure 6-(b), and concentric SRGs, Figure 6-(c) were simulated and the irradiance chart diagrams are shown in Figure 8. In comparison with experimental measurements, Figure 7-(c) and (d), it was found that the simulation results showed good agreement in terms of the line profiles, the intensity distribution along x or y axis, and diffraction robe locations.
Through these experimental and theoretical analyses, we could confirm that the proposed method to form SRGs on the optical fiber endfaces does have practical feasibility and can endow a new degree of freedom to design integrated optical systems compatible to fiber optics or waveguide optics.

4. Conclusion

By adapting azobenzene polymer layer, both linear and concentric surface relief gratings (SRGs) have been successfully inscribed over optical fiber endfaces based on developing maskless lithography technology. Utilizing unique advantage of the azo-polymer such as direct writing and multiple exposure capabilities, various SRGs were flexibly designed and fabricated. Two-beam interference patterns were applied for 2-D linear SRGs and SMF-CSF concatenated device was proposed to generate concentric interference patterns for circular SRGs, respectively. Theoretically, the diffraction pattern out of SRGs was investigated by utilizing a Monte Carlo ray tracing package. In comparison with experimental measurement, the simulated results showed good agreement in terms of irradiance beam patterns.

Through the proposed inscription technology based on polymer thin layer coating, micro/nano scale phase front control could be possible for manipulating the propagation properties of the light and it could be applicable in integrated optical components as well as sub-systems for optical communications.

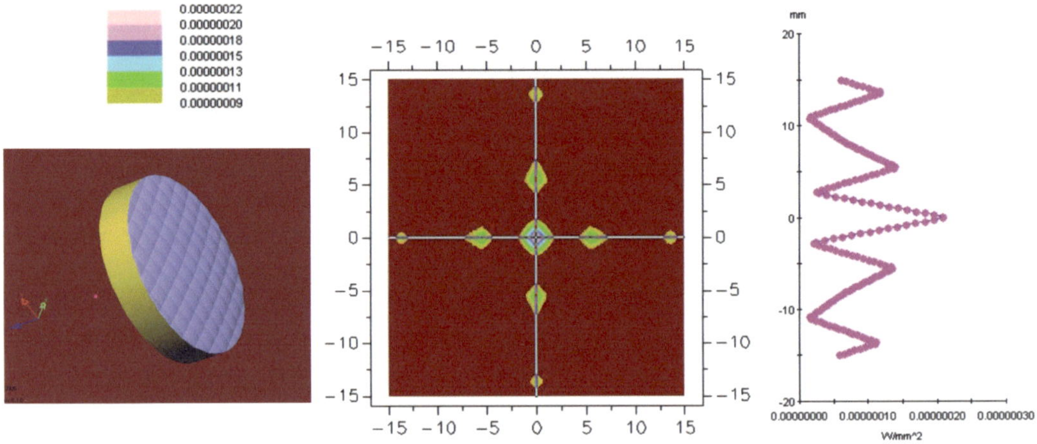

(a)

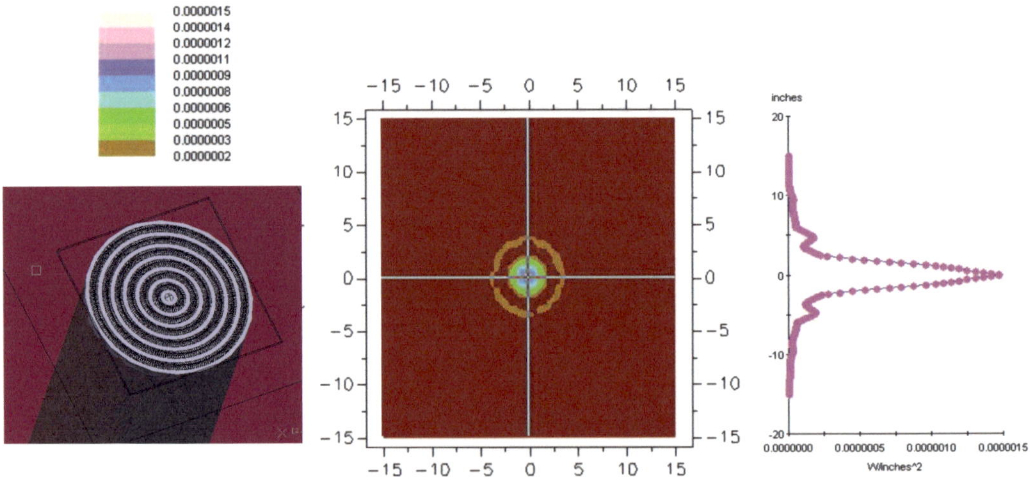

(b)

Fig. 8. The simulation results of the irradiance chart using LightTools™

Acknowledgement
This work was supported by the National Research Foundation of Korea Grant funded by the Korean Government (NRF-2009-352-C00042) and in part by the KOSEF (ROA-2008-000-20054-0, R15-2004-024-00000-0), in part by the KICOS (2009-8-1339, 2008-8-1893), in part by the ITEP (2008-8-1901, 2009-8-0809), and in part by the Brain Korea 21 Project of the KRF

5. References

A. M. Streltsov et al., "Fabrication and analysis of a directional coupler written in glass by nanojoule femtosecond laser pulses.", *Opt. Lett.*, Vol. 26, pp. 42 (2001)

C. J. Barrett, A. L. Natansohn, and P. L. Rochonm, *J. Phys. Chem.* Vol. 100, pp. 8836 (1996)

D. F. Cornwee, "Nonprojective transformation in optics," *Opt. Eng.*, Vol. 294, pp. 62-72 (1982)

D. K. Yi et al., "Surface-modulation-controlled three-dimensional colloidal crystals.", *Appl. Phys. Lett.* Vol. 80, pp. 225 (2002)

D. Y. Kim, L. Li, X. L. Jiang, V. Shivshankar, J. Kumar, and S. K. Tripathy, *Macromolecules* 28, pp.8835 (1995)

D. Y. Kim et al., "Laser-induced holographic surface relief gratings on nonlinear optical polymer films.", *Appl. Phys. Lett.*, 66, pp. 1166 (1995)

E. Hecht, "Optics," Addison Wesley, San Francisco (2002)

G. Eisenstein and D. Vitello, "Chemically etched conical microlenses for coupling single-mode lasers into single-mode fibers," *Appl. Opt.*, Vol. 21, pp. 3470-3474 (1982)

H. Ghafoori-Shiraz and T. Asano, "Microlens for coupling a semiconductor laser to a single-mode fiber," *Opt. Lett.*, Vol. 11, pp. 537-539 (1986)

H. Kogelnik, "On the propagation of Gaussian beams of light through lenslike media including those with a loss and gain variation," *Applied Optics*, Vol. 4, pp. 1562 (1965)

H. Rau, in Photochemistry and Photophysics; Vol. 2, edited by J. Rebek (CRC Press, Boca Raton, FL, 1990), pp. 119-141 (1990)

H. Sasaki, "Normalized power transmission in single mode optical branching waveguides," *Electron. Lett.*, Vol. 17, pp.136-138 (1981)

Hidehiko Yoda and Kazuo Shiraishi, "A new scheme of a lensed fiber employing a wedge-shaped graded index fiber tip for the coupling between high power laser diodes and single mode fibers," *J. Lightwave Technol*, Vol. 19, pp. 1910 (2001)

J. Kim, M. Han, Selee Chang, W. Lee and K. Oh, "Achievement of large spot size and long collimation length using UV curable self-assembled polymer lens on a beam expanding core-less silica fiber," *IEEE Photonics Technology Letters*, Vol. 16, paper no. 11, pp. 2499-2501 (2004)

Junichi Sakai and Tatsuya Kimura, "Design of a miniature lens for semiconductor laser to single-mode fiber coupling," *J. Quant. Elentron.* Vol.16, pp. 1059 (1980)

J. W. Ogland, "Mirror system for uniform transformation in high power annular laser," *Appl. Opt.*, Vol. 17, pp. 2917-2923 (1978)

Kyung-Rok Kim, Selee Chang, and K. Oh, "Refractive microlens on fiber using UV-curable fluorinated acrylate polymer by surface-tension," *IEEE Photonics Technology Letters*, 15, paper 8, pp. 1100-1102 (2003)

L. G. Cohen and M. V. Schneider, "Microlenses for coupling junction lasers to optical fibers," *Appl. Opt.*, Vol. 13, pp. 89-94 (1974)
Michael J. hayford et al., "Illumination module user's guide," Optical Research Associates (2003)
M. S. Ho, C. Barrett, J. Paterson, M. Esteghamatian, A. Natansohn, and P. Rochon, *Macromolecules* Vol. 29, pp. 4613 (1996)
N. C. R. Holme, L. Nikolova, P. S. Ramanujam, and S. Hvilsted, *Appl. Phys. Lett.* Vol. 70, pp. 1518 (1997)
P. Rochon, E. Batalla, and A. Natansohn, "Optically induced surface gratings on azoaromatic polymer films.", *Appl. Phys. Lett.* Vol. 66, pp. 136 (1995)
S. Choi, Kyung-Rok Kim and K. Oh, "Interferometric inscription of surface relief gratings on optical fiber using azo polymer film," Appl. Phys. Lett., Vol. 83, pp. 1080 (2003)
Shaoping Bian et al., "Photoinduced surface deformations on azobenzene polymer films", *J. Appl. Phys.*, Vol 86, pp. 4498 (1999)
Shojiro Kawakami, "Light Beam Redistribution using computer generated phase plates," *J. Lightwave Technol.*, Vol. 7, pp.1412-1418 (1989)
S. J. Zilker et al., "Holographic data storage in amorphous polymers", *Adv. Mater.*, Vol. 10, pp. 855 (1998)
T. S. Lee, D. Y. Kim, X. L. Jiang, L. Li, J. Kumar, and S. K. Tripathy, *Macromol. Chem. Phys.* Vol. 198, pp. 2270 (1997)
W. Y. Hung, "Novel design of wide angle single-mode symmetric Y-junctions," *Electron. Lett.*, Vol. 24, pp. 1184-1185 (1988)
Wen-Ching Chang, "A novel low-loss wide angle Y-branch with a diamond-like microprism," IEEE *Photon. Technol. Lett.*, Vol. 11, pp. 683-685 (1999)
X. Wang, L. Li, J. Chen, S. Marturunkakul, J. Kumar, and S. K. Tripathy, *Macromolecules* 30, pp.219 (1997)
X. L. Jiang, L. Li, J. Kumar, D. Y. Kim, and S. K. Tripathy, "Polarization dependent recordings of surface relief gratings on azobenzene containing polymer films.", Appl. Phys. Lett. Vol. 68, pp. 2618 (1996)
X. T. Li et al., "Photoinduced liquid crystal alignment based on a surface relief grating in an assembled cell.", *Appl. Phys. Lett.* Vol. 74, pp. 3791 (1999)

Organo-siloxane supramolecular polymers used in CO2 detection-Polymer thin film

Gabriela Telipan[1], Lucian Pislaru-Danescu[1], Mircea Ignat[1], Carmen Racles[2]
[1]*Research and Development National Institute for Electrical Engineering ICPE-CA*
[2]*Petru Poni Institute for Macromolecular Chemistry*

1. Introduction

Supramolecular polymers are macromolecules in which the monomer units are kept together by non-covalent interactions. In a broader sense, the term is also used for self-organized macromolecules of conventional polymers involving non-covalent interactions to determine their material properties (M. Zigon, G. Ambrozic, 2003). It is well known that the physical properties of linear polymers and organic molecules are strongly modified when they contain associating end groups. Supramolecular polymers combine the features of supramolecular species with polymeric properties
Self-organizing materials, which include liquid-crystalline polymers, block copolymers, hydrogen-bonded complexes, biological polymers, have great potential for various functional materials. Manipulation of supramolecular nanostructure in self-organizing materials is of critical importance for achieving desired functions and properties in solid state and liquid crystalline state molecular materials (M. Lee, B.K. Cho, et.al. 1999).
Although hydrogen bonds between neutral organic molecules are not very strong, they play a very important part in molecular recognition, self-assembling in bio-macromolecules, increasing miscibility in polymers blends, organization in liquid crystals. The directionality and versatility of the hydrogen bonds are the major keys for their implication in supramolecular polymers design. It has been observed that hydrogen bonding in supramolecular polymers is enhanced by liquid crystallinity and phase separation (. L. Brunsveld, B.J.B. Folmer, et.al., 2001).
The self-assembly of carboxylic acids as proton donors with pyridyl fragments as proton acceptors is most frequently used in the formation of H-bonded structures [4]. Such supramolecular polymers have been obtained and investigated, especially for liquid crystalline properties [M.Lee, B.K. Cho, et.al. 1999, M. Parra, P. Hidalgo et.al., 2005, H. Han, A.H. Molla et.al., 1995, Y.S. Kang, H. Kim et.al., 2001, M. Parra, P. Hidalgo et. al., 2005; P.K. Bhowmik, X.Wang et. al., 2003, 2, 4-8]. Siloxane-containing benzoic acids and bipyridine also form interesting supramolecular structures, with cubic mesophases (E. Nishikawa, E. T. Samulski, 2000). This paper presents the sensitivity of gases in specially CO_2 of the organo-siloxane supramolecular polymers.

2. Synthesis and characterization of organo-siloxane supramolecular polymers

2.1. Materials

4,4`-Bipyridyne, (Fluka), m.p. 109-112°C, was used as received.

1,3-Bis(3-carboxypropyl)tetramethyldisiloxane, $[HOOC(CH_3)_3(CH_3)_2Si]_2O$, CX, was synthesized by using the method described in ref. [10] (the hydrolysis of 1,3-bis(3-cyanopropyl)-tetramethyldisiloxane); m.p. = 50 °C. IR, cm^{-1}: 3000 cm^{+1} (OH), 1710 (-CO-); 1075 (Si-O-Si). ^{1}HNMR (CCl_4); □, ppm: 0.0-0.25 (Si-CH_3); 0.41-0.85 (t, -CH_2-Si); 1.4-2.0 (m, -CH_2-); 2.15-2.6 (-CH_2-COOH); 10.8-11.00 (COOH); 84% yield.
1,3-bis(carboxytrimellityliido-N-propylene)tetramethyldisiloxane, m.p. 200°C.

2.2. Methods for polymers synthesis

A series of organo-siloxane supramolecular polymers was obtained, starting from 4,4'-bipyridine as an acceptor and different silicon-containing carboxylic acids as hydrogen-donor molecules [10].
The standard procedure was applied, i.e. mixing of stoechiometric amounts of 4,4'-bipyridine and siloxane diacids, in a non-polar solvent, followed by distillation and vacuum drying. We also tested the contact method, which gave very good results, proving that such supramolecular polymers are very easily obtainable, due to the natural molecular recognition process.
In this study, two of these supramolecular polymers were tested for potential application as gas sensors. The structures of the studied compounds are presented in Fig. 1. The starting disiloxane diacids, 1,3-bis(carboxypropyl)tetramethyl-disiloxane and 1,3-bis(carboxytrimellitylimido-N-propylene)tetramethyldisiloxane were synthesized following the methods described in the literature (J.E. Mulvaney, C.S. Marvel 1961, A. Staubli, E. Ron, R. Langer, 1990); (M.p. 50 °C).

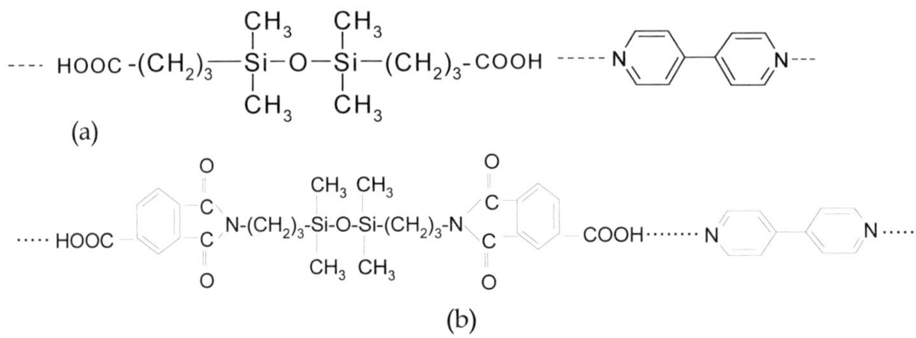

Fig. 1. Chemical structure of the supramolecular polymer.(a): CH2 polymer; (b). CH5 polymer

2.3. Methods for polymeric structural characterization of polymer

2.3.1. Thermal analysis

The thermal characterization by DSC showed for CH2 a melting temperature of 63°C, while by polarized optical microscopy a narrow smectic mesophase was observed, between 62 and 68°C. The heating rate didn't allow us to observe two endothermal peaks in DSC. For CH5, the DSC results confirmed the polarized optical microscopy observations, i.e. the presence of a mesophase between 130 and 160 °C, which were proved by the presence of two endothermal peaks on heating. In Fig. 2, the DSC curves of polymer CH5 are presented, at first (a) and second heating scan (b).

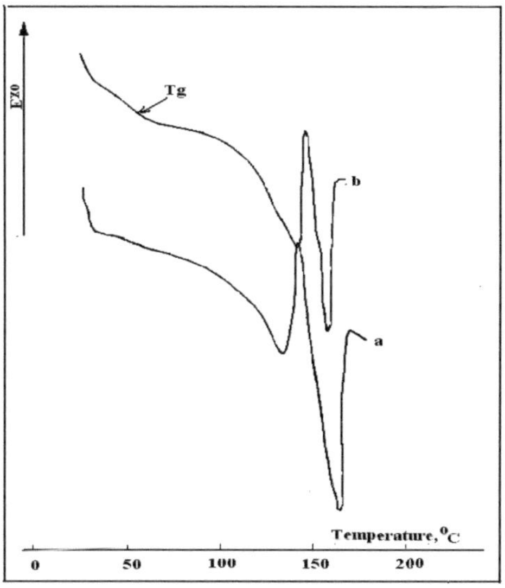

Fig. 2. The DSC curves on first heating (a) diacid and the second heating (b) for the supramolecular supramolecular polymer CH5

2.3.2. FT-IR spectra

The formation of the hydrogen bonds was primarily verified by FT-IR spectroscopy which showed two specific broad absorption bands centered at around 1923 and 2475 cm^{-1} for CH2 and at around 1940 and 2450 cm^{-1} for CH5 (H. Han, A.H. Molla, 1995), as can be observed in Fig. 3 and 4. All the other absorption bands, corresponding to the proposed structures were present: siloxane Si-O-Si asym. at 1030 cm^{-1}, CH_3 def. at 1250 cm^{-1}, CH_3 rock at 812 cm^{-1}, COOH at about 1700 cm^{-1}, aromatic at 1600 cm^{-1}. Nevertheless, a slight shift of the carboxylic C=O band did occur, reported to the starting diacids (see for example Fig. 3. All the other absorption bands, corresponding to the proposed structures were present in the polymers, with small shifts compared to the starting compounds, as can be observed in Table 1.

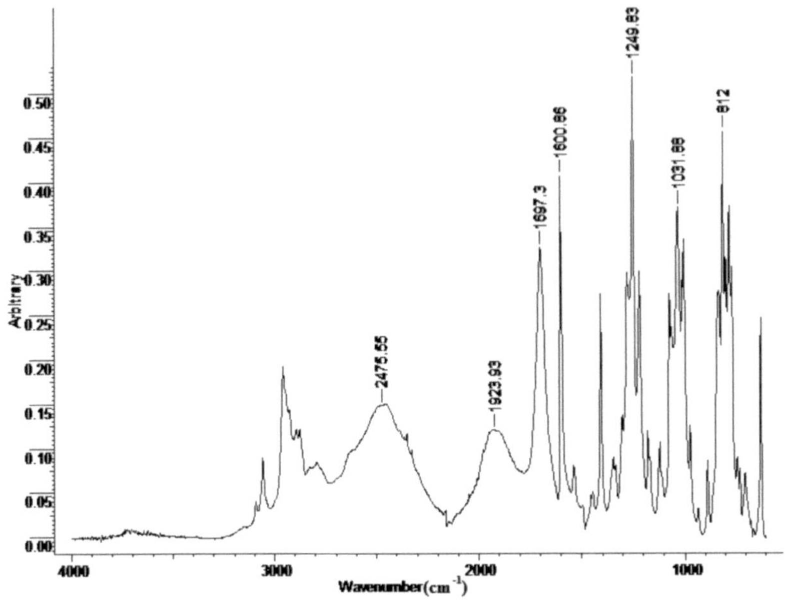

Fig. 3.FT-IR spectra for CH2 polymer of the supramolecular polymer

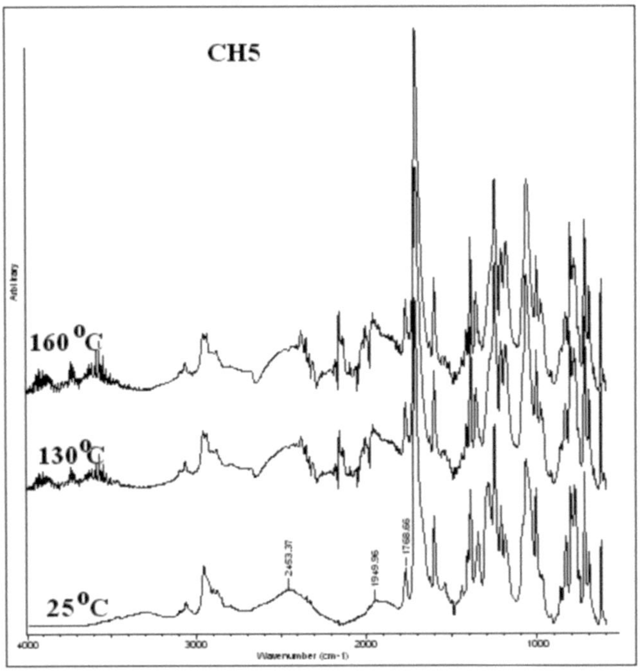

Fig. 4. FT-IR spectra for CH5 polymer

Compound	wavenumber, cm^{-1}	Assignment
4,4`-Bipyridine	1590	heteroaromatic
	812, 806	arC-H γ oop
1,3-bis(carboxytrimellityliido-N-propylene)tetramethyldisiloxane	787	arc-H 1,2,4 trisubstituted
	841, 1060, 1253	CH_3-Si; Si-O-Si
	1716	C=O acid
	1777	C=O imide
CH5	785	arc-H 1,2,4 trisubstituted
	807	arC-H γ oop
	838, 1067, 1252	CH_3-Si; Si-O-Si
	1600	aromatic
	1717	C=O acid
	1771	C=O imide
	1950, 2443	H-bonding

Table 1. The main FT-IR absorption bands in the starting compounds and supramolecular polymers

2.3.3. ^{1}H-NMR spectra

^{1}H-NMR spectra showed one set of signals and slightly modified chemical shifts compared to the starting compounds, indicating that the polymers were stable in chloroform, at least on the analysis time scale. The polymer spectrum is presented in Fig. 5.

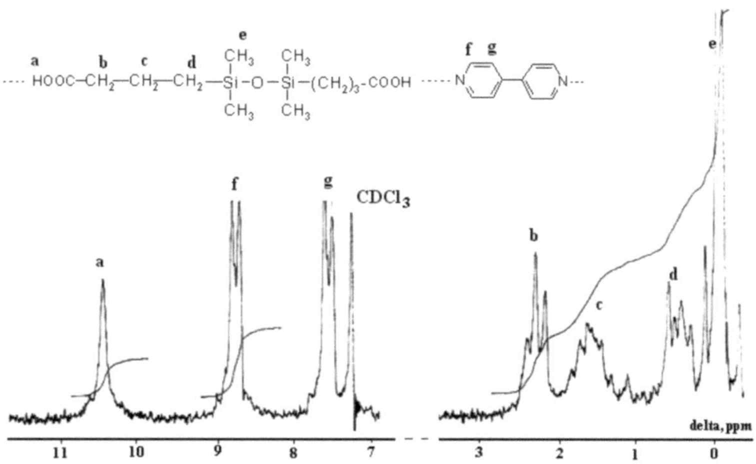

Fig. 5. ^{1}H-NMR Spectrum of the supramolecular polymer

3. Gas sensing mechanism of organo-siloxane supramolecular polymers

In order to analyse the influence of the testing gas - in this case CO_2 - on the chemical structure of the supramolecular polymers, we registered the IR spectrum of sample CH2, after maintaining it for an hour in a CO_2 atmosphere. As can be observed in Fig. 6, the bands corresponding to the H-bond didn't change and no other existing absorption bands suffered noticeable modification. This result shows that the compound is chemically stable at exposure to the testing gas.

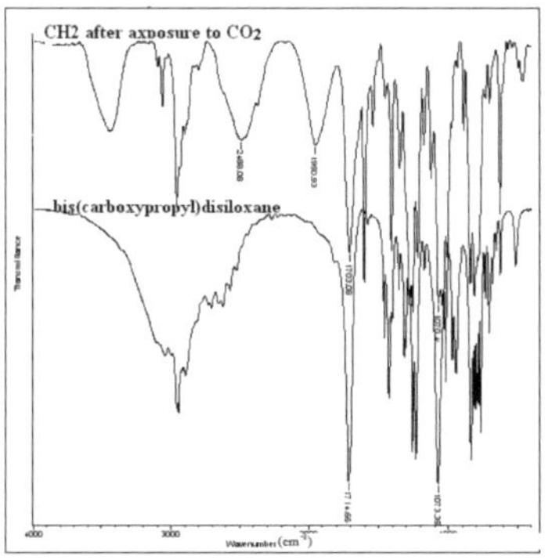

Fig. 6. FT-IR spectra of the starting supramolecular polymer CH5 before and after exposure to CO_2

4. CO_2 sensing properties of polymers

4.1. Thin and thick technology for microsensors fabrication

4.2. Device 1

Fig. 7 shows the structure of the sensor. An alumina subtrate 6 x 6 x 0.5 mm was used. On one side of the substrate was screen-printed an interdigitated electrode array using Au ink and heat treated at 950°C for 1 h. The sensitive layer obtained by dissolving the polymer in chloroform was deposed on the substrate with provided electrode by spin coating method. The sensor contains the pads of Pd – Ag conductive ink and the conductive layer of Ag ink heat treated at 750°C for 30 minutes.

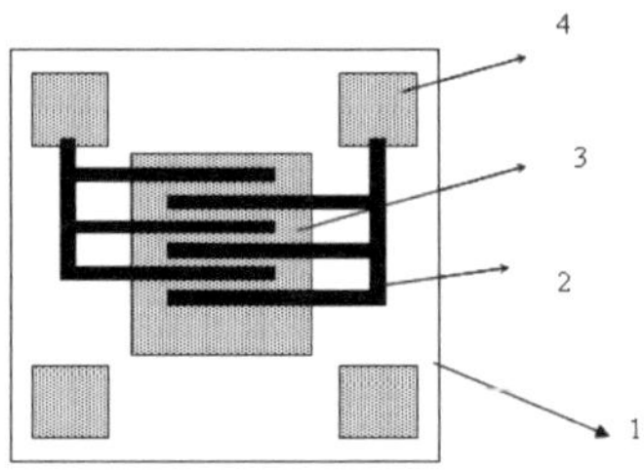

Fig. 7. Sensor configuration; 1 – alumina substrate; 2 – electrode; 3 –polymeric layer; 4. pad.

4.3. Device 2

The sensor is composed from an alumina substrate 5 x 5 x 0.6 mm, where on one side of the substrate was deposed by magnetron sputtering 2 plates gold electrodes distanced at 3 mm. The deposed conditions were: pressure argon 1,9 torr, voltage 500V and the deposed time was 30 minutes. The sensitive layer obtained by dissolving the polymer in chloroform was deposed on the substrate with provided electrodes by spin coating method in the 200 nm thickness. The sensor was mounted on the transistor ambase-Fig. 8.

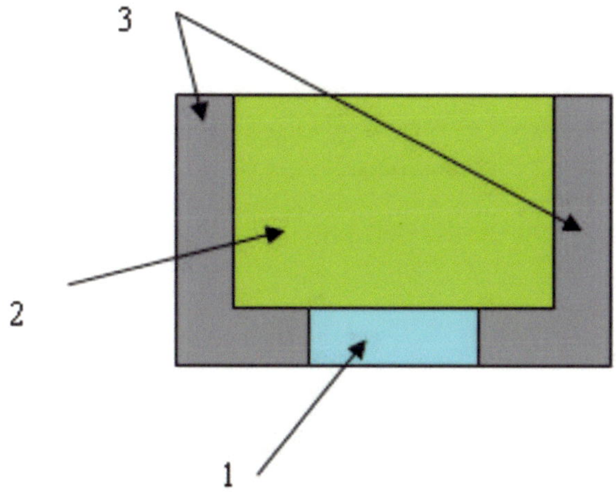

Fig. 8. The sensor configuration
1- alumina substrate; 2- sensitive layer; 3- electrodes

4.2. The characterization of microsensors

4.2.1. Device 1

The testing measurement

The detection testing of gases CO_2 and NO_x was performed in a 100 ppm concentration and the voltage measurements were effected with a multimeter type APPA 301 TRUE RMS.
The sensor was exposed in the CO_2 and NO_x atmospheres in 100 ppm concentration and was measured the voltage in function of the time. For the CH2 polymer were obtained for CO_2 the voltage value about 70 mV after 10 minutes after gas exposure and for NO_x was obtained the small value of the voltage by 4 mV. For the CH5 polymer were obtained 238 mV for NO_x and 92 mV for CO_2 atmospheres after 30 minutes exposure. Results that, the polymer CH5 is more sensitive to gases against CH2 polymer and the CH5 polymer is more sensitive to NO_x. CH2 is sensitive to the CO_2 and unsensitive to NO_x. Fig. 9 and 10 present the characteristics voltage - time for the CH2 and CH5 polymers.

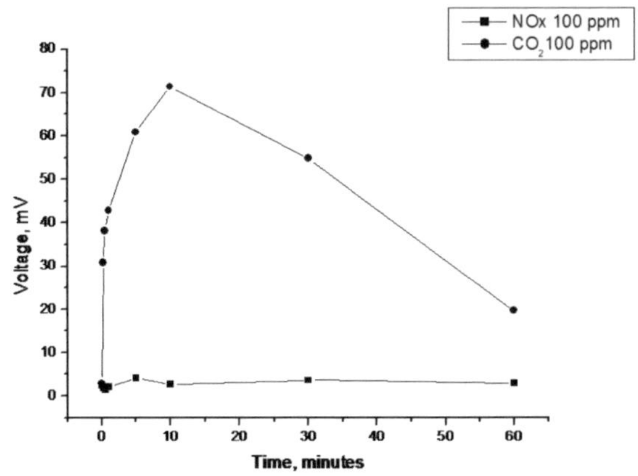

Fig. 9. The characteristic voltage time for the CH2 polymer

4.2.2. Device 2

The testing measurements

The detection testing of gas CO_2 was performed in a 100 and 1000 ppm gas concentration and the voltage measurements were effected by testing module, in automated process mode. An control panel, provides a lot of measuring value, by rate 1/10 second.

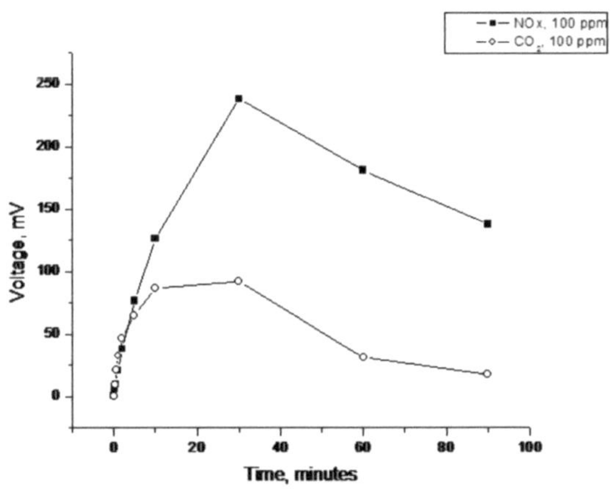

Fig. 10. The characteristic voltage-time for CH5 polymer

The bench of testing for the gas sensor consists in an enclosure where there are set the testing conditions of the sensor as well as in connected equipment. The whole process of testing is automated, being controlled by a programmable automaton. The gas for testing is introduced in a controlled way in the testing enclosure, through a mass debitmeter. In the testing enclosure is set a constant temperature, controlled by a temperature regulator-Fig. 11.

Fig. 11. The apparatus for the gas sensor testing

The gas sensing characteristics

The testing gas was CO_2 in 100 and 1000 ppm concentrations. Fig. 12 shows the characteristics voltage-time for 3 sensors exposed at 1000 ppm CO_2. It can be see that in the first 5 minutes gas exposure, the signals of the sensors are very weak with the voltage values of 660, 720 and 800 mV corresponding of sensor 1, sensor 2 and sensor 3. Fig. 13 presents the characteristics voltage-time for the sensor 3 exposed in 100 and 1000 ppm where were obtained the maximum voltage values of 92 mV for 100 ppm CO_2 and 970 mV for 1000 ppm CO_2 after 30 minutes gas exposure.

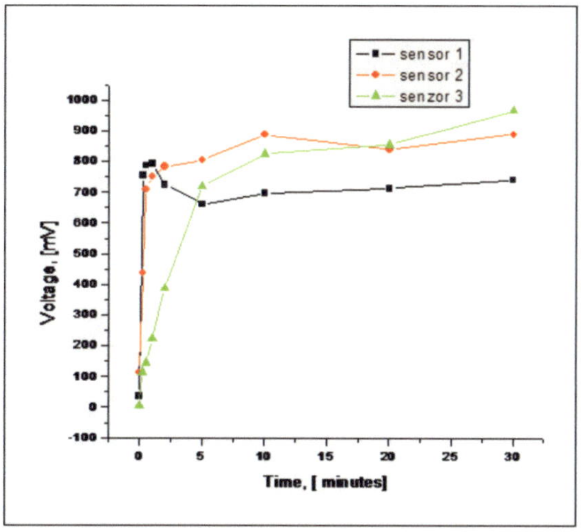

Fig. 12. The characteristics voltage-time for 3 sensors exposed at 1000 ppm CO_2

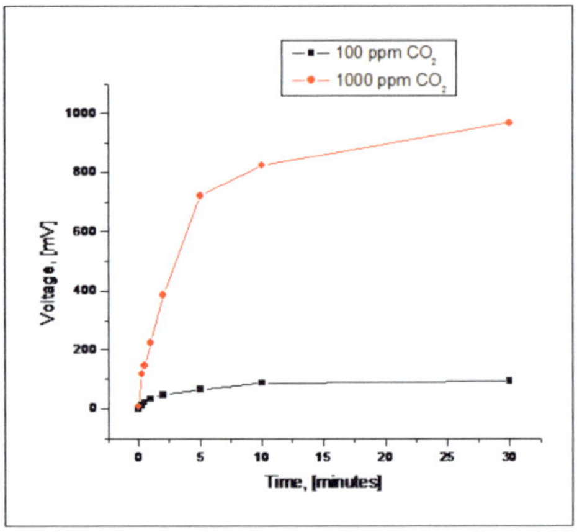

Fig. 13. The characteristics voltage-time for sensor 3 exposed at 100 and 1000 ppm CO_2

4.3. The signal conditioning of polymeric sensitive layer for device 2

The electronic design of conditioning detector presented in Fig. 14, contains two amplification stages, cascade connected and realized with integrated circuit operational amplifier with JFET transistors at input, LF 356 type. The input impedance of these is theoretically infinite, practically very large, around $Z \approx 10M\Omega$. Therefore, the polarizing currents, when the sensitive element of the transducer is connected, are extremely small and the measured value is very precise. The sensitive element of the sensor of CO_2 is connected to the operational amplifier U1 in common mode, on the reversing input. Every one of these amplification stages realizes an amplification factor A1 and A2, respectively, corresponding with the realized reversing input

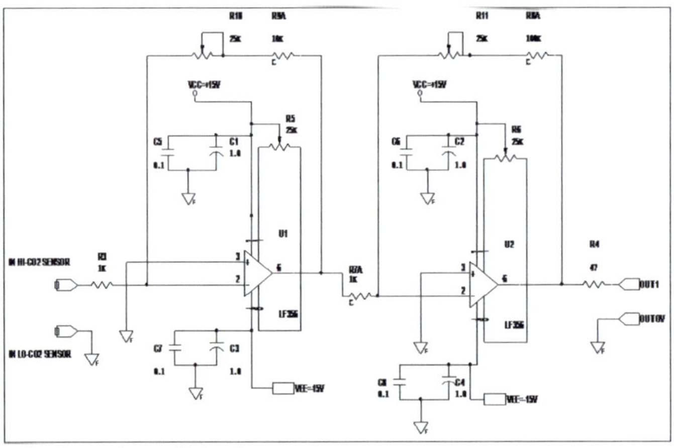

Fig. 14. Electronic design of conditioning

The global amplification factor can be written, if we count that every amplification stage represents a basic cvadripole and the interaction is made exclusively through the terminals,

$$A = A1 * A2, \tag{1}$$

where:

$$A1_{max} = -\frac{R9A + R10}{R3} = -\frac{10+25}{1} = -35, \tag{2}$$

and

$$A2_{max} = -\frac{R8A + R11}{R7A} = -\frac{100+25}{1} = -125. \tag{3}$$

We have obtained:

$$A_{max} = (-35)(-125) = 4375 \tag{4}$$

If the values of the potentiometers R10 and R11 are null, is obtained:

$$A_{min} = (-10)(-100) = 1000 \tag{5}$$

Obviously, if we would have wanted to use a single amplification stage for realizing the same maximal amplification value, most probably this would have transformed in an oscillator.

Of course, the final value of the global amplification is stabilized around the value $A \approx 2500$, being related with the resulted values in the transfer characteristic of the sensitive element.

The integrated circuit operational amplifiers U1 and U2, respectively, type LF 356, have capabilities of regulating the input offsets, through the R5 and R6 potentiometers. The regulating algorithm is:

With the reversing input of the operational amplifier U1 at the null potential, it is activating on the R5 potentiometer, until the output voltage, (pin 6 U1), is null, with R7A disconnected;

With the reversing input of the operational amplifier U2 at the null potential, it is activating on the R6 potentiometer, until the output voltage, (pin 6 U2), is null, with R7A disconnected;

In an iterative way, there are repeated the procedures in the stages 1 and 2, until the output voltage of the operational amplifier U2 is zero, when the reversing inputs of the operational amplifiers U1 and U2 are at null potential.

The output signal OUT1-Fig. 15, is a common mode voltage, picked at the output of the operational amplifier U2.

The supply voltage V_{cc} and V_{EE}, respectively, is dual polarity power supply, stabilized. This is differentiated, $\pm 15V_{cc}$, in comparison with the null potential bar.

The capacitor groups C_1 - C_5; C_3 - C_7; C_2 - C_6; C_4 - C_8; make a decoupling of the supply voltage, in the immediate closeness of the operational amplifiers.

Another important design consideration is how circuit gain affects error, such as noise. When the operational amplifier U1 and U2 is operating at higher gains, the gain of the input stage is increased. At high gain, the input stage errors dominate. That is: input noise = eni and output noise = eno. Total noise RTI and RTO are calculated, (Charles Kitchin Lew Counts, 2004).

$$RTI = \sqrt{(eni)^2 + (eno/Gain)^2} \tag{6}$$

$$RTO = \sqrt{(Gain(eni))^2 + (eno)^2} \tag{7}$$

Typical noise of the LF 356 is specified, (LF356/LF357 JFET, 2001), as $eni = 12nV / Hz$, and $eno = 80nV / Hz$. Total noise RTI and RTO of the LF 356 operating at a gain of 10 is equal to:

$$RTI = 14{,}42nV / \sqrt{Hz}\ ,\ RTO = 144nV / \sqrt{Hz}$$

Also, the noise input current of the LF 356 is very low, $0.01 pA / \sqrt{Hz}$.

Another design for this conditioning CO_2 transducer amplifier is showing in Fig. 15. The AD 8224 can operate on a ± 15 V dual supply. The most important parameter for this application is common - mode rejection ratio (CMRR). The AD 8224 has a high CMRR, of 94 dB for G = 10. In addition, slew rate of the AD 8224 is 2V / µs and the transfer function, (Charles Kitchin and Lew Counts, 2004).

$$G1 = 1 + \frac{49.4k\Omega}{R1A^*}\text{, respectively,} \quad G2 = 1 + \frac{49.4k\Omega}{R1B^*} \tag{8}$$

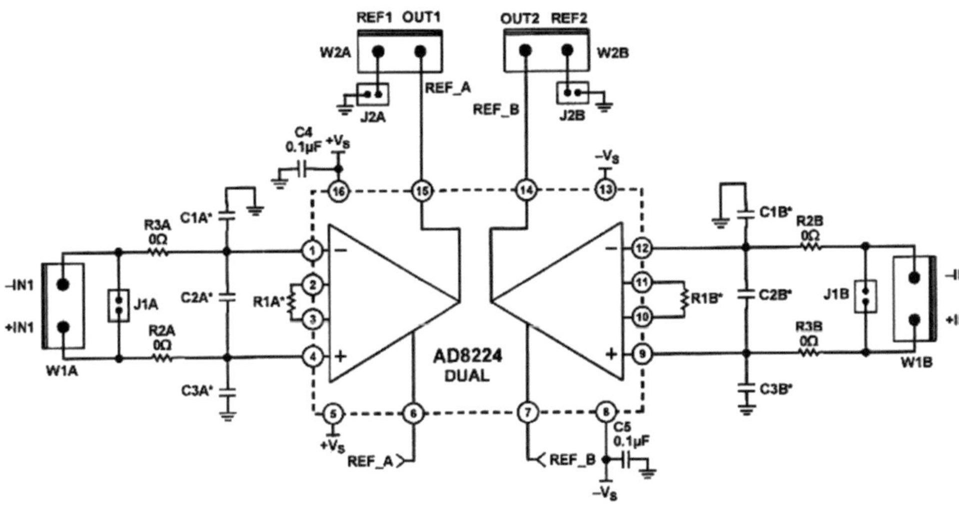

Fig. 15. Precizion, AD 8224 JFET Input instrumentation amplifier for CO_2 transducer interface.

Placing a resistor across the $R1A^*$ and $R1B^*$ terminals, sets the gain of the AD 8224. For best performance, by using standard value of 1% resistor $R1A^*$ = $R1B^*$ = 5.49 KΩ, calculated gain is 9.998.

For optimum performance, a low - pass R - C network can be placed at the input of the instrumentation amplifier. This filter limits the input signal bandwidth according to the following relationship, (Charles Kitchin and Lew Counts, 2004):

$$FilterFreq_{DIFF} = \frac{1}{2\pi R2A\left(2C2A^* + C1A^*\right)} \tag{9}$$

where:

$$C1A^{*} = C3A^{*} \text{ and } R2A = R3A \quad (10)$$

As shown, Fig.15, R2A=R3A=4.02kΩ can be metal film resistors and $C1A^{*} = C3A^{*}$ =1nF, $C2A^{*}$ = 10nF miniature size micas.

Fig. 16 presents the ansamble of image for CO_2 detector. As shown in Fig. 16, this application required a variable gain amplifier.

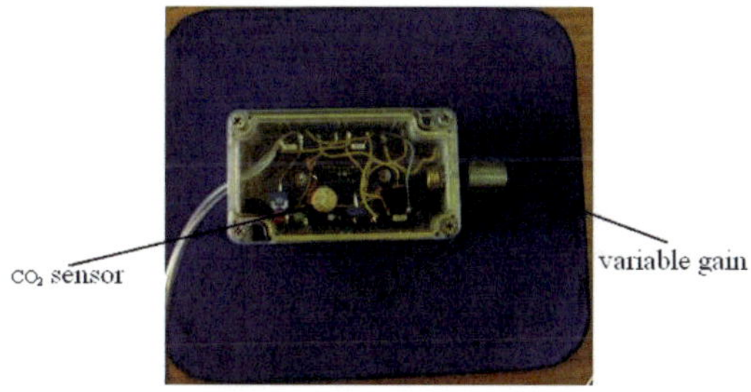

Fig. 16. The image of CO_2 detector

5. Theoretical aspects of equivalent scheme

5.1. General equivalent circuit of active sensor

As a first step in the direction to make clear the study of gas sensor is following classification;

a) Active sensor and represents a sensor which convert the gas effect of the sensible material in a microvoltage -Fig. 17. Example: gas electrochemical sensor.
b) Passive sensor represents a sensor which convert the gas effect of the sensible material in a change of an electric resistance or an electric capacitance-Fig. 18,.example; gas semiconductor sensor.

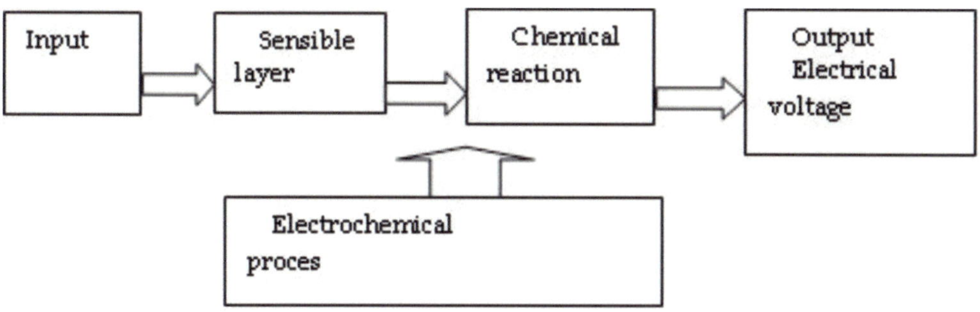

Fig. 17. Active sensor

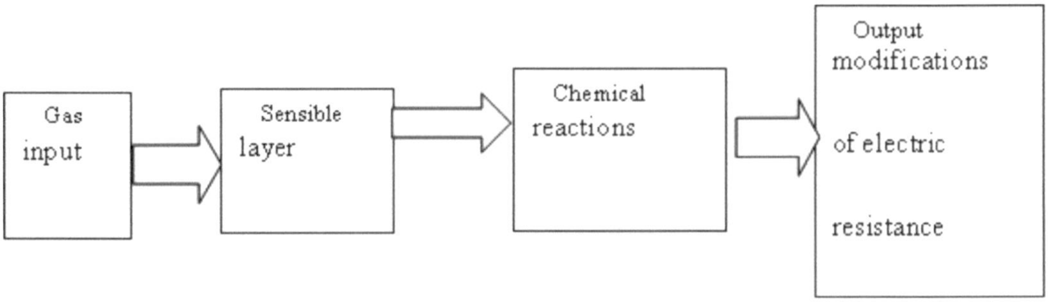

Fig. 18. Pasive sensor

By the electric characteristic Gopel (W.Gopel, 1996), proposes a classification:

- Sensor type resistor.
- Sensor type diode.
- Sensor type capacitor.

An active gas sensor (with polymeric sensitive layer) accepting as a converter with stimuli gas concentration (percentage, ppm) and a terminal or electric port:voltage (U) is presented in fig.19.

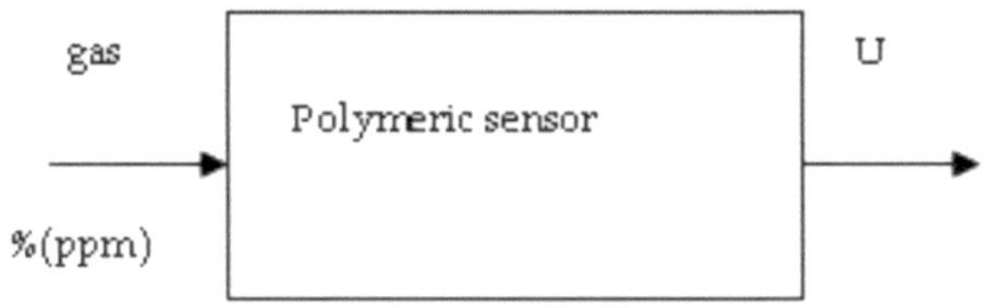

Fig. 19. The gas sensor as a converter.

5.2 The ideal characteristics of an equivalent voltage source.

In keeping with the theory of electric circuits the active sensors have the following equivalent circuit -Fig. 20.

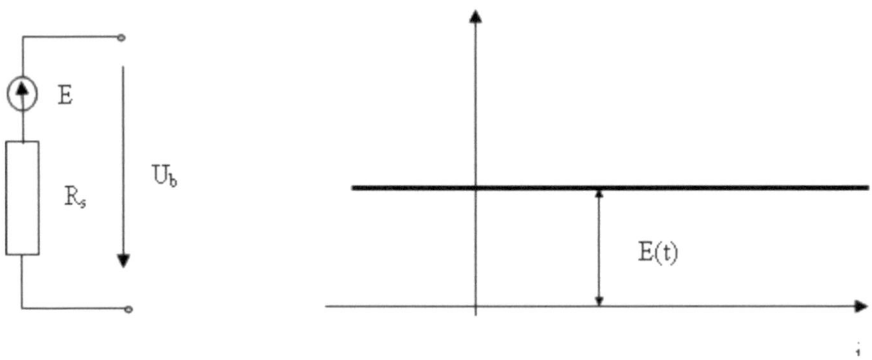

Fig. 20. General equivalent circuit of active sensor

This equivalent circuit is a general scheme which represents a voltage source with base equations (Joubert) (C.I.Mocanu, 1979, C.A. Desoer,E.S.Kuh, 1969):

$$U_b - E = R_s i \tag{11}$$

where E - the electromotive force, R_s – voltage source series internal resistance, U_b – terminal voltage of source, i – electric current.In Fig. 21 is presented and ideal characteristics for equivalent circuit.

5.3 Current source.Equivalent source and characteristics

Any voltage source consisting of an electromotive force E independent of the terminal current I , having an internal series resistances R_s can be replaced by a current source

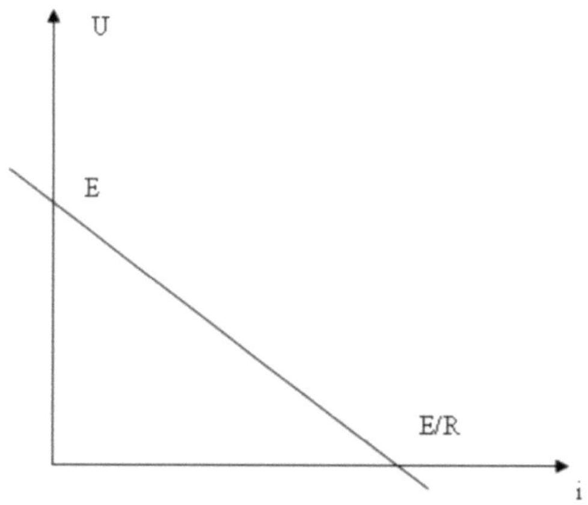

Fig. 21. The ideal characteristics of an equivalent voltage source.

consisting of a current I_g independent of the terminal voltage U_b, having an internal resistance R_g with equation:

$$I_g - I = \frac{U_b}{R_g} \tag{12}$$

The behaviour of sensor gas asimilated as the voltage source is not linear on different subdomains and the real characteristics $U(i)$ -Fig.22.

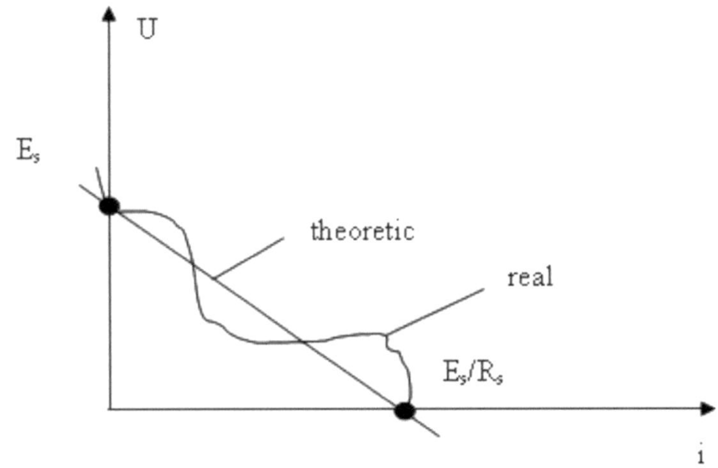

Fig. 22. The real and ideal characteristics $U(i)$ for a voltage source

5.4 Analog equivalent circuit of the active gas sensor

The theoretical aspect which was presented forward is a synthetic macroscopic treatment of active sensor gas electric analysis. We consider for a detail study a geometric model –Fig. 23. Gaseous substances can affect the performance of the device. The adsorbtion and desorbtion such as gas molecules on the surface of sensible sensor layer may cause signal amplitude or phase or frequency fluctuation. This kind of perturbation can be described by dynamic process. The surface inference, fluctuations, particles adsorbtion, desorbtion and random movement of particles are the important aspects for the knowlidge of activ gas sensors.

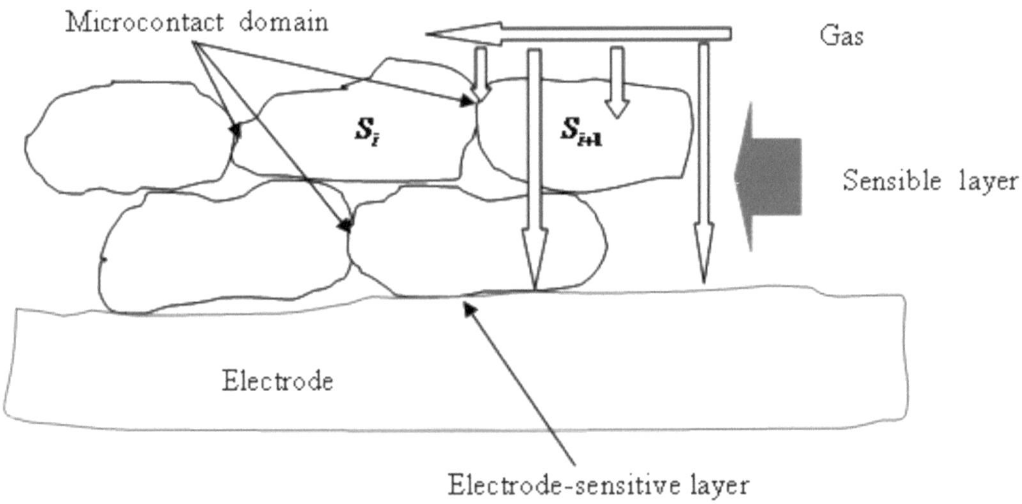

Fig. 23 Sensor geometry modelling

Is necessary to formulate the electrokinetics problem of the active gas sensor:

Either a known topology of active gas sensor domain. To be investigated local and general equivalent circuit of domain referring to la imprintelectric field and the density (distributions) of this fields with the goal of modelling and determination of output electric signal; voltage, current and the specific parameters; sensibility, response time, linearity of signal, resolution, hysterezis.

Δ-active gas sensor domain is n-connected space; $\Delta = \Delta_s \cup \Delta_e \cup \Delta_{ec} \cup \Delta_{slc}$ and is formed by conductors (I and II category) in contact and which generate different imprint electric fields (C.A. Desoer,E.S.Kuh, 1969, C.I.Mocanu 1984):

- electrolityc nature with gas reaction $\overline{E}_{egr}$,
- electrolytic nature $\overline{E}_{ei}$ (another than gas sensor),
- microcontact nature $\overline{E}_{ci}$,

or others (C.A. Desoer, E.S. Kuh, 1969, C.I.Mocanu 1984). The domains represent; Δ_{sl}-domain of sensible layer, Δ_{slc}-domain of microcontact between the different parts (subdomains) of sensible layers, Δ_e-domain of electrode, Δ_{ec}-domain of microcontact between sensible layer and electrodes of sensor.

Either Φ_g the molecular flux of the gas specie giving rise to the active sensor signal and referred to the geometrical model:

$$\Phi_g = \Phi_{react} + \Phi_{rm} = \Phi_{sl} + \Phi_e + \Phi_{ec} + \Phi_{slc} + \Phi_{rm} \quad (13)$$

where; Φ_{sl} - the molecular flux to sensible part (layer) of active sensor, Φ_e – the molecular flux to electrode surface, Φ_{ec} – the molecular flux to the microcontact between electrode - sensible layer,

Φ_{slc} – the molecular flux to the contact between two sensible subdomains (S_i, S_{i+1}), Φ_{react} – the reacted flux molecules, Φ_{rm} – the reflected molecular flux.

The sensible layer of sensor include $S_1,...,S_i,...S_n$ subdomains $\subset \Delta_{sl}$:

$$\Delta_{sl} = S_1 \cup S_2 ... \cup S_i \cup S_{i+1} \cup .... \cup S_n. \quad (14)$$

The assumption of sensible layer subdomains ($S_1,...,S_i,...S_n$) in microcontact is founded on the electron microscopie investigations (SEM) of sensible layer what indicates the existence of subdomains and the discontinuities in structure. We investigate especially the structure of NASICON ($Na_3Zr_2SiO_2PO_{12}$) disc, Na-super-ionic conductor/nitrite $TiO_2ZrO_2Y_2O_3$ and polymeric sensitive layers). We assume that each reacting molecules flux initiates a microvoltage source:

Φ_{sl} - E_{sl}, , $\Phi_e - E_e$, $\Phi_{ec} - E_{ec}$, $\Phi_{slc} - E_{slc}$ and the sensor domain can be modelled that in fig.24.

We assume that each reacting molecules flux initiates a microvoltage source:
Φ_{sl} - E_{sl}, , $\Phi_e - E_e$, $\Phi_{ec} - E_{ec}$, $\Phi_{slc} - E_{slc}$ and the sensor domain can be modelled that in fig.25.

In this circuit E, R represent the elements of voltage source; microelectromotive force (microvoltage) and electric resistance reported to the microsubdomains.
The detailed equivalent circuit includes a matrix structure (conform with the geometry and topology described - Fig.23) with elements of sensible subdomain in quadrature axis; $E_{sl_i}, R_{sl_i}, E_{sl_{i+1}}, R_{sl_{i+1}}$ and in longitudinal axis; $E'_{sl_i}, R'_{sl_i}, E'_{sl_{i+1}}, R'_{sl_{i+1}}$ .In diagram are evidenced the component possible noise signal of microcontact sensible layer- electrode(E_{ec}, R_{ec}) and reaction of gas with electrode (E_e, R_e) which have the longitudinal components and appear that effect of local microcorrosion.Essential signal for sensor is the microvoltage generated in sensible layer or sensible subdomain; E_{sl} .
Because the detailed equivalent circuit requires many informations about the subdomain is proposes a synthetic circuit in Fig.25.
The electrokinetic equations by Kirkkof rules, are -Fig.24:

$$I_{sl} + I_{slc} = I_s \tag{15}$$

For $(R_{sl}, E_{sl})\| (R_{slc}, E_{slc})$ the relations of microvoltage source and equivalent resistance are:

$$E_{eq} = \frac{E_{sl}R_{slc} + E_{slc}R_{sl}}{R_{sl} + R_{slc}} \tag{16}$$

$$R_{eq} = \frac{R_{sl}R_{slc}}{R_{sl} + R_{slc}} \tag{17}$$

and the equivalent circuit becomes that in Fig 26 with the general equation;

$$(E_{eq} + E_{ec} + E_c) - U_s = I_s(R_{eq} + R_{ec} + R_e) \tag{18}$$

Finnaly:

$$U_s = (E_{eq} + E_{ec} + E_e) - I_s(R_{eq} + R_{ec} + R_e) \tag{19}$$

or with equivalent relations of microvoltage source and resistance:

$$U_s = \frac{E_{sl}R_{slc}}{R_{sl} + R_{slc}} + \frac{E_{slc}R_{sl}}{R_{sl} + R_{slc}} + E_{ec} + E_e - I_s \cdot \frac{R_{sl}R_{slc} + (R_{ec} + R_e)(R_{slc} + R_{sl})}{R_{sl} + R_{slc}}$$

There are following cases:

I. $E_{slc} = E_{ec} = E_e = 0$ represents the ideal case when the are nor internal noise microvoltage sources:

$$U_s = E_{sl}k_1 - k_2 \tag{20}$$

where:

$$k_1 = \frac{R_{slc}}{R_{sl} + R_{slc}}, k_2 = \frac{R_{sl}R_{slc} + (R_{ec} + R_e)(R_{slc} + R_{sl})}{R_{sl} + R_{slc}} \cdot I_s \tag{21}$$

The characteristic $U_s(E_s)$ is presented in Fig.27.

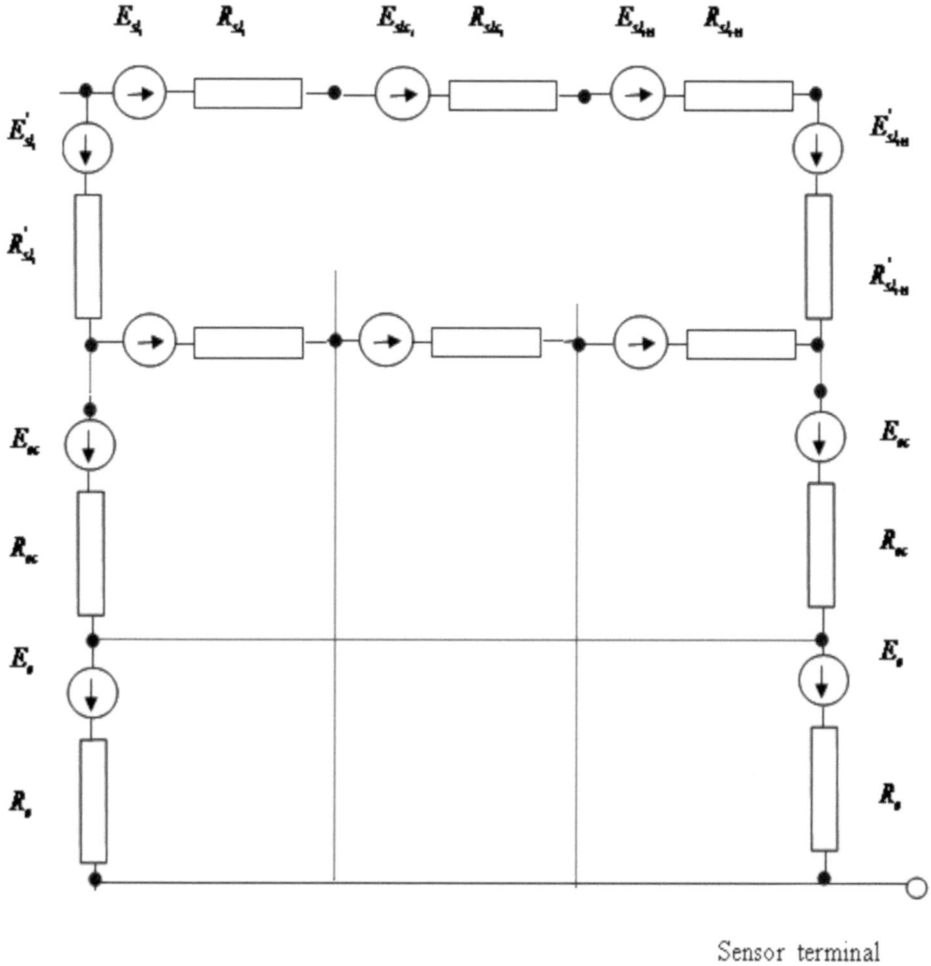

Fig. 24. Detailed equivalent circuit of active gas sensor

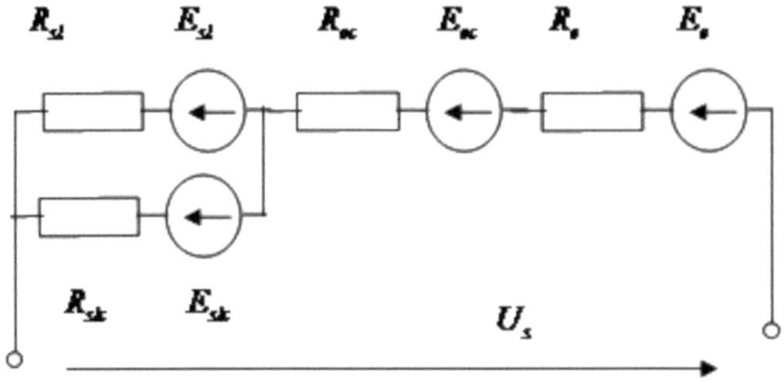

Fig. 25. An equivalent synthetic circuit of active gas

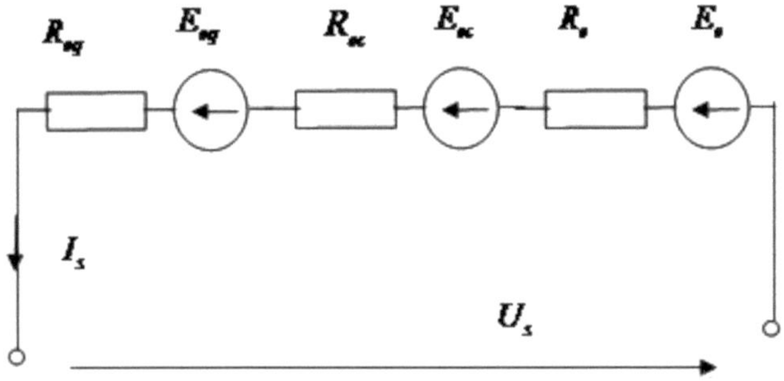

Fig. 26. The final equivalent synthetic circuit of active gas

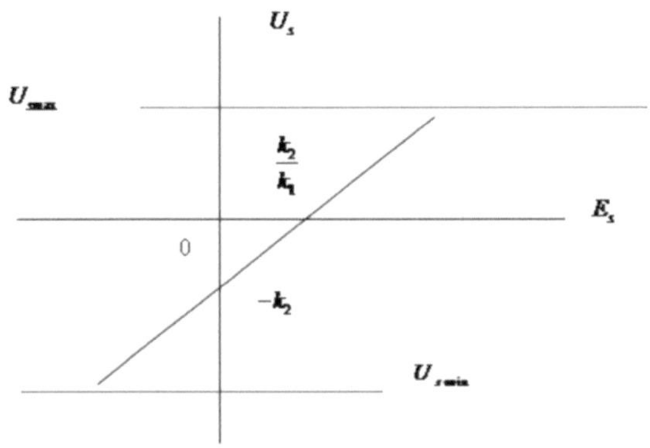

Fig. 27. Characteristic U_s (E_s) in ideal case without the noise microsources

Essential every problem of gas sensor generation; measurement, control, involves the conversion of the conversion of the chemical in electrical energy. In dealing, with active gas structure, chemical-electric coupled systems, it is very convenient at some times and essential at others to treat the combined gas sensor as an entity. The basic idea of an electro-chemical analogy comparison is the following:

	Gas circuit	Electric circuit
Potential	Concentration - c	Electric potential - V
	$\Delta c = c_2 - c_1$ Concentration difference	$\Delta U = V_2 - V_1$ Voltage (Potential difference)
Flux density	$\Phi_d = -DS\frac{\partial c}{\partial x}$, $\Phi_{dg} = \frac{\Phi_d}{S} = -\frac{\nabla c}{\frac{1}{D}}$ Fick's difussion flux dernsity S-surface of diffusion	$\bar{J} = \sigma \bar{E}, I = \frac{U}{R}$ Ohm's law $\bar{J}$ – electric density current, I – electric current, R – electric resistance
Constitutive parameter	D – diffusion coefficient	σ – electric conductivity

Table 2. Basic analogies

According to Fick's first diffusion law [5.5] the evolution of the gas in active sensor, the rate (flux) Φ_d depends on the concentation gradient c, on sensible material subdomain S_i:

$$\Phi_{S_i} = \Phi_d = -DS\left(\frac{dc}{dx}\right)_{x=\frac{d}{2}} \tag{22}$$

where S is the surface area and D the diffusion coefficient, d- the thickness of sensible subdomain. To obtain the gas evolution rate as a function of gas active sensor, the time dependence of dc/dx must be known. By Fick's second diffusion law:

$$\frac{\partial c}{\partial x} = D\frac{\partial^2 c}{\partial x^2} \tag{23}$$

If $dc = c_0 dp$ with c_0 – initial concentration and p – gas pression, according to Henry law equation may therefore be written as:

$$\Phi_p = DSc_o(\frac{dp}{dx}) \tag{24}$$

which represent another analog model with gas pression p -potential.

Is possible to determine the solute distributions $c(x,t)$ during transient states, (I. Lundsrom, 1996, W. Jost, 1952, H. Carslaw, J. C. Jaeger, 1959, V. S. Vladimirov, 1980, J. Crank, 1956).

In the case of uniform initial concentration equation (25) can be satisfied by the solution (V.S. Vladimirov, 1980):

$$c(x,t) = C_i\frac{4}{\pi}\sum_{n=0}^{\infty}\frac{(-1)^n}{2n+1}\cos\frac{\pi x(2n+1)}{d}\cdot\exp\left\{-\left[\pi(2n+1)\right]^2\frac{Dt}{d^2}\right\} \tag{25}$$

where t – time and C_i – initial gas concentration.

With the basic analogies which was described in Table 1, is presented in Fig. 28 the equivalent circuit of the active gas sensor as a chemical gas subsystem (input of sensor) coupled with an electric subsystem (output of the sensor).

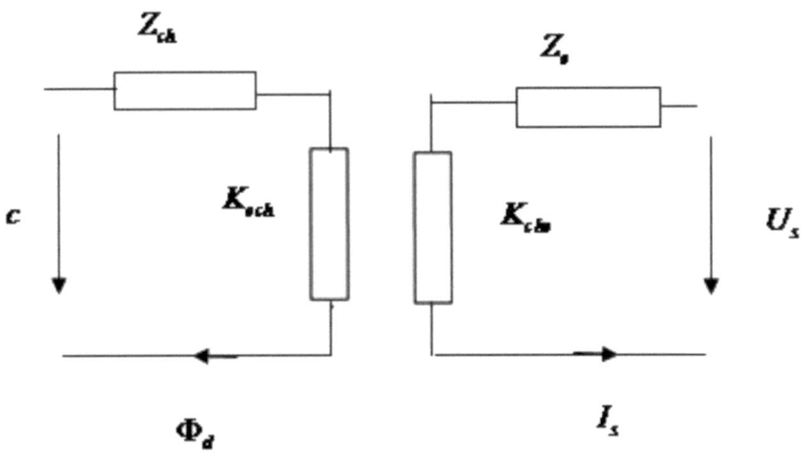

Fig. 28. Analog equivalent circuit of the active gas sensor

The equation of this coupled circuits are:
Input

$$c = K_{che} I_s + Z_{ch}^{I_s} \Phi_d \tag{26}$$

Output

$$U_s = Z_e^{\Phi_d} I_s + K_{ech} \Phi_d \tag{27}$$

where c – gas concentration, U_s – output voltage, Φ_d – flux gas, I_s – electric current sensor, $Z_{ch}^{I_s}$ – specific chemical gas sensor impedance to $I_s = 0$ (no load output):

$$Z_{ch}^{I_s} = \frac{c}{\Phi_{d\,(I_s=0)}} \tag{28}$$

$Z_e^{\Phi_d}$ – specific electric impedance to $\Phi_d = 0$ (no-load input):

$$Z_e^{\Phi_d} = \frac{U_s}{I_{s\,(\Phi_d=0)}} \tag{29}$$

K_{che} – coupled ratio or operator between chemical gas input and electric output:

$$K_{che} = \frac{c}{I_{s\,(\Phi_d=0)}} \tag{30}$$

K_{ech} - coupled ratio or operator between the electric output and chemical gas input:

$$K_{ech} = \frac{U_s}{\Phi_{d\,(I_s=0)}} \tag{31}$$

The coupled operators; K_{che} and K_{ech} represents the reflected impedances.

6. Conclusions

A organo-siloxane supramolecular polymer was obtained, starting from 4,4′-bipyridine as an acceptor and silicon-containing carboxylic acids as hydrogen-donor molecules. The sensors were made by thin and thick film technologies. The thickness of polymeric layer deposed by spin coating on the alumina substrate was 200 nm. For the device 1, the sensor was exposed in the CO_2 and NO_x atmospheres in 100 ppm concentration gas and was measured the voltage in function of the time. For the CH2 polymer were obtained for CO_2 the voltage value about 70 mV after 10 minutes after gas exposure and for NO_x was obtained the small value of the voltage by 4 mV. For the CH5 polymer were obtained 238 mV for NO_x and 92 mV for CO_2 atmospheres after 30 minutes exposure. For the 3 sensors of the device 2, exposed at 1000 ppm CO_2, were obtained in the first 5 minutes gas exposure very weak voltage signals of 660, 720 and 800 mV corresponding of sensor 1, sensor 2 and sensor 3. For the sensor 3 exposed in 100 and 1000 ppm CO_2 were obtained the maximum voltage values of 92 mV for 100 ppm CO_2 and 970 mV for 1000 ppm CO_2 after 30 minutes gas exposure. The apparatus for CO_2 detection is composed by the sensor and electronic device of signal conditioning. In the paper were discussed the theoretical aspects about equivalent scheme circuit of active sensors.

7. References

P.K. Bhowmik, X.Wang, H. Han, Main-chain, thermotropic, liquid-crystalline, hydrogen-bonded polymers of 4,4'-bipyridyl with aliphatic dicarboxylic acids *J. Polym. Sci.*, Part A: Polym. Chem., 2003, vol. **41 (9)**, pp. 1282-1295

L. Brunsveld, B.J.B. Folmer, E.W. Meijer, R.P. Sijbesma,Supramolecular polymers, *Chem. Rev.*2001, vol. **101**, pp.4071.

H.Carslaw, J.C.Jaeger "Conduction of Heat in Solids",Oxford University Press,Fair Law, 1959.

C.A. Desoer, E.S. Kuh "Basic Circuit Theory" New York, Editions,Mc.Graw Hill,1969.

W.Gopel "Ultimate limits in the miniaturization of chemical sensors", *Sensors and Actuators*, A56 (1996), pp. 83-102.

H. Han, A.H. Molla, P.K. Bhowmik, Hydrogen-Bonded Main-Chain Liquid Crystalline Polymers of Trans-1,2-Bis(4-pyridyl)ethylene with Aliphatic Dicarboxylic Acids. Polym. Prepr., *Am. Chem. Soc. Div. Polym. Chem.* **1995**, 36(2), 332-333.

Y.S. Kang, H.Kim, W.C. Zin, Phase behaviour of hydrogen-bonded liquid crystalline complexes of alkoxycinnamic acids with 4,4'-bipyridine, *Liquid Crystals,* Volume **28,** Number 5, 1 May 2001, pp. 709-715.

Charles Kitchin and Lew Counts, "A Designer's Guide to Instrumentation Amplifiers, Analog Devices, Inc. Printed in U.S.A, 2ND Edition, 2004, pp. 64-74.

W.Jost "Diffusion in Solids Liquids,Gases" Academic Press Inc., New York, 1952.

M.Lee, B.K. Cho, Y.S. Kang, W.C. Hydrogen-Bonding-Mediated Formation of Supramolecular Rod-Coil Copolymers Exhibiting Hexagonal Columnar and Bicontinuous Cubic Liquid Crystalline Assemblies. *Zin, Macromolecules*, 1999, vol. **2**, pp. 8531-8537.

LF356/LF357 JFET Input Operational Amplifiers, National Semiconductor Corporation, 2001.

I.Lundsrom "Approaches and mechanisms to solid state based sensing", *Sensor and Actuators* **B35-36** (1996), pp. 11-19.

C.I.Mocanu "Teoria circuitelor electrice"Ed.Didactica si Pedagogica, Bucuresti,1979. C.I.Mocanu"Teoria câmpului electromagnetic", Ed.Didactică şi Pedagogică, Bucureşti, 1984.

J.E. Mulvaney, C.S. Marvel, *J. Polym. Sci.*, 1961, Vol. **50**, pp. 41.

E. Nishikawa, E. T. Samulski, New mesogens with cubic phases: hydrogen-bonded bipyridines and siloxane-containing benzoic acids I. Preparation and phase behaviour, *Liq. Cryst.*, 2000, vol. **27(11)**, pp. 1457-1462.

M. Parra, P. Hidalgo, J. Alderete, New supramolecular liquid crystals induced by hydrogen bonding between pyridyl-1,2,4-oxadiazole derivatives and 2,5-thiophene dicarboxylic acid**,** *Liq. Cryst.*, 2005, Vol. **32(4),** pp. 449-455.

.M. Parra, P. Hidalgo, J. Barbera, J. Alderete, Properties of thermotropic liquid crystals induced by hydrogen bonding between pyridyl-1,2,4-oxadiazole derivatives and benzoic acid, 4-chlorobenzoic acid or 4-methylbenzoic acid, *Liq. Cryst.*, 2005, vol. **32(5)**, pp. 573-577.

A. Staubli, E. Ron, R. Langer, Hydrolytically Degradable Amino Acid Containing Polymers. *J. Am. Chem. Soc.*, 1990, Vol. **112**, pp. 4419.

V.S.Vladimirov "Ecuatiile fizicii matematice" Ed.Stiintifica si Enciclopedica, Bucuresti,1980.

M. Zigon, G. Ambrozic, Supramolecular polymers = Supramolekularni polimeri *Materiali in Tehnologije*, 2003, vol **37 (5),** pp. 231-236.

16

Planar Quarter Wave Stack Reflectors Prepared from Chalcogenide Ge-Se and Polymer Polystyrene Thin Films

Tomas Kohoutek, Jiri Orava, Martin Hrdlicka, Jan Prikryl,
Tomas Wagner and Miloslav Frumar
*Department of General and Inorganic Chemistry, Faculty of Chemical Technology,
University of Pardubice, Legion´s Sq. 565, 53210 Pardubice, Czech Republic*

1. Introduction

For many decades the multilayer structures consisting of alternating dielectric films with sufficient differences in the refractive index have played an important role in designing highly effective planar optical elements, namely mirrors and filters. The one-dimensional photonic bandgaps are formed in the dielectric multilayers if optical constants and thicknesses of films meet the Bragg resonance condition. These photonic bandgaps display themselves as high intensity reflection or transmission bands in the optical spectra of the dielectric multilayers. Dielectric multilayers are widely used as highly effective reflectors and filters in current optical devices. Their bandgap structure can be predicted according to the theory of the light propagation through stratified dielectric media (Yeh et al., 1977) and their optical properties then adjusted appropriately.

The interest in dielectric multilayers was renewed again when the omnidirectional bandgap (Fink et al., 1998) was reported for the reflectors having the refractive index difference of films higher than $\Delta n \geq 0.8$. The omnidirectional total reflection from such quarter wave stacks (QWS) allows the designing of the reflectors that can reflect the light at any angles and polarizations. A common goal in current QWS device design is then a maximization of a bandwidth of the omnidirectional bandgap by using materials having a high refractive index difference.

From material point of view, the first successful fabrications of planar omnidirectional reflectors with bandgaps in the visible, near-infrared and middle-infrared ranges were reported on materials such as SiO_2/TiO_2, Si/SiO_2 or GaAlN/GaN. These materials exhibited favourable optical properties and a high chemical, thermal, and mechanical stability. However, they brought at the same time high cost and some technological difficulties relating to their high temperature processing (annealing) which can lead to the structural defects of films, e.g., crystallization, cracking or delamination (Chen et al., 1999 and Deopura et al., 2001).

Recent development also showed a possibility of a low temperature fabrication of QWS devices from chalcogenide and polymer films. Chalcogenide glasses and organic polymers seem to be a good alternative to the above-mentioned dielectric materials and deposition methods because they require processing at temperatures below T < 200°C, offer a sufficiently high refractive index difference and also exhibit favourable mechanical properties, see already published results (Fink et al., 1998; Decorby et al., 2005; Clement et al., 2006; Kohoutek T. et al., 2007; Abouraddy et al., 2007 and Kohoutek T. et al., 2008[a]. As shown in the literature (Decorby et al., 2005 and Abouraddy et al., 2007), the omnidirectional QWS devices can be fabricated as planar or scrolled multilayers using thermal evaporation (TE) of high index As-Ge-Se or As-Se chalcogenide glasses and the spin-coating (SC) of PAI or PES polymers which form a low index films. A combination of highly transparent chalcogenide and polymer films enables these QWS elements to cover the wavelength ranges reaching from the visible to the infrared spectral regions including telecommunication (1.3-1.7 µm), security laser (~2.8 µm), night vision (3-5 µm) or even high-power laser (~5,4 and ~10.6 µm) wavelengths.

In this paper we present the preparation and determination of the optical properties of dielectric reflectors which consisted of eight chalcogenide $Ge_{25}Se_{75}$ films periodically alternating with seven polymer Polystyrene (PS) films. The reflectors were designed with the first order normal incidence bandgap at a telecommunication wavelength of $\lambda = 1.55$ µm and the refractive index difference between the used films was ~0.821. We also studied the optical properties of the Ge-Se/PS reflector deposited on ~40 nm gold film to show the effect of metal layer on the stopband of a newly formed metal/dielectric reflector.

Keywords
Polymers, Chalcogenide glasses, Dielectric multilayers, Optical properties

2. Experimental

Deposition of Reflectors

The Ge-Se/PS reflectors were fabricated as one-dimensional photonic crystals from alternating high index chalcogenide (eight layers) and low index polymer (seven layers) films as schematically illustrated in Figure 1. The film thickness values necessary to centre the first order normal incidence bandgap of the reflectors near the telecommunication C-band (λ ~1.55 µm) were calculated according to the QWS condition, $d_i = \lambda/4\ n_i$ (Kim S.-H. & Hwangbo C.K., 1998), knowing the appropriate values of film refractive index of individual Ge-Se and PS films.

Polystyrene films with refractive index $n_{\lambda\ =\ 1.55\ \mu m}$ ~1.531 were prepared toluene solutions where 0.72 g of polymer pellets (Sigma-Aldrich, $M_n = 130k$, $M_w = 240k$) were dissolved in 10 ml of pure toluene using a spin-coating technique. The demanded thickness of polystyrene films (d ~250 nm) was achieved by spinning a drop of the solution for 50 seconds at a spin speed of 5000 rpm. We carried out the calibration of film thicknesses as a function of solution concentration and spin speed before multilayer preparation. After deposition of each polymer PS film, the films and then the multilayers were thermally stabilized in a vacuum oven at T = 90 °C and p ~5 Pa for one hour to remove solvent residua.

Before the deposition of chalcogenide films, the bulk glass with a composition of $Ge_{25}Se_{75}$ was prepared by a standard synthesis from high purity (99.999 at. %) elements in sealed evacuated quartz ampoule placed in a rocking furnace (T = 990 °C, 24 hours) followed by air quenching of the melt. Pieces of bulk glass were placed in a corundum crucible and heated in a vacuum chamber. Chalcogenide Ge-Se films with a thickness of d ~165 nm and refractive indices $n_{\lambda = 1.55\ \mu m}$ ~2.352 were then thermally evaporated onto chemically pre-cleaned silica glass substrates (1 x 1 inch). The substrates were fixed to a planetary rotation system which improved the homogeneity of film thicknesses during evaporation. The deposition rate was ~5 nm/s and residual pressure in the vacuum chamber p ~10^{-4} Pa. The film thicknesses and evaporation rates were measured by a calibrated quartz crystal monitor.

Some of the multilayers were fabricated on substrates coated in advance with a thermally evaporated gold layer with a thickness of d ~40 nm to study the influence of metal on the optical properties of the newly formed metal/dielectric reflector.

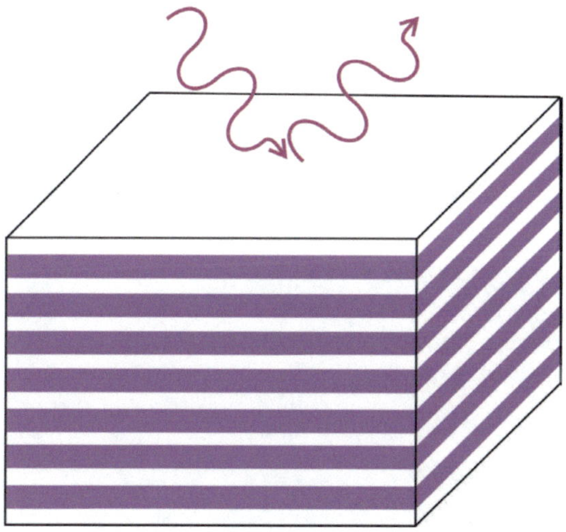

Fig. 1. A schematic sketch of the intended chalcogenide/polymer reflector.

Measurement of Optical Properties

The optical properties of the fabricated reflectors were investigated by means of an optical spectroscopy. The normal incidence reflectivity was recorded by a Jasco V-570 UV-vis-NIR spectrophotometer in the spectral range of 400-2300 nm with a resolution of 2 nm. The experimental ellipsometric data of individual Ge-Se and PS films were recorded using a spectral ellipsometer VASE, J. A. Woollam Co. Inc. with NIR at angles of incidences 60, 65 and 70 ° and in the spectral range of 300-2300 nm. The optical functions n(λ) and thicknesses of the studied chalcogenide and polymer films were evaluated by fitting the ellipsometric Ψ and Δ spectra using the Tauc-Lorentz dispersion formula in the entire spectral region (according to Jellison G.E & Kosine F.A., 1996). The Cauchy formula was used to fit the optical functions of the studied samples in the transparent region (Tompkins H.G & McGahan W.A., 1999). Knowing the optical parameters and thicknesses of films, we were able to generate the optical reflectivity spectra of the intended dielectric reflectors which consisted of eight Ge-Se + seven PS layers for both the normal and the oblique light incidence and for TE- and TM-polarizations, respectively.

3. Results

Dielectric and Metal/Dielectric Reflectors

The fabrication of Ge-Se/PS reflectors was followed by studying their cross-section to confirm if the film order and thickness accuracy in the multilayer were correct. The cross-section image of the multilayer was captured using the JEOL JSM-5500LV scanning electron microscope SEM after a rupture of the prepared reflector across a groove made by a diamond tip. The periodically alternating films made from chalcogenide glass (brighter stripes) and polystyrene (wider and darker stripes) can be then seen in Figure 2. Chalcogenide films appeared brighter in the image because of their low electron conductivity. The lower conductivity causes the films to be locally charged by the electrons gathered in the scanned area. The polymer films have slightly higher electron conductivity than amorphous chalcogenide films and therefore appear darker in the SEM image. Although the film order in the reflector is clearly apparent from the image, the resolution of our microscope did not allow us to focus on the details of film boundaries. To achieve a higher image resolution, it is necessary to observe the samples cross-sectionally using a high resolution transmission electron microscopy (HR-TEM) (Kohoutek et al.[b], 2008 and Kohoutek et al., 2009).

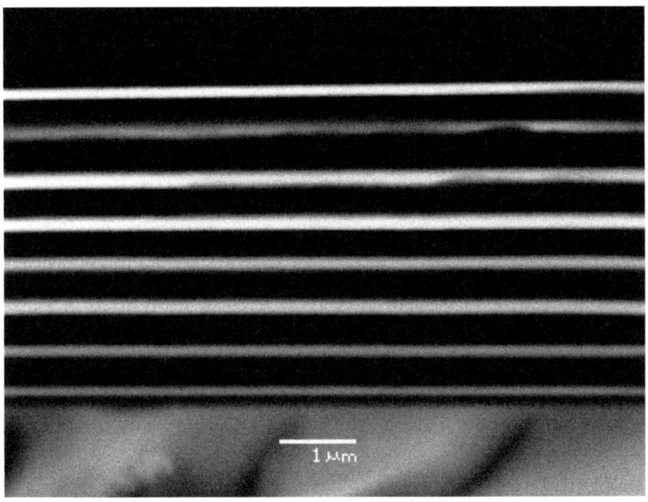

Fig. 2. The SEM image shows a cross-section of Ge-Se/PS reflector. The darker and thinner films belong to polystyrene while brighter and thicker layers to Ge-Se films.

The optical properties of chalcogenide Ge-Se and polymer PS films were studied using spectral ellipsometry. This method is very useful for the determination of the optical parameters of thin films and their thicknesses. Figure 3. shows the optical function $n(\lambda)$ of Ge-Se and PS films in the spectral range of 500-1700 nm. The difference in the index of refraction between Ge-Se ($n = 2.352$) and PS ($n = 1.531$) films was ~0.821 at $\lambda = 1.55$ μm. This wavelength was intended for the stopband of the prepared reflectors. As the absorption of films is expected to be in the visible part of the electromagnetic wavelengths it is very unlikely that it plays a major role to the reflector stopband in the near infrared range of ~1.55 μm. As mentioned above, the values of film thicknesses, $d_{Ge-Se} = 165$ nm and $d_{PS} = 250$

nm, were calculated according to the QWS condition before the reflector deposition. The thickness of each prepared individual film was then checked using spectral ellipsometry. The average inaccuracy in film thickness was found to be ± 7 nm (~4.2 %) for chalcogenide and ± 17 nm (~6.8 %) for polymer films. A similar inaccuracy in the film thicknesses was also reported in the literature (DeCorby et al., 2005).

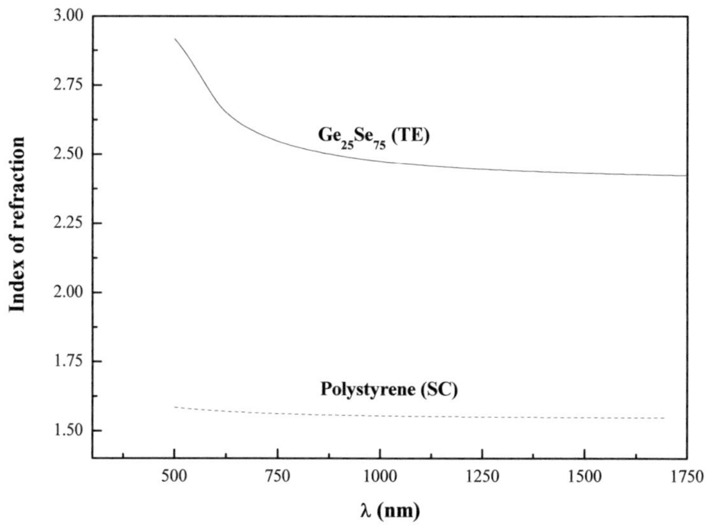

Fig. 3. The films refractive indices, i.e., $n(Ge\text{-}Se)_{\lambda = 1.55\ \mu m} \sim 2.352$ and $n(PS)_{\lambda = 1.55\ \mu m} \sim 1.531$, were determined from optical functions n(λ) of the films according to the ellipsometric measurements.

The normal incidence optical reflectivity of Ge-Se/PS and Au/Ge-Se/PS reflectors was studied using an UV/vis/NIR spectroscopy in the spectral range of 400-2300 nm. Figure 4. represents the comparison between the measured optical reflectivity of the dielectric reflector (a) and the metal/dielectric reflector (b) and theoretical optical reflectivity of both reflectors obtained by modelling ellipsometric parameters of a complex system of 15 or 16 layers and the substrate. The normal incidence stopband with the optical reflectivity higher than r ≥ 98.5 % was found to be in the range of 1527-1730 nm with a maximum at λ = 1588 nm, r = 99.09 %. In the same spectra is seen both near infrared first order stopbands of Ge-Se/PS reflector and also the second order stopbands near λ = 800 nm. The formation of the second order stopbands reflects a high order and a good accuracy of the fabricated reflectors. By adding a ~40 nm thick gold layer between the reflector and the substrate, the stopband splits into two parts separated by a ~30.55 % transmission band with a maximum at λ = 1772 nm. The stopband splitting was also accompanied by an increase in the maximum optical reflectivity of the metal/dielectric reflector up to r ≥ 99.41 % at λ = 1553 nm and the broadening of the stopband bandwidth in the range of 1464-1685 nm, where r ≥ 98.5 %. The effect of gold layer on the reflector stopband is apparent in Figure 5., where details of the above-mentioned increase in optical reflectivity and the band splitting of the first order stopbands of dielectric and metal/dielectric reflectors are shown.

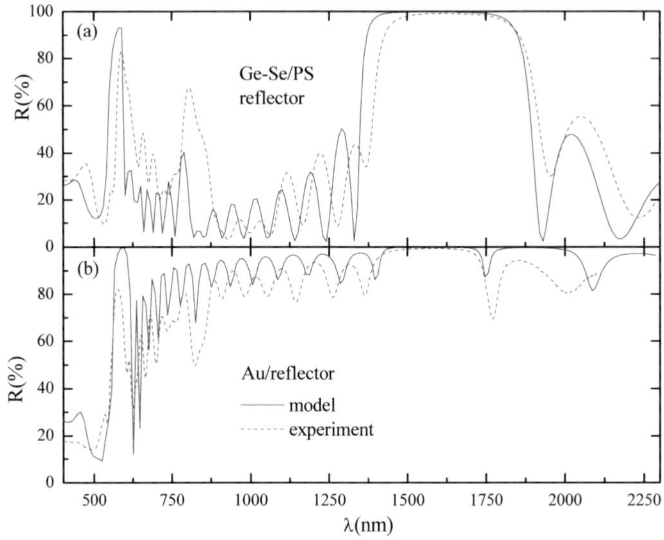

Fig. 4. The comparison of the experimental and modelled normal incidence optical reflectivity is shown for Ge-Se/PS reflector (a) and Au/Ge-Se/PS reflector (b).

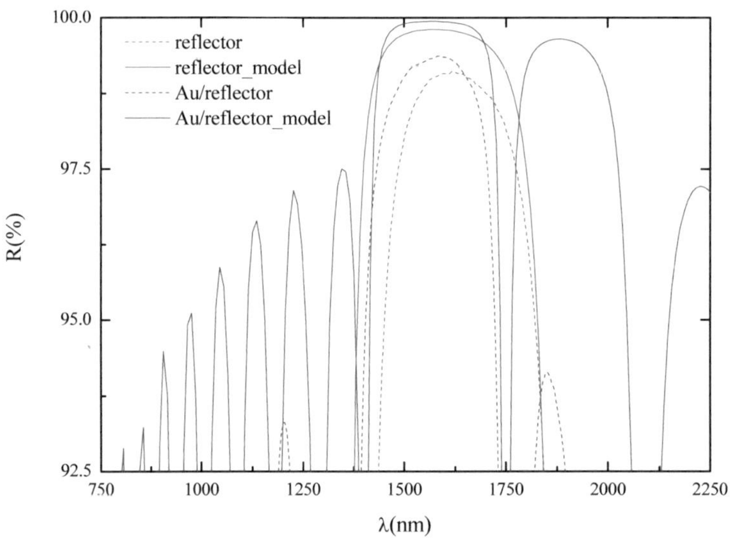

Fig. 5. The optical reflectivity spectra of Ge-Se/PS reflector and Au/Ge-Se/PS reflector show reflectivity maxima of normal incidence stopbands in detail, i.e., an increase in the optical reflectivity of metal/dielectric reflector compared to the maximum achieved for Ge-Se/PS reflector.

Based on our ellipsometric study of the parameters of individual chalcogenide and polymer films, we can demonstrate the behaviour of Ge-Se/PS and Au/Ge-Se/PS reflectors at oblique angles of incidence. The optical reflectivity spectra of the reflectors were modelled for oblique angles of 35°, 45° and 55° and transverse magnetic (TM) and transverse electric (TE) modes of the light. As apparent from Figure 6., the stopband position of the Ge-Se/PS reflector shifted towards lower wavelengths (a blue-shift) with an increasing angle of propagating light. The shift was more apparent in the reflectivity of the light in TM-mode, especially in its longer wavelength spectra, as TM-mode defines the low frequency edge of the stopband. Such a shift is also in an agreement with theoretical predictions (Yeh at al., 1977). The first order bandgap of the TM mode decreases with an increasing angle of the light incidence, i.e., $\Delta \upsilon$ = 265, 196 and 76 nm for $r \geq 98.5$ %. The bandgap of the TE mode increases at the same time, i.e., $\Delta \upsilon$ = 417, 442 and 468 nm. A similar angular dependence of the optical properties observed for the first order bandgap of the dielectric reflector can also be seen for the second order bandgaps near 750 nm. In comparison with Ge-Se/PS dielectric reflectors, modelled angular optical reflectivity spectra of metal-dielectric Au/Ge-Se/PS reflectors revealed a partially different behaviour, see Figure 7. An effect of the blue-shift of the bandgap and its broadening at higher oblique angles is similar. However, an addition of a gold layer to the dielectric reflector forms a new transmission band in the middle of the original bandgap which becomes highly intensive in the TM-mode as the angle of the light incidence increases, i.e., t = 28.19 % for λ = 1683 nm and 35°, t = 42.28 % for λ = 1640 nm and 45° and t = 64.65 % for λ = 1595 nm and 55°, respectively. In the TE-mode, the transmission band subsequently vanishes at higher oblique angles, i.e., t = 7.64 % at λ = 1691 nm and 35°, t = 4.57 % at λ = 1653 nm and 45° and t = 2.65 % at λ = 1613 nm and 55°, respectively.

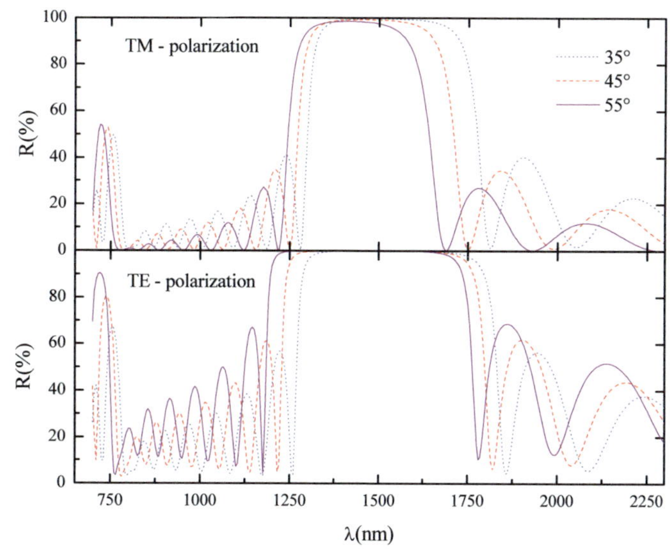

Fig. 6. Comparison of the modelled optical reflectivity of Ge-Se/PS reflector is shown for the light with TM- and TE-polarization at oblique angles 35°, 45° and 55°.

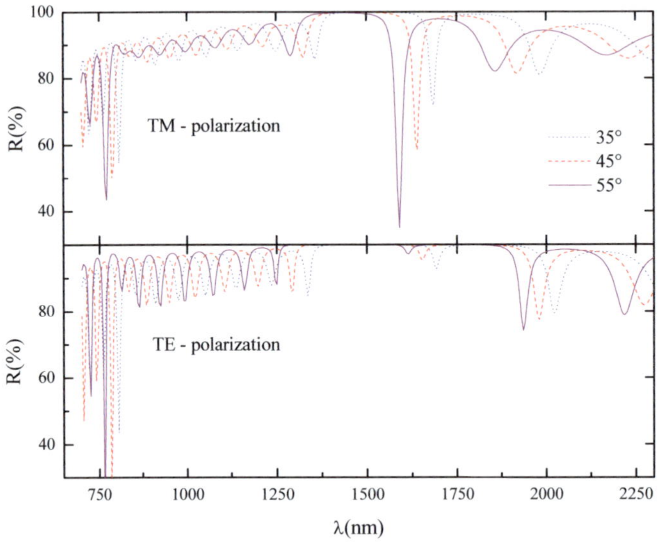

Fig. 7. The modelled optical reflectivity of Au/Ge-Se/PS reflector documents a broadening, blue-shift and stopband splitting of the bandgap of metal/dielectric reflector at oblique angles of 35°, 45° and 55° and TM- and TE-polarized light.

4. Discussion

Dielectric and Metal/dielectric Reflectors

In this study, we have shown that dielectric and metal/dielectric reflectors with the optical reflectivity higher than r > 99 % can be fabricated using low-temperature and technologically simple deposition techniques from chalcogenide glasses and polymer films. We aimed rather at an investigation of the optical properties of the used materials and dielectric and metal/dielectric reflectors, than at the fabrication of reflectors with the highest practically achievable optical reflectivity, which would of course demand a much higher number of deposited layers.

The factors determining the quality of Ge-Se/PS reflectors and dielectric reflectors in general are an inaccuracy in thin film deposition, homogeneity of film optical properties and their surface roughness. The inaccuracy in a film deposition was ± 7 nm for chalcogenide and ± 17 nm for polymer films according to ellipsometric results. It represents roughly a ~5 % deviation in thickness. In our opinion, this value can be further reduced maximally up to the level of 1-2% especially if we consider that the main issue remains the thickness homogeneity of spin-coated films across their entire surface. For comparison, the inaccuracy in thickness for crystalline films prepared using more sophisticated and more expensive vacuum deposition methods such as an e-beam or CVD is under 1% (Yao et al., 2007 and Perez et al., 2005). In case of 1D-photonic crystals, i.e., multilayers a film thickness inaccuracy in order of percent units has basically a low impact on the optical properties of such multilayers.

On the other hand, the optical properties of films and multilayers can be strongly influenced by their eventual compositional inhomogeneity across the entire film thickness. This factor is very important especially for thermally evaporated chalcogenide films. It is well known that an effect of phase separation often occurs in chalcogenide glasses and their thin films including Ge-Se system (see Borisova Z., 1981 and Orava J. et al., 2009). Regarding the fabricated reflectors, a phase separation in Ge-Se films was minimized using a relatively high evaporation rate ~5 nm/s during deposition. The higher deposition rate means higher energy brought on the bulk glass placed in a crucible and its quicker evaporation which prevents a phase separation. In our opinion, a compositional inaccuracy of amorphous films in order of several tenths of per cent is acceptable if we consider that properties, e.g., optical or electrical, of glassy and amorphous (disordered structure) materials change very slightly with such a change in composition. It is in sharp contrast to crystalline (ordered structure) materials where any slightest compositional change causes a considerable change in material properties. Freshly prepared (as-deposited) chalcogenide films also underwent an annealing procedure at T = 90 °C which is necessary for their stabilization. This temperature is, in fact, not appropriate (T_G of Ge-Se film is much higher) but limited by thermal stability of PS films in the reflector.

The root mean square surface roughness of individual Ge-Se and PS films, R_q, determined from atomic force microscopy (AFM) data was in the range of R_q ~ 2-5 nm. These values were lower than the surface roughness of oxide sol-gel films R_q > 10 nm (reported in Almeida & Rodrigues, 2003) and slightly higher than values achieved by well defined e-beam deposition R_q ~ 1 nm (Yao et al., 2007). Higher surface roughness of films in photonic devices can cause undesired scattering of the light and noises which reduce their efficiency. An important issue for the multilayers is the quality of film boundaries, especially, if the rmultilayer structure is made from materials with different mechanical properties. Good adhesion, chemical stability and thermal expansion coefficient of films are also required.

The optical reflectivity of the fabricated dielectric reflector and metal-dielectric reflector was compared with modelled optical reflectivity obtained from ellipsometric results (see Orava et al., 2008). A strong correlation between these values is apparent in Figs. 4. and 5. Modelled curves predicted higher optical reflectivity of the metal-dielectric reflector compared to the Ge-Se/PS reflector near λ = 1550 nm and also the behaviour of both reflectors at oblique angles. An increase in the optical reflectivity of metal-dielectric reflector was then confirmed by the reflectivity measurement performed on the fabricated reflectors. In general terms an addition of metal layer between the dielectric reflector and the substrate leads to an increase in the optical reflectivity, blue-shift and broadening of the stopband. In our case, an addition of a ~40 nm gold layer to the dielectric multilayer caused a 0.32 % increase in the optical reflectivity of reflectors (r = 99.09 → 99.41 %) and blue-shift of the stopband maximum (λ = 1588 → 1553 nm) and bandwidth broadening r ≥ 98.5 % (Δυ = 203 → 221 nm). We should notice that a 0.32 % increase in the stopband maximum of the reflector represents really significant value if we consider the reflectivity range r > 99 %. From the application point of view, an addition of one thin metal layer can substitute a deposition of several pairs of dielectric films necessary to achieve the same value of optical reflectivity. On the other hand, the optical reflectivity of the multilayer increases significantly in the entire wavelength range, which reduces the difference between reflectivity maxima and minima. The metal/dielectric reflectors are then suitable namely for the applications where the highest possible reflectivity at the lowest number of deposited films is required.

Future prospects of Ge-Se/PS multilayers could be in their deposition on flexible polymer substrates (as reported by DeCorby et al., 2007) or in designing more complex optical devices, including Bragg reflectors (see the review by Abouraddy et al., 2007). Exploiting high transparency of chalcogenide and polymer films, the optical properties of the described dielectric multilayers can be tuned from the visible up to λ = 3000 nm (above this wavelength, the absorption of the stretching vibrations of C-C, C-H bonds in polymers can be found). The near infrared (λ = 800-2500 nm) dielectric reflectors and filters can be then designed for applications in telecommunication (signal processing), moister or dust detection in the air (astronomy, meteorology), measurement of oxygen in hemoglobin (oximetry, medicine), soft-laser frequencies (security lasers), etc.

5. Conclusions

In this study, we fabricated Ge-Se/PS dielectric and Au/Ge-Se/PS metal/dielectric reflectors from amorphous chalcogenide and polymer films with the optical reflectivity higher than r > 99 % near λ ~1550 nm using low-temperature and inexpensive deposition techniques. Our attention was paid to the determination of the optical properties of films and multilayers to demonstrate the suitability of the selected materials for designing 1D-photonic crystals. The main conclusion of the study is that highly intensive one dimensional photonic bandgaps easily tunable at any wavelength range in the near infrared (λ ~800–2500 nm) can be achieved by using chalcogenide/polymer multilayers. It is mainly because of a sufficient difference in the refractive index of these materials and their high infrared transparency. Another important conclusion is that an addition of a metal layer to a dielectric reflector can further enhance the stopband maxima or reduce a number of necessary layer pairs in the reflector to reach the same optical reflectivity values. The QWS devices based on high refractive index chalcogenides and low refractive index polymers seem to be promising for applications demanding favorable performance ratio.

Acknowledgements

The authors thank the Research Centre (University of Pardubice, Project No. LC 523) and the Czech Republic Grant Agency (GA CR, Grant No. 203/08/P204) for their financial support of this study.

6. References

Abouraddy A.F.; Bayindrin M.; Bendit G.; Hart S.D.; Kuriki K.; Orf N.; Shapira O.; Sorin F.; Temelkuran B. & Fink Y., Towards Multimaterial Multifunctional Fibres that See, Hear, Ssense and Communicate, *Nat. Mater.*, Vol. 6, 2007, pp. 336-347, ISSN: 1476-1122.

Almeida R.M. & Rodrigues A.S., Photonic Bandgap Materials and Structures by Sol–gel Processing, *J. Non-Cryst. Solids*, Vol. 326&327, 2003, pp. 405-409, ISSN: 0022-3093.

Borisova Z.U. (1981), *Glassy Semiconductors*, pp. 104-122, Plenum, ISBN: 0-306-40609-8, New York.

Chen K.M.; Sparks A.W.; Luan H.-C.; Lim D.R.; Wada K. & Kimerling L.C., SiO_2/TiO_2 Omnidirectional Reflector and Microcavity Resonator via the Sol-Gel Method, *Appl. Phys. Lett.*, Vol. 75, 1999, pp. 3805-3807, ISSN: 0003-6951.

Clement T.J.; Ponnampalam N.; Nguyen H.T. & DeCorby R.G., Improved Omnidirectional Reflectors in Chalcogenide Glass and Polymer by using the Silver Doping Technique, *Opt. Express*, Vol. 14, 2006, pp. 1789-1796, ISSN: 1094-4087.

DeCorby R.G., Nguyen H.T., Dwivedi P. & Clement T.J., Planar Omnidirectional Reflectors in Chalcogenide Glass and Polymer, *Opt. Express*, Vol. 13, 2005, pp. 6228-6233, ISSN: 1094-4087.

DeCorby R.G.; Ponnampalam N.; Nguyen H. T. & Clement T. J., Robust and Flexible Free-Standing All-Dielectric Omnidirectional Reflectors, *Adv. Mater.*, Vol. 19, 2007, pp. 193-196, ISSN: 0935-9648.

Deopura M.; Ullal C.K.; Temelkuran B. & Fink Y., Dielectric Omnidirectional Visible Reflector, *Opt. Lett.*, Vol. 26, 2001, pp. 1197-1199. ISSN: 0146-9592.

Fink Y.; Winn J.N.; Fan S.; Chen C.; Michel J.; Joannopoulos J.D. & Thomas E.L., A Dielectric Omnidirectional Reflector, *Science*, Vol. 282. No. 5394, 1998, pp. 1679-1682, ISSN: 0036-8075.

Jellison G.E. & Modine F.A., Parametrization of the optical functions of amorphous materials in the interband region, *Appl. Phys. Lett.*, Vol. 69, 1996, pp. 371-373, ISSN: 0003-6951.

Kim S.-H. & Hwangbo C.K., Design of Omnidirectional High Reflectors with Quarter-Wave Dielectric Stacks for Optical Telecommunication Bands, *Opt. Lett.*, Vol. 23, 1998, pp. 1573-1575, ISSN: 0146-9592.

Kohoutek T.; Orava J.; Hrdlicka M.; Wagner T.; Vlcek Mil. & Frumar M., Planar Quarter Wave Stacks Prepared from Chalcogenide Ge-Se and Polymer Polystyrene Thin Films, *J. Phys. Chem. Solids*, Vol. 68, 2007, pp. 2376-2380, ISSN: 0022-3697.

[a]Kohoutek T., Wagner T., Orava J., Krbal M., Vlcek Mir., Ilavsky J., Vlcek Mil. & Frumar M., Multilayer Planar Structures Prepared from Chalcogenide Thin Films of As-Se and Ge-Se Systems and Polymer Thin Films using Thermal Evaporation and Spin-coating Techniques, *J. Non-Cryst. Solids*, Vol. 354, 2008, pp. 529-532. ISSN: 0022-3093.

[b]Kohoutek T.; Orava J.; Wagner T.; Hrdlicka M.; Vlcek Mil. & M. Frumar, Preparation of Dielectric Mirrors from High-Refractive Index Contrast Amorphous Chalcogenide Films, *J. Phys. Chem. Solids*, Vol. 69, 2008, pp. 2070-2074, ISSN: 0022-3697.

Kohoutek T.; Orava J.; Přikryl J.; Mistrik J.; Wagner T. & Frumar M., Near Infrared Quasi-Omnidirectional Reflektor in Chalcogenide Glasses, *Opt. Mater.*, Vol. 32, 2009, pp. 154-158. ISSN: 0925-3467.

Orava J.; Sik J.; Wagner T. & Frumar M., Optical Properties of $As_{33}S_{67-x}Se_x$ Bulk Glasses Studied by Spectroscopic Ellipsometry, *J. Appl. Phys.*, Vol. 103, 2008, pp. 083512-1-8, ISSN: 0021-8979.

Orava J.; Kohoutek T.; Wagner T.; Cerna Z.; Vlcek Mil.; Benes L.; Frumarova B. & Frumar M., Optical and Structural Properties of Ge-Se Bulk Glasses and Ag-Ge-Se Thin Films, *J. Non-Cryst. Solids*, Vol. 355, 2009, pp. 1951-1954, ISSN: 0022-3093.

Perez G.; Bernal-Oliva A.M.; Marquez E.; Gonzalez-Leal J.M.; Morant C.; Genova I.; Trigo J.F. & Sanz J.M., Optical and Structural Characterisation of Single and Multilayer Germanium/Silicon Monoxide Systems, *Thin Solid Films*, Vol. 485, 2005, pp. 274-283, ISSN: 0040-6090.

Tompkins H.G. & McGahan W.A. (1999), *Spectroscopic Ellipsometry and Reflectometry*, pp. 23-34, Wiley Inter-Science, ISBN: 0-471-18172-2, New York.

Yao J.; Shao J.; He & H. Fan Z., Effects of Annealing on Laser-Induced Damage Threshold of TiO_2/SiO_2 High Reflectors, *Appl. Surf. Sci.*, Vol. 253, 2007, pp. 8911–8914, ISSN: 0169-4332.

Yeh P.; Yariv A. & Hong C.-S., Electromagnetic Propagation in Periodic Stratified Media, *J. Opt. Soc. Am.*, Vol. 67, 1977, pp. 423-448, ISSN: 0740-3224.

17

Polymeric thin film technology for neural interfaces: Review and perspectives

Cristina Riggio, Gianni Ciofani, Vittoria Raffa, Silvia Bossi, Silvestro Micera and Alfred Cuschieri
Scuola Superiore di Studi Universitari e Perfezionamento Sant'Anna
Pisa Italy

1. Introduction: Thin polymeric film in nanomedicine

An important and exciting direction of research in nanomedicine would be to gain a fundamental understanding of how living cells respond to nanostructures. At this aim, thin film technology plays a key role in helping to understand the cell-surface interactions.
Generally, thin films are deposited onto bulk materials to achieve properties unattainable or not easily attainable in the substrate alone. In particularly, in biomedicine, polymeric thin films are used such as coating to improve the properties of biocompatibility, thus avoiding typical inflammatory response of immunitary system, especially when the system have to be permanently implanted (Jeong *et al.*, 1986).
Various biodegradable polymeric drug delivery devices have been developed for the sustained release of a variety of drugs, which include micro and nanoparticles, films, foams, wafers, discs, and micro- and nanofibers (Jain, 2000) Among them, films have gained growing interest in various applications. (Dorta *et al.*, 2002; Jackson *et al.*, 2002; Perugini *et al.*, 2003; Dhanikula *et al.* 2004; Jackson *et al.* 2004; Grant *et al.*, 2005; Alexis *et al.* 2005; Westedt *et al.*, 2006; Heller *et al.*1980). For example, films for stent applications have been devoleped to prevent early and late complications such as thrombotic closure and restenosis that have been reported with all current metallic stent devices. (Westedt *et al.* 2006; Drachman *et al.* 2000; Alexis *et al* 2004, Hanefeld *et al.* 2006). The surfaces of most metals are electropositively charged and, therefore, are thrombogenic because blood elements are negatively charged. So the drawbacks of metallic stents have encouraged significant efforts to explore other materials as possible stent matrices. Alexis and colleagues studied the *in vitro* release kinetics of two important antirestenosis drugs: paclitaxel and rapamycin from biodegradable stent matrices. The poly(lactic-co-glycolic acid) (PLGA) and poly-DL-lactic acid (PDLLA) were selected in view of their relatively fast degradation rates (Alexis *et al.*, 2006).
Many polymer-based implantable film formulations were developed to provide controlled, local release of drug for the treatment of tumors (Ho *et al,* 2005). Local administration of chemotherapeutic agents has been investigated for the treatment of various cancers, such as brain, prostate, esophageal, head and neck, ovarian, and breast cancer (Jeong *et al.*, 1986; Webber *et al.*, 1998; Zhou *et al.*, 1998; McCarron *et al.*, 2000). The implantation of a polymer-

based device containing anti-cancer drug can provide a high-dose of chemotherapy to a specific area for a prolonged period of time (McCarron *et al.* 2000). The goal of Grant and collaborators, for example, was to develop a delivery system that would benefit from the properties of chitosan and egg phosphatidylcholine lipids, while providing a sustained release of therapeutically effective levels of a hydrophobic agent, paclitaxel, over several months (Grant *et al.*, 2005).

The specific combination of chitosan and egg phosphatidylcholine was found to produce films with a minimal degree of swelling and high stability. The chitosan lipid blend serves as the matrix for the delivery system, and the drug is incorporated into nanoparticles that are dispersed throughout the film. This film with high degree of biocompatibility was found to be a promising system for the localized delivery of the highly lipophilic anticancer agent.

Another typical application of thin polymeric film in biomedicine is to prevent post surgical adhesions. Postsurgical adhesions are abnormal tissue attachments that may result from tissue abrasion during surgical procedures. Shi *et al.* studied biodegradable polymeric film formulations for the controlled delivery of paclitaxel as an effective inhibitor of the formation of postsurgical adhesions (Shi *et al.*, 2004). The films were intended to provide both a barrier effect to separate traumatized tissues and also to release paclitaxel to inhibit adhesion processes.

2. Neural interfaces and drug delivery system

From the rapid growth in biotechnology, neural engineering has emerged as a new field. The merging of neurophysiology and engineering has resulted in approaches to link brain activity with man-made devices to replace lost sensory and motor function (McGee *et al.*, 2009). The excitement in this field is based not only on the prospect of helping a wide range of patients with neural disorders, but also on the certainty that this new technology will make possible to gain scientific insight into the way populations of neurons interact the complex, distributed systems that generate behaviour.

After analytical demonstration of nerve excitation and associated action potential generation by Alan Hodgkin and Andrew Huxley in 1952, the possibility of using electrical phenomenon to cure diseases and to understand the principles of electrical nerve signals have been investigated through implementation of neural interfaces.

Neural prosthetics are devices that link machines to the nervous system with the purpose of restoring lost functions. Two broad approaches are used in this field: neurons are stimulated or inhibited by applied current, or their activity is recorded to intercept motor intention. Stimulation can be used for its therapeutic efficacy, as in deep brain stimulation, to ameliorate the symptoms of Parkinson's disease or to communicate input to the nervous system (for example by transforming sound to neural input with cochlear prosthetics). In contrast, recordings are used to decode ongoing activity for use as a command or input signal to an external device (Schwartz *et al.*, 2006).

A great number of neuro-prostheses use interfaces with peripheral nerves or muscles for neuromuscular stimulation and signal recording. Only in this way, these new artificial devices can be easily incorporated into the natural control strategies of the subjects and can be felt as parts of their own body thanks to the sensory feedback which can be delivered from the sensors embedded in the robotic artefact to the nervous system.

According to Navarro *et al.* 2005, a neural interface can be defined, from an engineering point of view, as a bidirectional transducer that establishes a contact between an artificial device and a neural structure within the body. On the other hand from a biological point of view, an interface is just a "foreign body". These two aspects (the engineering and the biological) must be both taken in high consideration and an optimized invasive neural interface is a compromise between the two requirements.
The biocompatibility is the first and most important requirement for a neural interface. In general, the compatibility between a technical and a biological system can be divided into the structural biocompatibility and the surface biocompatibility (Bronzino *et al.* 1985).

- The structural biocompatibility depends on the mechanical properties of the surrounding tissue and on the adaptation capability of the artificial material structure to the tissue. Device design and material properties should mimicry the biological structure of the target tissue.
- The surface biocompatibility deals with the interaction of the chemical, physical, biological and morphological surface properties of the foreign material and the target tissue with the desired interaction.

Several interfaces have been developed during the past years by many groups, applied both to the central nervous system (CNS, Schwartz 2004) and to the peripheral nervous system (PNS, Navarro *et al.*, 2005).
Notwithstanding the efforts carried out by several groups, the solutions developed so far suffer from several limitations which make very difficult the development of effective bionic systems. For example, cuff – and epineural - electrodes (Tarler *et al.*, 2004) are reliable and robust, imply a reduced invasiveness, can be used to extract interesting information but suffer from a limited selectivity which is a significant drawback especially to deliver a sensory feedback. Intraneural PNS interfaces characterized by needles to be inserted longitudinally (LIFE electrodes, Lawrence *et al.*, 2004) or transversally (USEA electrodes, McDonnall *et al.*,2004) into the PNS are very interesting combining a reduced (even if not absent) invasiveness with a good selectivity but their "bandwidth" (*i.e.*, the amount of information which can be exchanged between the natural and artificial systems) is still limited.
Moreover, the use of these devices, and in particular microfabricated neural prostheses, is limited due to an inflammatory tissue reaction following implantation, which quickly degrades the stimulation and recording features of the device (Polikov *et al.*, 2005).
To avoid this phenomenon, several groups addressed their study to novel neuronal interfaces able of promoting neural regeneration. Anyway, the reactive tissue response of the brain to chronically implanted materials remains a formidable obstacle to stable recording from implanted microelectrodes.
Over the course of days and weeks following implantation, a sheath of activated glia forms around the electrode, isolating the implanted device from the neuronal tissue (Szarowski *et al.*, 2003). Studies have reported a "kill zone" around implanted devices where neuronal cell bodies and fibers become severely reduced in density up to 200 μm around the implantation site (Biran *et al.*, 2005). The reactive tissue response includes both an acute phase, caused by implantation injury, and a chronic response of the tissue to the implanted material itself (Polikov *et al.*, 2005). The tissue response results in a decline in availability of viable neurons

for recording, and a reduction in signal to noise ratio over time, which reduces or eliminates the ability to effectively collect recordings (Vetter *et al.*, 2004).
Two strategies are possible for improving the tissue response to invasive neural interfaces: to increase the structure biocompatibility and/or the surface biocompatibility.
The first aspect can be achieved by the introduction of flexible interfaces. The inflammation at the implantation sites is thought to be aggravated by the mechanical mismatch between the stiff interface and the soft biological tissue. Their different mechanical properties induce relative drifts between the interface and the tissue encouraging the formation of a glial scar, which can encapsulate or break the probe with time (Cheung *et al.* 2006). Among the other materials, the polyimide has been having a great success for the development of neural interfaces as substrate for metal active sites and tracks. Because of its high flexibility (Young's modulus between 4 and 10 GPa). If compared with silicon and silicon insulation (SiO2, Si3N4), it got a similar insulation resistance and dielectric strength at a lower. Polyimide is proved as a non-toxic material in biomedicine (Richardson et al. 1993) In addition, polyimide-based neural interfaces can be processed by thin film technologies which ensure high precision and repeatability, allows a variety of designs suitable for implantation in different nerves and anatomical regions, and a higher number of active sites can be positioned within a small surface (Stieglitz *et al.* 2005). As an example, polyimide-based regenerative electrodes with a thin film structure have been allowed for much better regeneration than silicon dice (Lago et al. 2005) while the tf-LIFE (thin film LIFE base on polyimide) demonstrated a good compromise between invasiveness and selectivity (Lago *et al.* 2007).
From the other hand, alternative strategies are focusing also on the enhancement of surface biocompatibility. Local drug delivery to the region surrounding the implant as a means to avoid tissue response phenomena is actively pursued by several groups. The basic idea is to release bioactive components from a polymer coating on the electrode to the damaged neurons in order to induce adhesion and regeneration.
Cue *et al.* used electrochemical polymerization to optimize the surface of the metal electrode sites. Poly(3,4-ethylenedioxythiophene) (PEDOT)/poly(styrene sulfonate) (PSS) was electrochemically deposited on the microelectrodes of neural probes (Cue *et al.*, 2003). The PEDOT/PSS coating decreases the impedance modulus and, moreover, in comparison to a previous study on polypyrrole, PEDOT demonstrated much better electrochemical stability (Cui *et al*, 2001). A bioactive peptide was incorporated in the PEDOT film during the electrochemical polymerization to be released during the *in vitro* tests. The coated probes were cultured with rat glial cells that grew on the coating area of the probes and high quality acute neural recordings were obtained through the coated electrodes.
One of the limits of the use of polymer surface coating is the increasing of dimensions of the device, potentially adding a layer to the surface that could range from tens to hundreds of micrometers in thickness. The high surface area-to-volume ratio of the coatings may reduce the ability for sustained delivery over long periods of time, which could minimize its potential for limiting the chronic neuroinflammatory response.
To overcome these disadvantages, Williams *et al.*, presented a method for targeted microscale drug delivery, based on micromachined "wells" fabricated into the substrate of a MEMS (micro-electro-mechanical systems)- based chronic implant (Williams *et al.*, 2005). These wells are holes that extend through the device thickness and allow for the integration of matrices (*i.e.*, hydrogels) infused with bioactive substances.

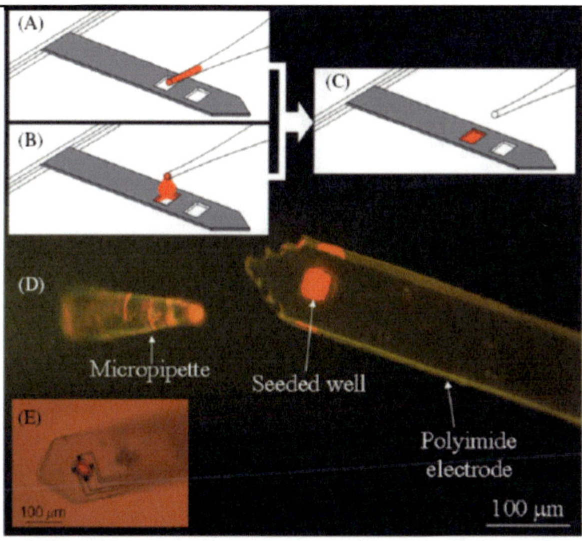

Fig. 1. Illustration of electrode well seeding technique. Schematics of the device and pipette depicting the injection and the droplet deposition methods prior to gel loading are given in (A) and (B), respectively, (C) schematic of a device with a filled well, (D) fluorescent image of probe immediately after the gel loading and subsequent withdrawal of pipette tip. The edges of the probe appear red due to internal reflection of the fluorescent DiI emanating from the well through the polyimide substrate. The volume of gel contained within the well is approximately 9 pL based on the well dimensions and (E) shows a polyimide based electrode with a seeded well in the middle of a tetrode arrangement of recording sites.

Since the matrices replace the substrate instead of adding to it, the device footprint is not increased. The lower surface area-to-volume ratio lends the technique to extended time release periods. Another benefit arises from the possibility of inducing cellular growth through the holes, which, if achievable, may serve to physically anchor the implant into the surrounding tissue. Particularly for electrically-based probes, any reduction of device movement post-implantation would aid in the localization of electrical signals and their subsequent assignment to individual cells.

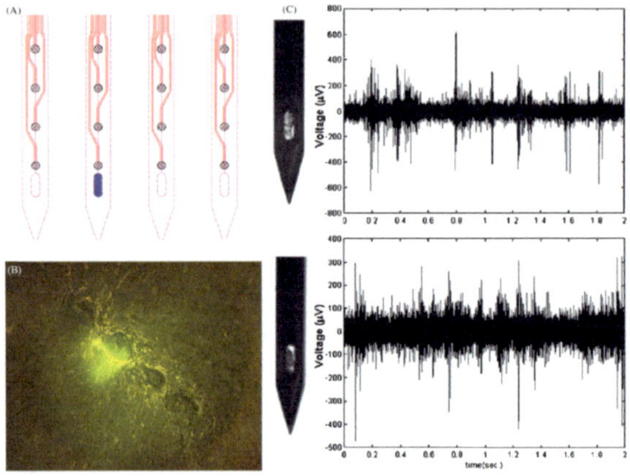

Fig. 2. Wells seeded with NGF. The upper left schematic (A) illustrates the layout of an implanted microelectrode array, showing the relative position of the NGF-seeded well in relation to a recording site. In this four-shank implant, only one of the shanks has an NGF-seeded well. The implanted animal was sacrificed after 40 days, sectioned at the level of the NGF-seeded well, and immunohistochemically stained with antibodies against GFAP and the extracellular matrix protein, laminin. (B) is a histological image from this implant. The site that contained the NGF-seeded well (second from the left in this image) showed increased levels of matrix deposition versus the unseeded electrode shank sites. (C) shows the electrophysiological recordings taken from two individual electrode sites closest to two different seeded wells, each at 40-days post-implantation.

Qualitative observations depicted in Fig. 2B show a notable tissue response in the form of laminin deposition, seen as a green "cloud" over and adjacent to the site of NGF release from the gel. Additionally, an increased level of GFAP expression (yellow stained cells) can be seen around the implant sites. These results are typical for the sites of NGF seeding, but noticeably different from the well sites without NGF, which all showed a relatively normal response. A two-second sample of the extracellular recordings taken from two different recording sites, each located 15 µm from two different seeded wells is given in Fig. 2 C. These recordings demonstrate that the sites near the NGF-seeded wells remain electrically viable following implantation. The recordings are typical for the sites adjacent to the NGF-seeded wells, although not statistically different from the other sites in the array. In all animals implanted with functioning electrodes (n = 4, 48 hours and 40 days), neural signals were recorded from the sites adjacent to the wells that had been seeded with bioactive molecules. Qualitatively, the electrodes were able to record unit activity with acceptable signal amplitude.

Winter *et al.,* have investigated the use of biodegradable, neurotrophin-eluting hydrogels *(i.e.,* poly(ethylene glycol)-poly(lactic acid), PEGPLA) as a means of attracting neurites to the surface of stimulating electrodes. PEGPLA hydrogels with release rates ranging from 1.5 to 3 weeks were synthesized (Winter *et al,* 2007). PEGPLA was selected because of the unique properties of the individual components. PEG has been shown to improve immune response to implanted elements and has been used to increase the circulation time of drugs. Thus PEG as a hydrogel component may reduce immune reaction to the implanted prosthesis and increase the circulation time of released neurotrophins associated with degrading PEG molecules. The PLA portion of the polymer provides biodegradability, allowing for eventual elimination of the polymer from the array. Additionally, the degradation rate can be used to regulate neurotrophin release through variations of the PEGPLA ratio. These hydrogels were applied to multielectrode arrays with sputtered iridium oxide charge-injection sites. PC12 cell cultures were exposed to NGF-releasing boluses, suspended in trans well inserts for 5/14 days, and compared to a positive control receiving 50 ng/mL of NGF, to a negative control receiving no NGF, and a to sham receiving BSA-releasing boluses.

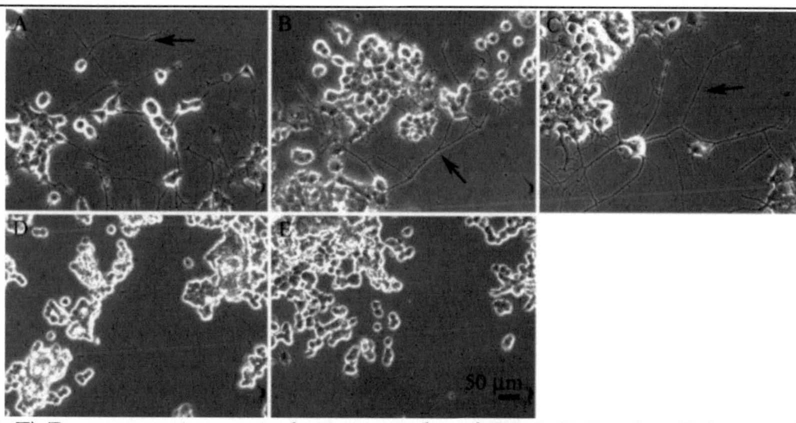

Fig. 3. (A-E) Representative optical micrographs of PC12 Cells after 5 days with boluses in culture. (A) Positive control receiving 50 ng/mL NGF, (B) and (C) substrates display large numbers of neurites (arrows), whereas (D) negative control receiving 0 ng/mL NGF and (E) sham, BSA-releasing bolus display only few, short neurites.

The results of this work show that PEGPLA polymers can produce controlled, sustained release of neurotrophins, and that these neurotrophins produce neurite extension in a neuronal cell culture model.

It was also verified that these hydrogels can be applied to electrically stimulating neural prostheses, and do not impede the function of these devices. Thus, neurotrophin-eluting hydrogels provide one possible means to attract neurites to an electrical prosthesis surface, potentially lowering electrical stimulation thresholds. If lower thresholds can be achieved, larger numbers of electrodes could be driven with the same power requirements, increasing pixel density. These high density devices will lead to greatly improved fidelity in electrically evoked sensory and functional responses.

Recently, Pierce *et al* used another approach to mitigate the reactive tissue response of the brain to chronically implanted materials. A bioactive coating in the form of an ultra-porous silica sol–gel was applied on the surface of silicon-based microelectrodes (Pierce *et al.*, 2009). The sol–gel technique allows for application of thin films of ultraporous silica, the morphology of which is characterized as a flat surface with pore size and surface features <25nm (Jedlicka *et al.*, 2005). The films have a thickness of approximately 100 nm. These ultra-porous sol–gel thin films have been shown to be excellent substrates for neuronal growth (Jedlicka *et al.*, 2006, 2007a).

In Pierce's study the attention was centred on two features of the microelectrode: the adherence of the coating and the resultant changes in electrical properties. Sol–gel presence and uniformity on the probes were assessed by the preparation of a fluorescently labelled coating that was revealed by fluorescence micrographs.

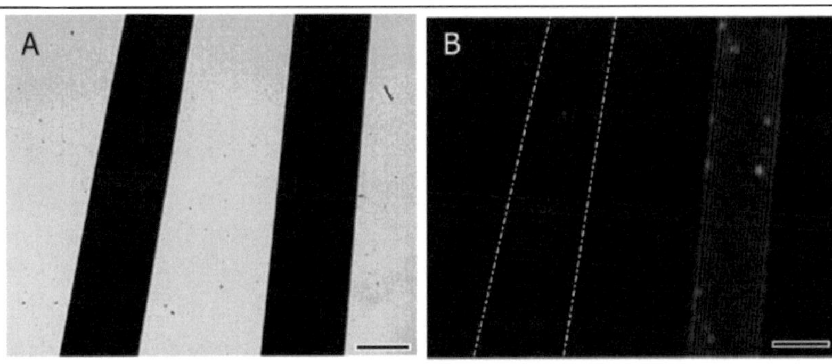

Fig. 4. Microscopy confirms the adherence of sol–gel coatings to microelectrodes. (A) Bright field image of probes shows their location. (B) Fluorescent micrograph of coated probes. For the right probe, fluorescein salt was added to the precursor sol and was encapsulated throughout the coating. The left probe was coated in an unlabeled sol-gel as a control. The fluorescence of the right probe confirms coating adhesion. These images demonstrate that the silica sol–gels can be successfully applied to silicon-based microelectrodes and Winter confirm the ability to encapsulate molecules in the TMOS precursor to enable drug delivery to the implantation site.

To evaluate the stability of the coating an established agarose brain model was used: the coating appeared unaltered after insertion into the model, demonstrating adherence of the coating to the probe that experiences similar mechanical stresses during surgical insertion.

Another important evaluation concerns the electrical properties of the microelectrode. Without favourable electrical properties, a coating designed to mitigate the reactive tissue response would be of little value. The cyclic voltammetry study demonstrated that the dip-coated, ambient-dried thin film silica sol–gel coatings do not adversely affect the electrical characteristics of the implantable electrodes, and it may in fact provide beneficial electrical properties. A major advantage of the sol–gel produced silica is a wide versatility for biofunctionalization. All together, these features will allow the sol–gel silica coatings to be used as a platform material for the mitigation of the chronic phase of the reactive tissue response, without causing significant detriment to the functionality of the probes.

Green *et al* led a study on modified laminin peptides DEDEDYFQRYLI and DCDPGYIGSR used to dope poly(3,4-ethylenedioxythiophene) (PEDOT) electrodeposited on platinum (Pt) electrodes (Green *et al,* 2008).

DCDPGYIGSR was chosen due to positive results reported by Cui *et al.* (Cui *et al,* 2003) DEDEDYFQRYLI, a peptide which has not previously been incorporated into conducting polymers, was also assessed. This peptide contains the active sequence YFQRYLI, identified by Tashiro *et al.* and reported to mediate cell attachment and promote neurite outgrowth in both PC12 cells and cerebellar microexplant cultures (Tashiro *et al,* 1998). The effect of a large biomolecule dopant on conducting polymer physico- chemical properties was established with comparison to conventionally doped PEDOT/paratoluene sulfonate (pTS).

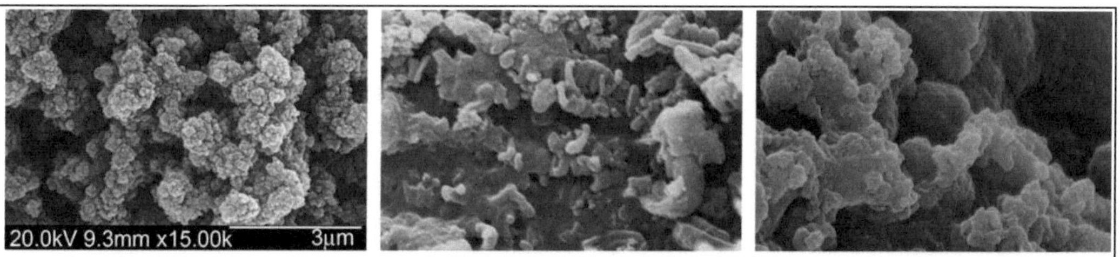

A. PEDOT/pTS B. PEDOT/DEDEDYFQRYLI C. PEDOT/DCDPGYIGSR

Fig. 5. SEMs at 15,000x magnification: (A) PEDOT/pTS; (B) PEDOT/DEDEDYFQRYLI; (C) PEDOT/DCDPGYIGSR

The resulting film was less homogeneous in microscopic appearance than the control PEDOT/pTS film. Doping PEDOT with laminin peptides, DEDEDYFQRYLI and DCDPGYIGSR, resulted in altered formation of the polymer during electrodeposition.

The use of large synthetic peptides as anionic dopants produced a softer interface with improved impedance characteristics, especially in the low frequency, biologically significant, region. However, reduced polymer electrochemical stability and lower adherence of the films to the Pt electrode was observed when compared to the conventionally doped pTS control. The larger DEDEDYFQRYLI dopant was shown to reduce the efficiency of electropolymerisation and consequently this polymer presented lower mechanical adherence than the DCDPGYIGSR-doped PEDOT.

The effect of peptide doped polymers on mammalian cell interactions and the cell response to peptides containing specific bioactive ligands was assessed using the neural-like PC12 cell line.

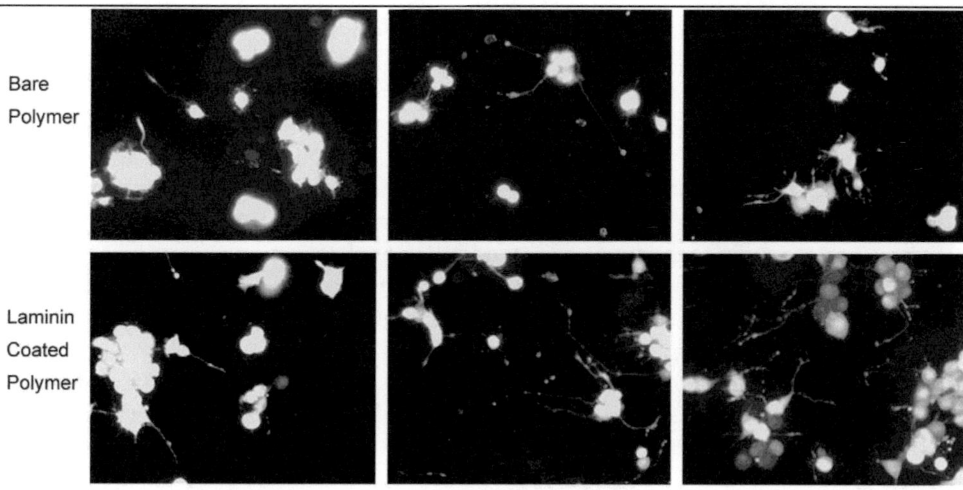

A. PEDOT/pTS B. PEDOT/DEDEDYFQRYLI C. PEDOT/DCDPGYIGSR

Fig. 6. Sample images of PC12 neurite outgrowth on PEDOT at 96 h post-plating with bare polymer (top) and adsorbed whole laminin coated polymer (bottom): A. PEDOT/pTS; B. PEDOT/DEDEDYFQRYLI; C. PEDOT/DCDPGYIGSR.

Uncoated DEDEDYFQRYLI-doped PEDOT appears to support higher neurite outgrowth per cell than both PEDOT/pTS and PEDOT/DCDPGYIGSR despite having the lowest density of attached cells. PEDOT/ DCDPGYIGSR also presented a greater neurite outgrowth per cell than the PEDOT/pTS control. All polymers showed a significant improvement in cell attachment and growth compared to the conventional Pt electrode.

3. A case study: alginate coated CNT array

Carbon nanotubes (CNTs) are molecular-scale tubes of graphitic carbon with outstanding properties. They are among the stiffest and strongest fibers known, and have remarkable electronic properties and many other unique characteristics. For these reasons they have attracted huge academic and industrial interest, with thousands of papers on nanotubes being published every year (Ciofani *et al.*, 2009).

More recently, nanotubes formed from conductive polymers have been applied to electrode sites, both reducing site impedance and providing a platform for controlled drug release (Abidian *et al.*, 2006; Abidian and Martin, 2008). Although many of these methods are promising, none have emerged as a definitive solution to the problem of reactive tissue response.

The case study presented in this Section is a combination of a polymer technology and carbon nanotubes array for the development of a drug delivery system at cellular level. Starting from these preliminary results, it is possible obtain innovative neural interface that combine the electrical properties of carbon nanotubes with the advantage of a drug delivery system, to avoid the typical response of the immunitary system.

Recently, the use of carbon nanotubes (Tasis *et al,* 2006) attracted significant attention of several groups for the development of novel neuronal interfaces (Nguyen *et al.*, 2007 Nguyen *et al.*, 2006, Gabay *et al.*, 2007). Composite materials containing multi-walled CNTs have shown the ability to limit astrocyte production of glial scarring, while maintaining good neuronal connections (McKenzie *et al.*, 2004).

Nguyen and collaborators found that PC12 cells cultured on PPy-coated CNT arrays (treated with a thin layer of collagen to promote cell adhesion) can form extended neural network upon differentiation. Starting from these considerations, in our laboratory we have proposed a combination of drug delivery system with such CNT array, exploiting a thin film of calcium alginate as drug reservoir embedded into the platform.

Two *in vitro* assays have been performed to validate the protein release on two cell lines: CrFK and PC12 cell, a cell line derived from a transplantable rat pheochromocytoma that respond reversibly to NGF by inducing a neuronal phenotype. In its presence, these cells undergo a dramatic change in phenotype whereby they acquire most of the characteristic properties of sympathetic neurons. The polymeric film embedded in the CNT array is described and characterized in terms of release kinetics using bovine serum albumin as drug model.

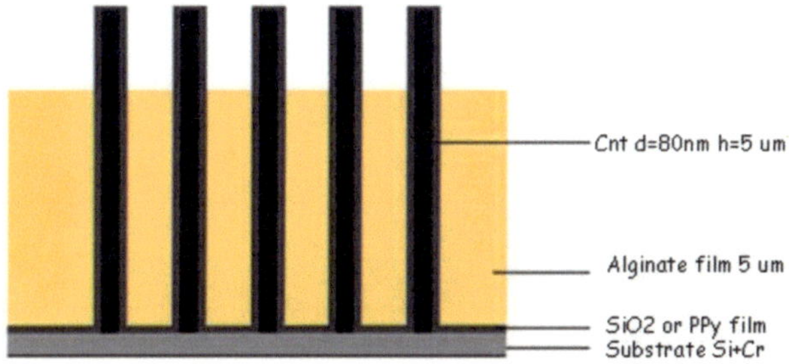

Fig. 7. Schematic illustration of the proposed CNT based system

Figure 7 represents the scheme of the proposed system. The main structure is composed by the CNT array, embedded with a thin film of alginate entrapping the drug.

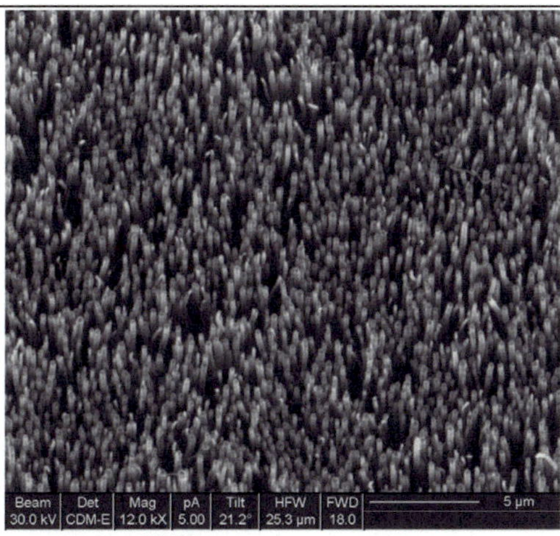

Fig. 8. FIB image of the as-grown CNT array

The as-grown CNT array is not stable when treated in liquid environments: during the drying process, CNTs irreversibly stick together to form microbundles, driven by the capillary force of water droplets (Figure 8). In order to avoid this phenomenon, a double approach was followed: SiO_2 film was deposited *via* sputtering in order to prevent the CNT sticking in a liquid environment, and to improve mechanical features of CNTs. A thin layer of a conductive polymer, polypirrole, was deposited by electrochemical deposition in order to reduce the impedance of the system and to improve the recording.
This phenomenon is completely avoided by performing a SiO_2 coating. The SiO_2 thin film, in fact, improves CNT mechanical features against the capillary force of water droplets during the drying process, thus preserving vertically alignment (Fig. 9b). The thin film of polypirrole partially avoided the problem, but it was not found satisfactory.

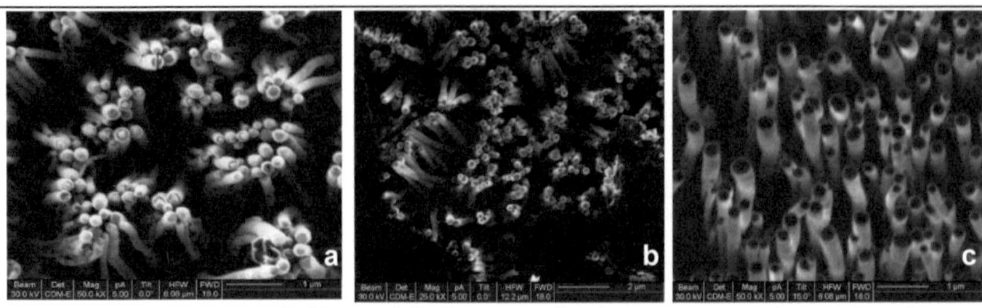

Fig. 9. Bare (a), PPy-coated (b), and SiO_2-coated (c) CNT array (samples dipped in water and dried in air before the imaging)

Among polymers, alginate has several unique properties that have allowed using it as a matrix for the entrapment and/or delivery of a variety of biological agents (Chretien *et al.*, 2005). It is a co-polymer extracted from some types of brown algae, and it is made up of two uronic acids: D-mannuronic acid and L-guluronic acid. Polyvalent cations are responsible for interchain and intrachain reticulations because they are tied to the polymer when two guluronic acid residuals are close (Mikkelsen *et al.*, 1995). The reticulation process consists of the simple substitution of sodium ions with calcium ions (Gombotz *et al.*, 1998). The relatively mild gelation process has enabled not only proteins (Ciofani *et al.*, 2008) but also cells (Murtas *et al.*, 2005) and DNA (Kimberly *et al.*, 2006) to be incorporated into alginate matrices with full retention of the biological activity.

For drug release kinetics investigation, BSA entrapped in the film was used as "protein model", as its molecular weight is similar to that one of NGF and its concentration can be much more easily evaluated concentration in the release bulk *via* spectrophotometry.

Crosslinking was thus performed with a 30% $CaCl_2$ solution, gently stirred and quickly removed (Simpson *et al.*, 2004).

In order to define a thickness of the film polymer comparable to the height of CNTs, different alginate solutions at several concentrations were tested, producing films on Si clean surface. Subsequently, *via* FIB analysis, the film thickness for the different conditions was measured, and finally the alginate concentration corresponding to a film thickness of approximately 5 μm was chosen.

The typical temporal trend of the protein release from the alginate thin film is reported in Figure 10.

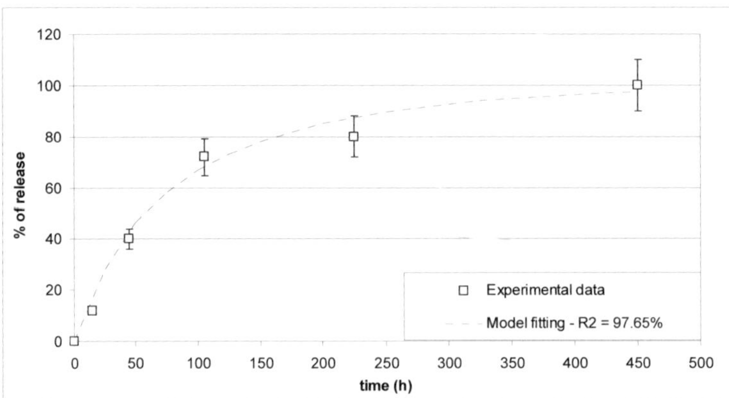

Fig. 10. Alginate release profile: experimental data and model fitting (*n* =3).

Protein amount is given as percentage of the initial amount entrapped into the film (200 μg per cm³ of film). The trend is well fitted (R^2 =97,65%) with a bi-exponential curve as already reported for alginate fibers (Ciofani *et al.* 2008a) and microspheres (Ciofani *et al.*, 2008b) and described by the following expression:

$$C_2(t) = \frac{C_{10}}{1+\frac{V_2}{V_1}} \cdot (1-e^{-h \cdot S \cdot (\frac{1}{V_1}+\frac{1}{V_2}) \cdot t}) + \frac{S \cdot C_{s0}}{V_2} \cdot (1-e^{-2 \cdot k_S \cdot t}) \qquad (1)$$

where C_2 is the protein in the bulk, C_{10} is the concentration inside the gel, S and V_1 are, respectively, the surface and the volume of film, V_2 is the volume of the bulk, h is the massive exchange coefficient, C_{s0} is the protein concentration on the surface of the film and finally k_s is the desorption rate constant.
Substituting known values and by fitting the experimental data with the mathematical model of equation (1), the h value results 10^{-9} m/s, in agreement with data given in the literature for alginate microsphere (Laca *et al.*, 1999).
In vitro experiments were carried out on two cell lines: in preliminary assays feline nephritic fibroblasts (CrFK) were used just to validate the diffusion of the drug in the bulk and the up-take by the cells. Other salient responses to NGF include cessation of proliferation, generation of long neurites, acquisition of electrical excitability, hypertrophy and a number of changes in composition associated with acquisition of a neuronal phenotype (Greene *et al.*, 1998)

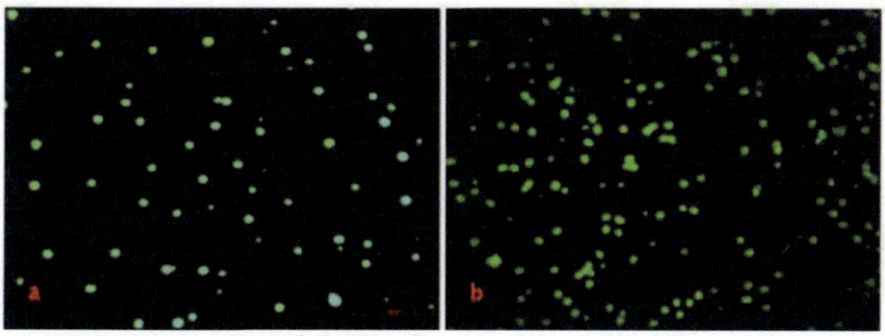

Fig. 11. Fluorescent microscopy of CrFK cells onto SiO_2 *(left)* of PPy *(right)* coated CNT interface entrapping fluoscent BSA coniugate after six hours of incubation

Figure 11 shows fluorescent microscopy images of CrFK cells after six hours of incubation onto CNT array coated with SiO_2 and PPy, respectively. These results demonstrate that the FITC-BSA, embedded in the alginate film, is efficiently released and internalized by the cells.
Alginate film coated on the CNT array and entrapping NGF was finally tested on PC12 cells monitoring their differentiation. An alginate solution entrapping 2 nM of NGF was casted on the CNT array and thereafter crosslinked with a 30% $CaCl_2$ solution as previously reported. PC12 cells were seeded on an *ad hoc* polystyrene substrate, fabricated with high

precision milling machine, at a density of 50,000/cm2. The substrate was thereafter placed on the CNT array system and the cells were grown in differentiating medium.
Figure 12 shows clearly differentiated PC12 cells after incubation on the CNT array coated with the releasing film. The microscope analysis was carried out up to three days of incubation, and, specifically, after 8 (Figure 12a), 24 (Figure 12b), 48 (Figure 12c) and finally after 72 hours (Figure 12d). Number of differentiated cells incremented during the time: at the third day of culture, the PC12 cells generate a neural network, that is a clear demonstration that the NGF is released from the film and still maintained its bioactivity.

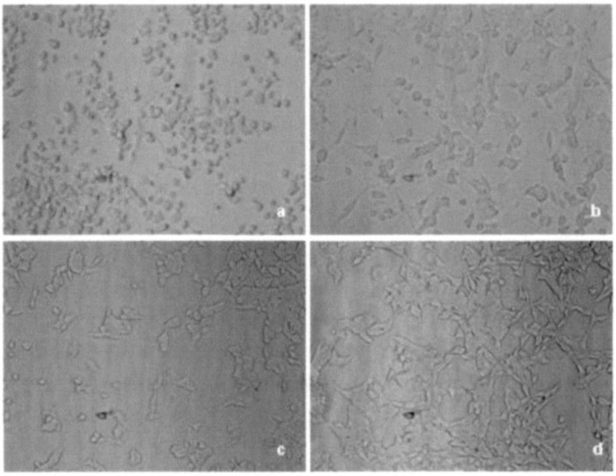

Fig. 12. Differentiating PC12 cells after 8 (a), 24 (b), 48 (c) and 72 h (d)

Fig. 13a and 13b show, respectively, the percentage of differentiated cells and the neurite length at the different time points. Figure 6a shows that already after eight hour, a non negligible number of cells, about the 10%, are differentiated. After 24 h there is a spread of the number of differentiated cells, about the 85% of the total cells. In the second day, the number increased up to 90% and, in the third day, about 96% of the cells presented well developed neurites. Figure 6b reports the trend of neurite length in the time: already after 24 hours, the mean length of the neurite is 30 ±17.9 μm and after 72 hours the length increases up to 28 ±15.9 μm.

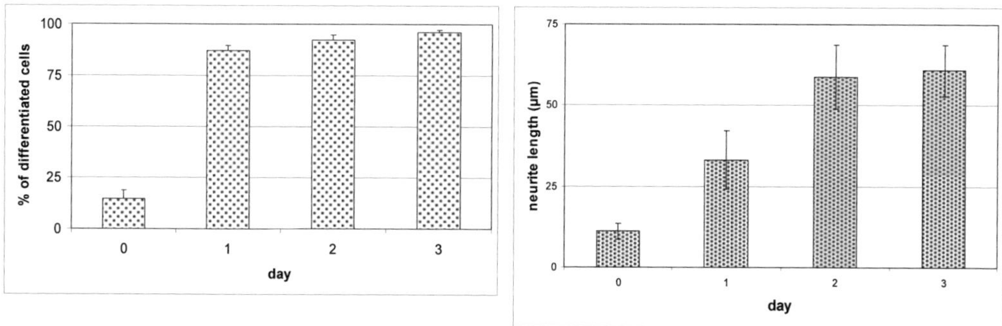

Fig. 13. Percentage of differentiated cells (a) and neurite length (b) *vs.* time ($n = 3$)

These data do not significantly differ ($p > 0.1$, Student's *t*-test) from control tests performed with "free" NGF (80 ng/ml in the culture medium) where, after three days of incubation, almost 95% of cell were differentiated with an average neurite length of about 30 µm.

4. Conclusion

Drug delivery to the central nervous system (CNS) remains a challenge despite advances in understanding the mechanisms involved in the development of neurodegenerative disorders and the actions of neuroactive agents. Drug accessibility to the CNS is limited by the blood-brain barrier; moreover the systemic administration of neuroactive biomolecules to support neuronal regeneration has several intrinsic problems, including the toxicity and poor stability associated with many bioactive factors (Maysinger *et al.*, 1997).

A variety of techniques to deliver therapeutics to the CNS has been established, including osmotic pumps (Lewin *et al.*, 1997) and silicone reservoirs. However, pumps frequently become clogged, thus limiting their ability to sustain effective concentrations; moreover these methods are often associated to highly invasive drawbacks, including device failure and higher potentials for inflammation and infection due to their non degradable components. Polymeric delivery systems have the potential to maintain therapeutic levels of a drug, to reduce side effects and to facilitate the delivery of drugs with short in vivo half-lives.

In this Chapter we have outlined different applications of polymers, usually intended as agent acted to improve the biodistribution of the desired drugs. Namely, we have highlighted the huge possibilities offered by the thin film technology applied to the concepts of "drug delivery" and "drug targeting". After a summary of the most relevant examples in the literature, we described the ongoing work in our lab, aimed at obtaining an efficient combination of physical and chemical features of an innovative neuronal interface based on a carbon nanotubes array.

The achieved results indicate that polymer technology could be efficiently embedded in CNT array (Hind *et al.*, 2004) acting as drug delivery system at cellular level. The implication of this study open several perspectives in particular in the field of neurointerfaces, combining several functions into a single platform (Navarro *et al.*, 1998; Navarro *et al.*, 2005).

5. References

Abidian MR.; Kim DH.; Martin DC. (2006). Conducting-polymer nanotubes for controlled drug release. *Advanced Materials*; vol.18, pp. 405–9.

Abidian MR.; Martin DC. (2008). Experimental and theoretical characterization of implantable neural microelectrodes modified with conducting polymer nanotubes. *Biomaterials* vol. 29, pp. 1273–83.

Alexis F.; Venkatraman SS.; Rath SK.; Boey F. (2004). In vitro study of release mechanisms of paclitaxel and rapamycin from drug-incorporated biodegradabile stent matrices. *Journal of Control release* vol. 98, pp. 67–74.

Biran R.; Martin DC.; Tresco PA. (2005). Neuronal cell loss accompanies the brain tissue response to chronically implanted silicon microelectrode arrays. *Experimental Neurology* vol. 195 pp. 115–26.

Bronzino JD.; (1995). Biomedical Engineering Handbook, CRC Press, pp., 2000.IEEE Press, Boca Raton.

Chen J. (1997). A multichannel neural probe for selective chemical delivery at the cellular level *IEEE Trans. Biomed.* vol. 44 pp. 760–69.

Cheung KC.; Renaud P.; Tanila H.; Djupsund K. (2007). Flexible polyimide microelectrode array for in vivo recordings and current source density analysis. *Biosensors and Bioelectronics* 22, 1783-1790

Chretien C., Chaumeil. JC. (2005). Release of a macromolecular drug from alginate-impregnated particles. *International Journal of Pharmaceutics* vol. 304, pp. 18–28.

Ciofani G.; Cascone MG.; Serino LP.; Lazzeri L. (2008). Urease loaded alginate microspheres for blood purification. *Journal of Microencapsulation* vol. 25, pp. 569–576.

Ciofani G.; Raffa V.; Pizzorusso T.; Menciassi A; Dario P. (2008). Characterization of an alginate based drug delivery system for neurological applications. *Medical Engeenering and Physics* vol. 30, pp. 848–855.

Ciofani G.; Raffa V. (2009). Mini-Reviews in *Medicinal Chemistry*, in press Chemically Functionalized Carbon Nanotubes: Emerging Vectors for Cell Therapy.

Cui XY.; (2001). Surface modification of neural recording electrodes with conducting polymer/biomolecule blends *Journal of Biomedical Materarials Research,* vol. 56, pp. 261–72.

Cui XY.; Hetke JF.; Wiler JA.; Anderson DJ.; Martin DC. (2001). Electrochemical deposition and characterization of conducting polymer polypyrrole/PSS on multichannel neural probes. *Sensor and Actuators A: Physics,* vol. 93, pp. 8–18.

Cui X.; Lee VA.; Raphael Y.; Wiler JA.; Hetke JF.; Anderson DJ. (2001). Surface modification of neural recording electrodes with conducting polymer/biomolecule blends. *Journal of Biomedical Materials Research*; vol. 56, pp. 261–72.

Cui X.; Wiler J.; Dzaman M.; Altschuler RA.; Martin DC.; (2003) In vivo studies of polypyrrole/ peptide coated neural probes. *Biomaterials*; 24:777–87.

Cui X.; David C.; Martin. (2003). Electrochemical deposition and characterization of poly(3,4-ethylenedioxythiophene) on neural microelectrode array *Sensors and actuators*. vol. 89, pp. 92-102.

Cui X.; (2003).In vivo studies of polypyrrole/peptide coated neural probes *Biomaterials,* vol. 24, pp. 777–87.

Dhanikula AB.; Panchagnula R. (2004). Developmentand characterization of biodegradable chitosan films for local delivery of paclitaxel. *AAPS J* vol. 6, pp. 1–12.

Dorta, MJ.; Oliva, A.; Mungula O., Llabres M.; Farina JB. (2002). In-vitro release of fluoropyrimidines fromPLGA film implants. *Journal of Pharmacy and Pharmacology* vol. 54, pp. 757–763.

Drachman DE.; Edelman ER.; Seifert P.; Groothuis AR.; Bornstein DA.; Kamath KR.; Palasis M.; Yang D.; Nott SH.; Rogers C. (2000). Neointimal thickening after stent delivery of paclitaxel: Change in compositionand arrest of growth over six months. *Journal of the American College of Cardiology*, vol. 36, pp. 2325–2332.

Hanefeld P.; Westedt U.; Wombacher R.; Kissel T.; Schaper A.; Wendorff JH.; Greiner A. (2006). Coating of poly(p-xylylene) by PLA-PEO-PLA triblock copolymers with excellent polymer-polymer adhesion for stent applications. *Biomacromolecules* vol. 7, pp. 2086–2090.

Heller PF.; Jenkins GM.; Cheng L.; Leong K.; Shao W.; Brinker J.; Hope E.; Nater C.; Lakatta E.; Froelich JP. (1995). The sustained release of galardin and taxol from gelatin chondroitin sulfate coacervate films. *Materials Research Society Symposia Proceedings* vol. 395, pp. 55–60.

Hinds; Chopra N.; Rantell T.; Andrews R.; Gavalas V.; Bachas LG. (2004). Aligned multiwalled carbon nanotube membranes, *Science*, vol. 303, pp. 62-65.

Ho EA.; Vassileva V.; Allen C.; Piquette-Miller M. (2005). In vitro and in vivo characterizations of a novel biocompatible polymer-lipid implant system for the sustained delivery of paclitaxel. *Journal of Control Release* vol. 104, pp. 181–191.

Gabay T Ben-David M.; Kalifa I.; Sorkin R.; Abrams Z.R.; Ben-Jacob E.; Hanein Y. (2007). Electro-chemical and biological properties of carbon nanotube-based multi-electrode arrays. *Nanotechnology* col. 6, pp.1–6.

Gombotz WR.; Wee SF. (1998). Protein release from alginate matrices. *Advanced Drug Delivery Reviews* vol. 31, pp. 267–285.

Grant J.; Blicker.; Piquette-Miller M.; Allen C. (2005). Hybrid films from blends of chitosan and egg phosphatidylcholine for localized delivery of paclitaxel. *Journal of Pharmaceutical Science* vol. 94, pp. 1512–1527.

Greene LA.; Farinelli SE.; Cunningham ME.; Park DS. (1998). in Culture and experimental use of the PC12 rat pheochromocytoma cell line *Culturing Nerve Cells.*

Green A.; Lovell NH.; Poole-Warren LA.; (2009) Cell attachment functionality of bioactive conducting polymers for neural interfaces *Biomaterials* vol. 30, pp. 3637–3644.

Jackson JK.; Skinner JC.; Burgess L.; Sun T.; Hunter WL.; Burt HM. (2002). Paclitaxel-loaded crosslinked hyaluronic acid films for the prevention of post surgical adhesions. *Pharmaceutical Research* vol. 19, pp. 411–417.

Jackson JK.; Smith J.; Letchford K.; Babiuk KA.; Machan L.; Signore P.;Hunter WL.; Wang K.; Burt HM. (2004). Characterization of perivascular poly (lactic-co-glycolic acid) films containing paclitaxel. *International Journal of Pharmaceutics* vol. 283, pp.97–109.

Jedlicka S.; Leavesley SJ.; Little KM.; Robinson JP.; Nivens DE.; Rickus JL.; (2006) Interactions between chemical functionality and nanoscale surface topography impact fibronectin conformation and neuronal differentiation on model sol–gel silica substrates. *Materials Research Society Proceedings.*

Jedlicka SS.; Little KM.; Nivens DE.; Zemlyanov D.; Rickus JL.; (2007) Peptide ormosils as cellular substrates. *Journal of Materials Chemistry*; 17:5058–67.

Jedlicka SS.; McKenzie JL.; Leavesley SJ.; Little KM.; Webster TJ.; Robinson JP,(2006). Sol-gel derived materials as substrates for neuronal differentiation: effects of surface features and protein conformation. *Journal of Materials Chemistry*; 16:3221–30.
Jeong, S.Y.; Kim, S.W.; (1986). Biodegradable polymeric drug delivery systems. *Arch. Pharm. Res.* 9, 63-73.
Jones LL.; Tuszynski MH. (2001). Chronic intrathecal infusions after spinal cord injury cause scarring and compression, *Microscopy Research and Technique*, pp. 317–324.
Kato Y.; Nishino M.; I Saito, Suzuki T.; Mabuchi K. (2006). Flexible Intracortical Neural Probe with Biodegradable Polymer for Delivering Bioactive *Components Proceedings of International Conference on Microtechnologies in Medicine and Biology*.
Kimberly L.; Douglas CA.; Piccirillo; Tabrizian M. (2006). Effects of alginate inclusion on the vector properties of chitosan-based nanoparticles. *Journal of Control Release* vol. 115, pp. 354–361.
Laca A.; Garcia LA.; Argueso F.; Diaz M. (1999). Protein diffusion in alginate beads monitored by confocal microscopy, The application of wavelets for data reconstruction and analysis. *Journal of Industrial Microbioogy and. Biotechnology* vol. 23, pp. 155–165.
Lago N.; Ceballos D.; Rodrıguez FJ.; Stieglitz T.; Navarro X. (2005). Long term assessment of axonal regeneration through polyimide regenerative electrodes to interface the peripheral nerve.. *Biomaterials,* 26, 2021-2031
Langer R.; (1998) Drug delivery and targeting, *Nature,* vol. 392, pp. 5–10.
Lewin SL.; Utley DS.; Cheng ET.; Verity AN.; Terris DJ. (1997). Simultaneous treatment with BDNF and CNTF after peripheral nerve transection and repair enhances rate of functional recovery compared with BDNF treatment alone, *Laryngoscope,* vol. 107, pp. 992–999.
Maysinger D.; Morinville A. (1997). Drug delivery to thc nervous system, *Trends Biotechnology* vol.15, pp. 410–18.
McCarron PA.; Woolfson AD.; Keating SM.; (2000). Sustained release of 5-fluorouracil from polymeric nanoparticles. *Journal of Pharmacy and Pharmacology* vol. 52, pp. 1451-1459.
McKenzie JL.; Waid MC.; Shi R.; Webster TJ.; (2004). Decreased functions of astrocytes on carbon nanofiber materials, *Biomaterials*, 25, 1309-1317,
Mikkelsen A., Eigsaeter A., (1995) Density distribution of calcium induces alginate gels: a numerical study *Biopolymers* vol. 36, pp. 17–41.
Murtas G.; Capuani M.; Dentini; Manetti C.; Masci G.; Massimi M.; Miccheli A.; Crescenzi V.; (2005). Alginate beads as immobilization matrix for hepatocytes perfused in a bioreactor. *Journal of Biomaterial Science Polymer Edition*. vol. 16, pp. 829–846.
Navarro CS.; Rodriguez CA.; Blau C.; Buti M.; Valderrama E.; Meyer JU; Stieglitz T. (1998). Stimulation and recording from regenerated peripheral nerves through polyimide sieve electrodes, *Journal of Peripheral Nervous System*, vol. 3, pp. 91-101.
Navarro X.; Krueger, TB.; Lago N.; Micera S.; Stieglitz T.; Dario P.; (2005) A critical review of interfaces with the peripheral nervous system for the control of neuroprostheses and hybrid bionic systems *Journal of the Peripheral Nervous System.*
Nguyen T.D. Vu; H. Chen; A.M. Cassell; Andrews R.J.; Meyyappan M.; J. Li, (2007) Vertically aligned carbon nanofiber architecture as a multifunctional 3-D neural electrical interface. *IEEE Transaction Biomedical Engeenering* vol. 54, pp. 1121–1128.

Nguyen-Vu T.D., Chen H.; Cassell AM.; Andrews R.; Meyyappan M.; Li J.; (2006). Vertically aligned carbon nanofiber arrays: an advance toward electrical-neural interfaces. *Small* vol. 2, pp. 89–94.

Perugini P.; Genta I.; Conti B.; Modena T.; Pavanetto F. (2003). Periodontal delivery of ipriflavone: Newchitosan/PLGA film delivery system for a lipophilicdrug. *International Journal of Pharm*aceutics vol. 252, pp. 1–9.

Piercea A L.; Sommakiaa S.; Rickus JL.; Ottoa KJ. (2009). Thin-film silica sol–gel coatings for neural microelectrodes, *Journal of Neuroscience Methods.*

Polikov VS, Tresco PA, Reichert WM. (2005) Response of brain tissue to chronically implanted neural electrodes. *Journal of Neuroscience Methods;*vol. 148, pp. 1–18.

Radin S.; Ducheyne P. (2007). Controlled release of vancomycin from thin sol–gel films on titanium alloy fracture plate. *Biomaterials* vol. 28, pp. 1721–9.

Rajeev AJ. (2000). The manufacturing techniques of various drug loaded biodegradable poly(lactide-*co*-glycolide) (PLGA) devices *Biomaterials*, vol. 21, pp. 2475-2490.

Retterer ST; (2004), Model neural prostheses with integrated microfluidics: a potential intervention strategy for controlling reactive cell and tissue responses *IEEE Transaction of Biomedical Engeneering*, vol. 51, pp. 2063–73.

Richardson RR.; Miller JA.; Reichert WM. (1993). Polyimides as Biomaterials: Preliminary Biocompatibility Testing. *Journal of Biomaterials.* 14, 627-63.

Santos X.; Rodrigo J.; Hontanilla B.; Bilbao G. (1998). Evaluation of peripheral nerve regeneration by nerve growth factor locally administered with a novel system, *Journal of Neuroscience Method*, vol. 85, pp.119–27.

Schmidt CE.; Leach LB. (2003). Neural tissue engineering: Strategies for Repair and Regeneration, *Annual Review of Biomedical Engeneering*, vol. 5, pp. 293–347.

Schwartz AB. (2004). Cortical Neural Prosthetics, *Annual Review of Neuroscience*, vol. 27, pp. 487-507.

Schwartz AB.; Cui X.; Douglas J.; Weber and Daniel W. Moran (2006). Brain-Controlled Interfaces: Movement Restoration with Neural Prosthetics *Neuron.*

Shi R.; Burt HM.; (2004). Amphiphilic dextran-graftpoly(epsilon-caprolactone) films for the controlledrelease of paclitaxel. *Int J Pharm* 271:167–179.

Simpson N.E.; Stabler CL.; Simpson, C.P.; Sambanis A. I. Constantinidis, (2004) The role of the CaCl2-guluronic acid interaction on alginate encapsulated betaTC3 cells. *Biomaterials* 25, 2603–2610.

Stieglitz T.; Scuettler M.; Koch KP. (2005). Implantable Biomedical Microsystems for Neural Prostheses. *IEEE Biomimetic Systems* 58 -64

Szarowski DH.; (2003). Brain Responses to micro-machined silicon devices, *Brain Research*, vol. 983, pp. 23-35.

Szarowski DH.; Andersen MD.; Retterer S.; Spence AJ.; Isaacson M.; Craighead HG. (2003). Brain responses to micro-machined silicon devices. *Brain Research;* vol. 983, pp. 23–35.

Tashiro K, Nagata I.; Yamashita N.; Okazaki K.; Ogomori K.; Tashiro N. (1994). A synthetic peptide deduced from the sequence in the cross-region of laminin A chain mediates neurite outgrowth, cell attachment and heparin binding. *Biochemical Journal;* vol. 302, pp. 39–45.

Tasis D.; N. Tagmatarchis, A. Bianco, M. Prato (2006). Chemistry of carbon nanotubes. *Chemical Review* vol. 106, pp. 1105–1136.

Vadim S.; Polikov PA.; Tresco and William M. Reicher; (2009). Bioelectronics and Implanted Devices EM McGee Medical Enhancement and Posthumanity, *Journal of Neuroscience Methods*, pp 1-18

Vetter RJ.; Williams JC., Hetke JF.; Nunamaker EA.; Kipke DR. (2004) Chronic neural recording using silicon-substrate microelectrode arrays implanted in cerebral cortex. *IEEE Transactions on Biomedical Engineering;* vol. 51, pp. 896–904.

Wand He.; Bellamkonda RV.; (2005) Nanoscale neuro-integrative coatings for neural implants *Biomaterials* vol. 26, pp. 2983–90.

Westedt U.; Wittmar M, Hellwig M, Hanefield P, Greiner A, Schaper AK, Kissel T. (2006). Coating of poly(p-xylylene) by PLA-PEO-PLA triblock copolymers with excellent polymer-polymer adhesion for *Journal of Control Release* vol. 111, pp. 235–245.

Webber.; W.L., Lago, F., Thanos, C., Mathiowitz, E., (1998). Characterization of soluble, salt-loaded degradable PLGA films and their release of tetracycline. *J. Biomed. Res.* 41, 18_/29

Williams JC.; Holecko MM; Massia SP; (2005). Neurotrophin-eluting hydrogel coatings for neural stimulating electrodes *Journal of Neural Enginnering,*

Winter JO.; Cogan SF.; Rizzo JF.; (2007) III - Journal of Biomedical Materials Research Part B.

Zhou, T.; Lewis, H., Foster, RE., Schwendeman, SP., (1998). Development of a multiple-drug delivery-implant for intraocular management of proliferative vitreoretinopathy. *Journal of Control Release* vol. 55, pp. 281-295

18

Determinations of Optical Field Induced Nonlinearities in Azo Dye Doped Polymer Film

Chong Hoon Kwak*, Hye Ri Yang
Department of Physics, Yeungnam University, South Korea
**Corresponding author: chkwak@ynu.ac.kr*

1. Introduction

Saturable dye doped polymer films have been widely used for optical data storage (Pham et al., 1997), nonlinear optical wave mixing processes (Sharma et al., 1994) such as degenerate and non-degenerate multiwave mixing (Fragnito et al., 1987), optical phase conjugation (Fujiwara et al., 1985), two beam coupling and polarization holograms (Huang et al., 1993). The optical field induced (photo-structural) changes of azo dye molecules embedded in polymer matrix lead to remarkable changes of nonlinear optical properties, including scalar effects like nonlinear refractive index and nonlinear absorption, and vector effects like photoinduced birefringence (PIB) and photoinduced dichroism (PID). Nonlinear optical materials exhibiting PIB and PID, such as azo dye doped polymers, amorphous chalcogenide semiconductor films and others have been intensively investigated due to their potential applications in many fields, particularly for polarization (or vector) holograms (Kwak et al., 1988 & Nikolova et al., 1988), nonlinear optical image processing (Kwak et al., 1989), and multispot arrays generation for digital optical computing systems (Kwak et al., 1992). It has been now widely accepted that the physical origin of the PIB and PID in azo dye doped polymer is attributed to the *trans-cis* photoisomerization of azo molecules via photoinduced molecular reorientation (Nikolova et al., 1988 & Sekkat et al., 1995 & Egami et al., 1997 & Yaroshchuk et al., 2003).
In this Chapter, we report on the determinations of the optical nonlinearities of azo dye doped polymer film by means of the holographic gratings as a scalar effect and the photoinduced birefringence as a vector effect, and present on the transient behaviors of the diffraction efficiency and the PIB, revealing the stretched exponential responses, which cannot be explained by the conventional photo-process model (Nikolova et al., 1988). A stretched exponential time response, which is also known as Kohlrausch-Williams-Watts (KWW) response, is frequently observed in amorphous polymers, glasses and other disordered materials. Although several authors have proposed some of the empirical models to explain the stretched exponential relaxation phenomena in amorphous materials (Johanson et al., 2007 & Benatar et al., 1993 & Dureiko et al., 1998 & Palmer et al., 1984), those models cannot be directly applied to the diffraction efficiency and the PIB kinetics observed in azo dye doped polymer.

The purpose of this work is to propose a novel approach to explain the transient behaviors of the diffraction efficiency and the photoinduced birefringence and to compare with the experimental data. A three state model for photoisomerization is presented to analyse the stretched exponential kinetic behaviors by employing an empirical stretched exponential time response in the *trans-cis* photoisomerization of azo molecules. Finally, we have conducted the Z-scan experiments to determine the sign of the optical nonlinearities.

2. Theory

2.1 Three state model for photoisomerization with stretched exponential kinetics

In this section, we will present a three state model for photoisomerization, describing the transient behaviors of both the diffraction efficiency of the holographic gratings and the photoinduced birefringence. We consider a three state system consisting of *trans*, *trans** and *cis* states, which is illuminated by a linearly (say, x-) polarized single pump beam for PIB, or coherent two pump beams for holographic gratings, as shown in Fig. 1. It is assumed that under the influence of the amorphous natures of azo dyes in polymeric environment, the dye molecules embedded in a polymer matrix exhibit the stretched exponential kinetics below the glass transition temperature (Huang et al., 1993). We empirically introduce a relaxation time dispersion function or stretched exponential time (KWW) response function, $g(t) = \beta(t/\tau)^{\beta-1}$, where τ is a characteristic time of azo molecules with polymer matrix and the parameter $\beta(0 \le \beta \le 1)$ is a stretched exponent that quantifies the extent of the deviation from pure exponentiality (i.e., $\beta = 1$), or equivalently, the degree of the relaxation time dispersion. The stretched exponential kinetics of glasses or amorphous materials has been related to the high degree of disorder, which leads to the existence of a large number of metastable states (Palmer et al., 1984).

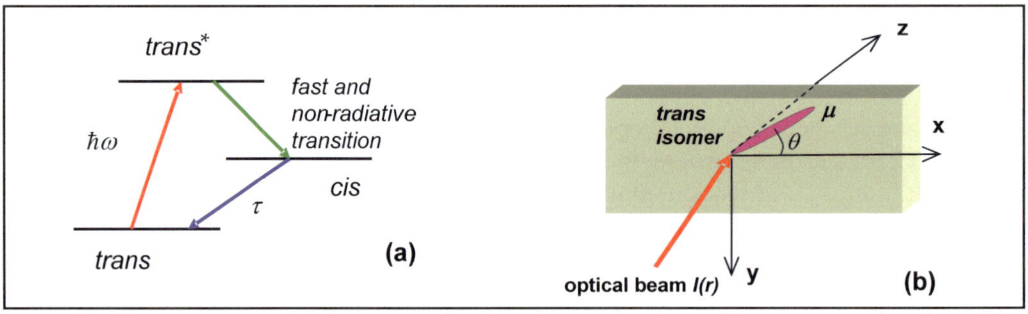

Fig. 1. (a) Three state model for *trans-cis* photoisomerizations, (b) geometry for describing the interactions of the *trans* molecules with the polarized optical beam intensity $I(r)$.

The optical transition rate from *trans* to *trans** state and thermal relaxation rate from *cis* to *trans* states are assumed to have a form of the stretched exponential time response and can be written as $W_{T\to C} = q\sigma_T I g(t) = q\sigma_T I\beta(t/\tau)^{\beta-1}$ and $A_{C\to T} = g(t)/\tau = \beta t^{\beta-1}/\tau^{\beta}$, respectively. Here, $q = \Phi/\hbar\omega$ is a constant depending on the molecular surroundings, where Φ is the quantum yield for *trans* to *cis* state and $\hbar\omega$ is the photon energy, so the dimension of q is the inverse of energy (i.e. J^{-1}). The anisotropic molecular absorption cross section is

$\sigma_T = \sigma_T^{\parallel}\cos^2\theta + \sigma_T^{\perp}\sin^2\theta$, where $\sigma_T^{\parallel}$ and $\sigma_T^{\perp}$ are the absorption cross sections of the *trans* molecule for the light polarized parallel and perpendicular to the molecular axis $\vec{\mu}$, respectively, θ is the angle between the *trans* molecular axis and the direction of the polarized light, I is the single pump beam intensity for PIB, or the intensity interference patterns for holographic gratings. To simplify the discussion, we assumed that only the *trans* molecules have an axial symmetry and that $\sigma_T^{\perp} = 0$ because of a rod-like *trans* molecular structure. Assuming a very rapid intersystem crossing from *trans** to *cis* with non-radiative transition, the *trans** state becomes a quasi-steady state, so the three state system reduces to the two state system by eliminating the number density of the *trans** state. Taking into account of the relaxation time dispersion of each isomer states, we readily have the rate equations for *trans* and *cis* states as;

$$\frac{dN_T}{dt} = -W_{T\to C}N_T + A_{C\to T}N_C, \quad (1a)$$

$$\frac{dN_C}{dt} = W_{T\to C}N_T - A_{C\to T}N_C, \quad (1b)$$

where N_T and N_C are the number densities of *trans* and *cis* states. For PIB, caused by a linearly polarized single beam, the optical transition rate $W_{T\to C}$ has a strongly angular dependency, whereas for the case of the diffraction gratings, the angular dependence of the optical transition rate $W_{T\to C}$ is negligible and should be ensemble averaged for all θ $(0 \le \theta \le 2\pi)$, because the grating spacing of about a few μm is very much larger than the *trans* or *cis* molecular sizes. With this in mind, the optical transition rates are given by $W_{T\to C} = \beta(t/\tau)^{\beta-1} q\sigma_T^{\parallel} I\cos^2\theta$ for PIB, and $W_{T\to C} = \beta(t/\tau)^{\beta-1} q\sigma_T^{\parallel} I/2$ for intensity gratings.

2.2 Diffraction efficiencies of holographic gratings

We present the transient diffraction efficiency of holographic gratings, which is constructed by a spatially modulated intensity grating, $I(x)$. The solutions of Eqs. (1) are given by

$$N_T(t) = N_0\frac{1}{1+S(x)}\left[1 + S(x)\exp\left\{-(1+S(x))\left(\frac{t}{\tau}\right)^{\beta}\right\}\right], \quad (2a)$$

$$N_C(t) = N_0\frac{S(x)}{1+S(x)}\left[1 - \exp\left\{-(1+S(x))\left(\frac{t}{\tau}\right)^{\beta}\right\}\right], \quad (2b)$$

which has been made use of the initial conditions of $N_T(0) = N_0$, $N_C(0) = 0$ and $N_0(= N_T(t) + N_C(t))$ is the total number density of isomers. Here, $S(x) = I(x)/I_S$ is the saturation parameter, and $I_S = 2/\tau q\sigma_T^{\parallel} = 2\hbar\omega/\Phi\tau\sigma_T^{\parallel}$ is the saturation intensity for the holographic grating, $I(x)$ is the intensity grating. The saturation parameter $S(x)$ represents the spatially modulated intensity distribution of two writing beams, and is expressed as (Fragnito et al., 1987; Kwak et al., 1988)

$$S(x) = S_0 + S_1\cos Kx, \quad (3)$$

where $S_0 = (I_a + I_b)/I_S$, $S_1 = 2\sqrt{I_a I_b}/I_S$ with the two writing beam intensities I_a and I_b, $K = 2\pi/\Lambda$ is the magnitude of the grating wave vector and Λ is the grating period. The refractive index change is assumed to be proportional to the changes of population density of Eqs.(2) and then is given by

$$n(x,t) = n_0 + \Delta n(x,t) = n_0 + \delta n \frac{S(x)}{1+S(x)}\left(1 - \exp\left[-(1+S(x))\left(\frac{t}{\tau}\right)^{\beta}\right]\right), \tag{4}$$

where n_0 is the linear refractive index, Δn is the nonlinear refractive index, and δn is the maximum nonlinear refractive index change. Following the same procedures of the literatures (Fragnito et al., 1987; Kwak et al., 1988) and using the residue calculus in the complex plane, we have the nonsinusoidal phase gratings of the form:

$$\Delta n(x,t) = \sum_{h=-\infty}^{\infty} \Delta n_h(t) \exp[ihKx], \tag{5a}$$

where $\Delta n_h(t)$ is the hth diffraction amplitudes of the phase gratings and are given by

$$\Delta n_0(t) = \delta n\left[1 - C_0 + \exp\left[-(1+S_o)\left(\frac{t}{\tau}\right)^{\beta}\right]\left(-D_0(t) + \sum_{n=-\infty}^{\infty} C_n D_{-n}(t)\right)\right], \tag{5b}$$

$$\Delta n_h(t) = \delta n\left[-C_h + \exp\left[-(1+S_o)\left(\frac{t}{\tau}\right)^{\beta}\right]\left(-D_h(t) + \sum_{n=-\infty}^{\infty} C_n D_{h-n}(t)\right)\right], \ (h \neq 0), \tag{5c}$$

Here, the Fourier coefficients, C_n and D_h are given by

$$C_n = \frac{\left(\sqrt{(1+S_0)^2 - S_1^2} - 1 - S_0\right)^{|n|}}{S_1^{|n|}\sqrt{(1+S_0)^2 - S_1^2}}, \tag{6a}$$

$$D_0(t) = I_0\left[S_1\left(\frac{t}{\tau}\right)^{\beta}\right], \ D_h(t) = 2(-1)^h I_h\left[S_1\left(\frac{t}{\tau}\right)^{\beta}\right], \ (h \neq 0), \tag{6b}$$

where $I_h(\bullet)$ denotes hth-order modified Bessel function of the first kind. The Fourier coefficients have the properties of $C_n = C_{-n}$, $D_h = D_{-h}$ for $I_h(\bullet) = I_{-h}(\bullet)$. These high-order Fourier coefficients result from the nonsinusoidal spatial modulation of the population grating due to the saturation effect. A precise knowledge of the amplitudes of the gratings, particularly for the first-order diffraction, is of fundamental significance for understanding the physical mechanism of the grating formations in saturable absorber. Henceforth we only consider the time evolution of the first-order spatial harmonic grating. Furthermore, we have numerically confirmed that the contribution of the high-order coefficients greater than 2 (i.e., $C_{|n|\geq 2}$ and $D_{|n|\geq 2}(t)$) to the resultant grating amplitude can be negligibly small for all time regime, within less than 10%. The diffraction efficiency of the first-order harmonic grating for Bragg condition is then approximately given by (Kogelnik et al., 1969)

$$\eta \approx \sin^2\left(-\left(\frac{\pi\delta nL}{\lambda'\cos\theta}\right)\left(C_1 + \exp\left[-(1+S_0)\left(\frac{t}{\tau}\right)^{\beta}\right](D_1(t) - C_0D_1(t) - C_1D_0(t))\right)\right) \tag{7}$$

where L is the thickness of the sample, λ' is the wavelength of read-out beam, θ is the incident half-angle between the two writing beams, τ is the characteristic lifetime of the material in grating formation, and β is the stretched exponent.

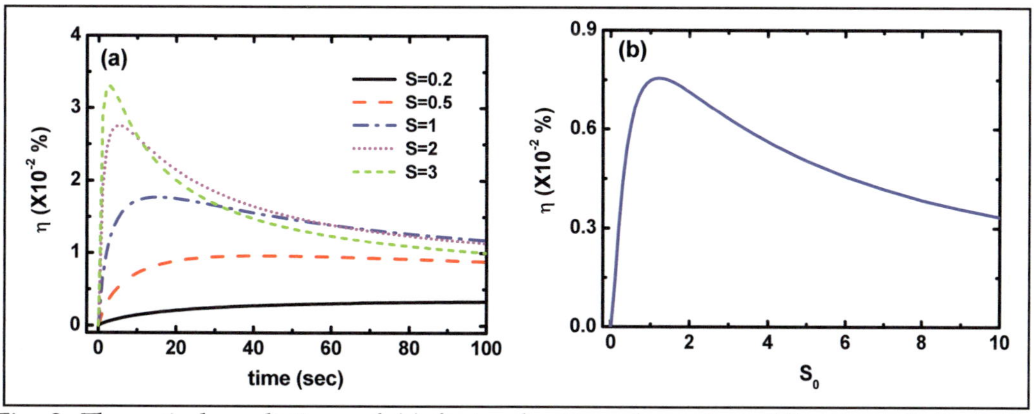

Fig. 2. Theoretical predictions of (a) first-order time dependent diffraction efficiency for several saturation parameters with $\beta = 0.35$, $\tau = 35\,\text{sec}$ and $\delta n = 5\times10^{-4}$, and (b) steady-state value of diffraction efficiency against saturation parameter, which is independent of the stretched exponent.

Fig. 2(a) represents the theoretical predictions of time-dependent diffraction efficiency for several saturation parameters S_0. It is shown that as increasing the saturation parameters the transient peaks of the diffraction efficiency reveal much higher than the steady state values. Figure 2(b) depicts the steady-state values of diffraction efficiency as a function of the saturation parameter S_0. As increasing the saturation parameter, the diffraction efficiency rapidly increases to a maximum value and then gradually decreases and eventually fades away.

2.3 Photoinduced birefringence

We will develop a theory describing the transient behaviors of the photoinduced birefringence (PIB) observed in azo dye doped polymer film by adopting a stretched exponential response, which cannot be explained by the conventional photo-process (Nikolova et al., 1988). For a linearly polarized (say, x-polarized) beam, we have the number densities of the states from the rate equation of Eqs.(1):

$$N_T(\theta,t)=\frac{N_0}{2\pi}\frac{1}{1+S\cos^2\theta}\left\{1+S\cos^2\theta\exp\left[-\left(1+S\cos^2\theta\right)\left(\frac{t}{\tau}\right)^{\beta}\right]\right\}, \tag{8a}$$

$$N_C(\theta,t)=\frac{N_0}{2\pi}\frac{S\cos^2\theta}{1+S\cos^2\theta}\left\{1-\exp\left[-\left(1+S\cos^2\theta\right)\left(\frac{t}{\tau}\right)^{\beta}\right]\right\}. \tag{8b}$$

It is noted that the saturation parameter, $S=I/I_S$ for PIB is a spatially uniform, and the total number density at polar angle θ, $N_T(\theta,t)+N_C(\theta,t)=N_0/2\pi$ is strongly angular dependent and the saturation intensity, $I_S=1/\tau q\sigma_T^{\parallel}=1\hbar\omega/\Phi\tau\sigma_T^{\parallel}$ is different from that of the holographic gratings. For an axially symmetric configuration about the polarization axis as in Fig. 1(b), the populations of *trans* and *cis* molecules are strongly anisotropic in space and only a function of the polar angle θ. In the laboratory frame, the macroscopic susceptibility of oriented molecules is given by

$$\chi_{ij}(t)=\chi_{polym}+\int d\chi_{\mu\mu}\cos_{\mu i}\cos_{\mu j}\equiv\chi_L+\delta\chi_{ij}(t),\ (i,j=x,y) \tag{9}$$

where $\cos_{\mu i}$ and $\cos_{\mu j}$ are direction cosines of the electric dipole moment vector $\vec{\mu}$ relative to the $i, j(=x,y)$ axes of the laboratory frame as in Fig. 1(b), and $d\chi_{\mu\mu}(\theta,t)=\left[\alpha_T N_T(\theta,t)+\alpha_C N_C(\theta,t)\right]d\theta$ is the total susceptibility of the group of molecules oriented in an angle $d\theta$. Since the host polymer is optically isotropic, the linear susceptibility for the polymer χ_{polym} is included. Here, the macroscopic (isotropic) linear susceptibility is $\chi_L=\chi_{ij}^{(1)}\delta_{ij}=\chi_{polym}+\alpha_T N_0$ with the isotropic condition of $\chi_{xx}^{(1)}=\chi_{yy}^{(1)}$ due to amorphous nature of the sample, where $\alpha_X\ (X=T,C)$ is the complex linear polarizability of the *trans* or *cis* isomers, and δ_{ij} is the Kronecker delta function. The photoinduced nonlinear susceptibility $\delta\chi_{ij}(t)$ is given by

$$\delta\chi_{ij}(t)=\frac{\delta\chi_o}{N_0}\int_0^{2\pi}N_C(\theta,t)\cos_{\mu i}\cos_{\mu j}\,d\theta\ , \tag{10}$$

where $\delta\chi_o=(\alpha_C-\alpha_T)N_0$, $\cos_{\mu x}=\cos\theta$ and $\cos_{\mu y}=\sin\theta$. Substituting Eq. (8b) into Eq. (10) and using the residue calculus of the complex plane, after some calculations, we have the photoinduced nonlinear susceptibilities for directions parallel and perpendicular to the direction of linearly polarized pump beam as

$$\delta\chi_{xx}(t)\equiv\delta\chi_o\left\{\begin{array}{l}A_{0,x}\left(1-\exp\left[-\left(1+\frac{S}{2}\right)\left(\frac{t}{\tau}\right)^{\beta}\right]I_0\left[\frac{S}{2}\left(\frac{t}{\tau}\right)^{\beta}\right]\right)\\ +2\exp\left[-\left(1+\frac{S}{2}\right)\left(\frac{t}{\tau}\right)^{\beta}\right]\cdot\sum_{m=1}^{\infty}A_{m,x}I_m\left[\frac{S}{2}\left(\frac{t}{\tau}\right)^{\beta}\right]\end{array}\right\}, \tag{11a}$$

$$\delta\chi_{yy}(t) \equiv \delta\chi_o \left\{ \begin{array}{l} A_{0,y}\left(1-\exp\left[-\left(1+\frac{S}{2}\right)\left(\frac{t}{\tau}\right)^{\beta}\right] I_0\left[\frac{S}{2}\left(\frac{t}{\tau}\right)^{\beta}\right]\right) \\ -2\exp\left[-\left(1+\frac{S}{2}\right)\left(\frac{t}{\tau}\right)^{\beta}\right]\cdot\sum_{m=1}^{\infty} A_{m,y} I_m\left[\frac{S}{2}\left(\frac{t}{\tau}\right)^{\beta}\right] \end{array} \right\}, \tag{11b}$$

where $I_m[\cdot]$ $(m = 1,2,3,\cdots)$ is the modified Bessel function of the mth order of first kind. Here, we present some of coefficients $A_{m,i}$ ($m = 0,1,2,3,\cdots$ and $i = x, y$) as follows:

$$\begin{array}{l} A_{1,x} = \frac{(8+S^2)\sqrt{1+S}-4(2+S)}{8S^2\sqrt{1+S}}, \quad A_{1,y} = \frac{(8+8S+S^2)\sqrt{1+S}-4(2+3S+S^2)}{8S^2\sqrt{1+S}} \\ A_{2,x} = \frac{4(2+S)\sqrt{1+S}-(8+8S+S^2)}{2S^3\sqrt{1+S}}, \quad A_{2,y} = \sqrt{1+S}\,\frac{4(2+S)\sqrt{1+S}-(8+8S+S^2)}{2S^3}, \\ A_{3,x} = \frac{2(16+16S+3S^2)\sqrt{1+S}-(32+48S+18S^2+S^3)}{2S^4\sqrt{1+S}}, \\ A_{3,y} = \sqrt{1+S}\,\frac{2(16+16S+3S^2)\sqrt{1+S}-(32+48S+18S^2+S^3)}{2S^4}, \ldots \end{array} \tag{12}$$

The complex refractive index changes can be written as $\tilde{n}_i(\omega') = \tilde{n}_0(\omega') + \delta\tilde{n}_i(\omega')$, where $\tilde{n}_0(\omega') = \sqrt{1+\chi_{polym}+\alpha_T N_0}$ is the complex linear refractive index including the background refractive index of the polymer matrix and $\delta\tilde{n}_i(\omega') = \delta\chi_{ii}/2\tilde{n}_0(\omega')$ is the complex nonlinear refractive index changes. The complex refractive indices are also written as $\tilde{n}_i = n_i - i\kappa_i$, in which $n_i = \mathrm{Re}[\tilde{n}_i]$ represents the anisotropic refractive index change, and $\kappa_i = \mathrm{Im}[\tilde{n}_i] = \lambda'\alpha_i/4\pi$ depicts the anisotropic absorption index change, where $\alpha_i = 4\pi\kappa_i/\lambda'$ is the anisotropic absorption coefficient and λ' is the wavelength of the probe beam. The photoinduced birefringence (PIB) is given by $\Delta n(t) = \mathrm{Re}[\delta\chi_{xx} - \delta\chi_{yy}]/2n_0(\lambda') \equiv \delta n_x(t) - \delta n_y(t)$ and the photoinduced dichroism (PID) is expressed as $\Delta\kappa(t) = \mathrm{Im}[\delta\chi_{xx} - \delta\chi_{yy}]/2n_0(\lambda') \equiv \delta\kappa_x(t) - \delta\kappa_y(t)$. Using Eqs. (11) and (12) the PIB kinetics can be approximately written as

$$\Delta n(t) \cong \delta\,\mathrm{n} \left(\begin{array}{l} B_0\left\{1 - I_0(\bullet)\exp\left[-\left(1+\frac{S}{2}\right)\left(\frac{t}{\tau}\right)^{\beta}\right]\right\} \\ +[B_1 I_1(\bullet) + B_2 I_2(\bullet) + B_3 I_3(\bullet)]\exp\left[-\left(1+\frac{S}{2}\right)\left(\frac{t}{\tau}\right)^{\beta}\right] \end{array} \right) \tag{13}$$

where $\delta n = \mathrm{Re}[\delta\chi_o]/2n_0$ is the maximum PIB change and the coefficients B_m are given by

$$B_0 = \frac{2+S-2\sqrt{1+S}}{2S\sqrt{1+S}},\ B_1 = \frac{\sqrt{1+S}\left(16+8S+2S^2\right)-4\left(4+4S+S^2\right)}{4S^2\sqrt{1+S}},$$
$$B_2 = \frac{(2+S)\left[4\sqrt{1+S}(2+S)-\left(8+8S+S^2\right)\right]}{S^3\sqrt{1+S}}, \tag{14}$$
$$B_3 = \frac{(2+S)\left[2\sqrt{1+S}\left(16+16S+3S^2\right)-\left(32+48S+18S^2+S^3\right)\right]}{S^4\sqrt{1+S}},\cdots$$

Since the contribution of high-order coefficients larger than m=3 to the PIB kinetics can be negligibly small, in what follows, we will use the approximated analytic formula, Eq. (13) with Eq. (14). Figure 3(a) represents the theoretical kinetics of normalized PIB divided by the maximum PIB change δn and compares the PIB kinetics of stretched exponent $(\beta = 0.35)$ with the pure exponentiality $(\beta = 1)$ for several saturation parameters, which reveals quite distinct transient behaviors at early time. It follows that as time goes to infinity the steady state value of PIB comes together with that of the pure exponentiality whatever one may take the stretched exponents. Figure 3(b) depicts the normalized steady state value of PIB divided by δn as a function of the saturation parameter. Using Eq. (13) with Eq. (14) the steady state value of PIB is uniquely determined by $\Delta n_{ss} = B_0\delta n = \delta n\left(2+S-2\sqrt{1+S}\right)/\left(2S\sqrt{1+S}\right)$. As increasing the saturation parameter, the PIB rapidly increases to a maximum value and then gradually decreases.

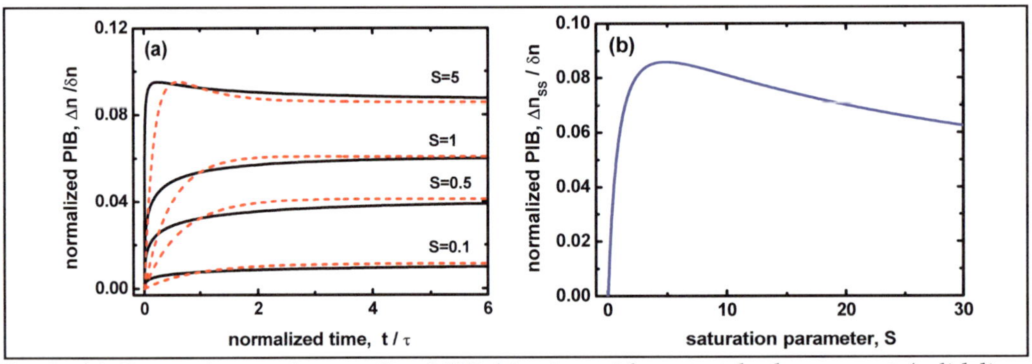

Fig. 3. (a) Comparisons of normalized PIB kinetics with a stretched exponent (solid lines: $\beta = 0.35$) and the pure exponentiality (dotted lines: $\beta = 1$) for several saturation parameters S, and (b) normalized steady state PIB divided by the maximum PIB change against saturation parameter. Steady state values of PIB is independent of the stretched exponent.

3. Experimental Results and Discussions

3.1 Sample preparation of azo dye doped polymer films

Methylorange doped PVA (MO/PVA) films are fabricated and are used as nonlinear media for investigating the transient behaviors of the holographic gratings and the photoinduced birefringence. PVA of 6wt% was melted by distilled water by means of double boiler processing. Small amount of azo dye was doped into PVA solution and is thoroughly mixed by agitator for about 24 hours. The MO/PVA mixture was coated on glass substrates by gravity deposition technique and baked at 50°C for about 1 hour in a heating oven. We fabricated several MO/PVA films for various MO concentrations of 0.01wt%, 0.02wt%, 0.05wt%, 0.08wt%, 0.12wt% and 0.14wt%.

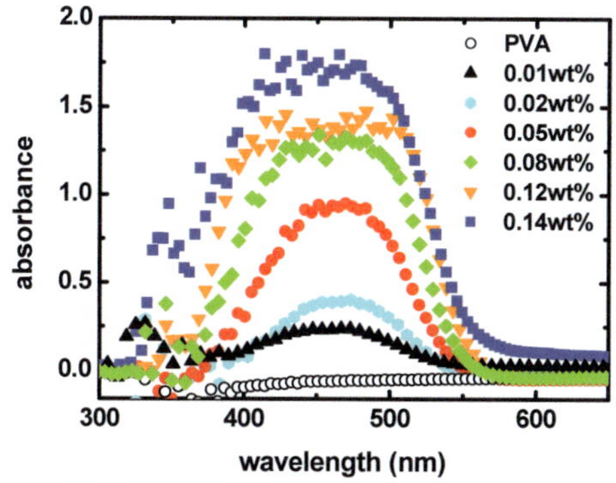

Fig. 4. Absorbance of MO/PVA films against wavelength for various MO concentrations.

The thickness of the film was approximately $20\mu m$. The absorption spectra of MO/PVA films for various MO concentrations are measured by using a spectrophotometer and are shown in Fig. 4. The linear absorbance has the maximum values for the wavelength region of blue-green light, while for the red wavelength region it shows nearly transparent, irrespective of MO concentrations. The pure PVA film without azo dye reveals no absorption for visible lights.

3.2 Determinations of optical nonlinearity by holographic gratings

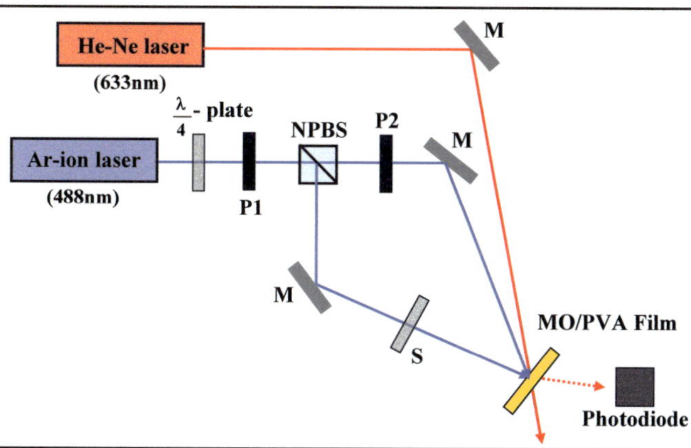

Fig. 5. Experimental setup for recording holographic gratings and for measuring the diffraction efficiencies. (NPBS: non-polarization beam splitter, M: mirror, P1, P2: polarizers, S: shutter).

Figure 5 shows the experimental setup measuring the real-time diffraction efficiency of the holographic gratings. Two coherent Ar-ion laser beams with the same linear polarization and the wavelength of 488 nm were used to construct the holographic gratings, and a He-Ne laser beam of 633nm wavelength was used for measuring the diffracted efficiencies. The incident half-angle between the two writing beams was approximately $\theta = 12^{\circ}$ and the beam intensity ratio of the two writing beams was kept to be unity. The read-out beam was incident by Bragg angle and the real-time first-order diffraction efficiencies were measured for various writing beam intensities and MO concentrations. The intensity of read-out beam was very small compared to the writing beam intensity, not to affect the grating formations.

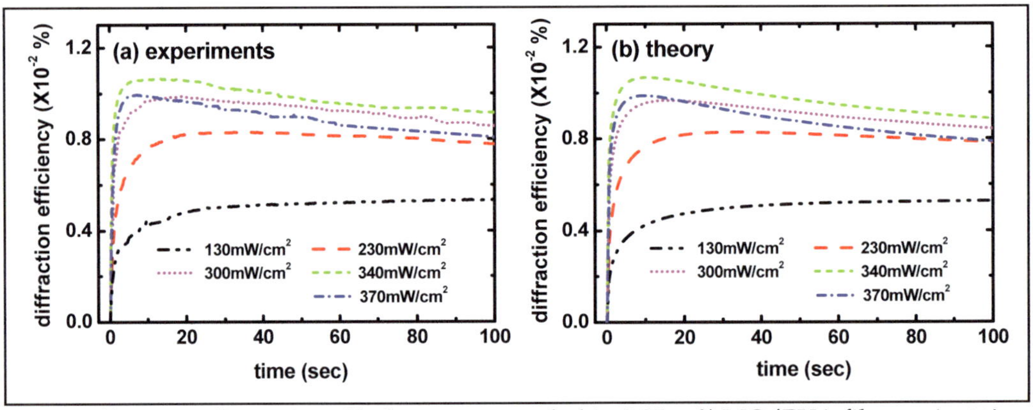

Fig. 6. Diffraction efficiencies of holograms recorded in 0.05wt% MO/PVA film against time: (a) experimental results, (b) theoretical curves for writing beam intensities with the stretched exponent of $\beta = 0.3 \pm 0.02$ and $\tau = 31.5 \pm 1.5\,\text{sec}$.

Figure 6 represents the real-time first-order diffraction efficiencies of holographic gratings for the concentration of 0.05wt% MO/PVA film with the theoretical predications according to Eq. (7). As clearly seen Fig. 6, theoretical curves are in good agreements with the experimental data. It is also found that as increasing the writing beam intensity the transient peak of the diffraction efficiency at early time, which is higher than the steady-state value, was observed, as theoretically predicted. Figure 7(a) represents the steady state diffraction efficiency as a function of total writing beam intensity at several MO concentrations with the theoretical predictions of Eq. (7), whose steady-state value is determined by $\eta \approx \left(C_1\, \pi \delta n L / \lambda \cos\theta\right)^2$. It is also clear that the maximum nonlinear refractive index change δn is linearly proportional to the MO concentration, as shown in Fig. 7(b).

From the best curve fitting to the data, we estimated the following physical parameters as: $\delta n = -(4.8 \pm 0.5) \times 10^{-4}$ of the maximum nonlinear refractive index change, $I_S = 500 \pm 20\,\text{mW/cm}^2$ of the saturation intensity, $\tau = 31.5 \pm 1.5\,\text{sec}$ of the characteristic lifetime and $\beta = 0.3 \pm 0.02$ of the stretched exponent in holographic gratings. It should be emphasized that the nonlinear refractive index chang has the negative sign, which is experimentally confirmed by the Z-scan experiment.

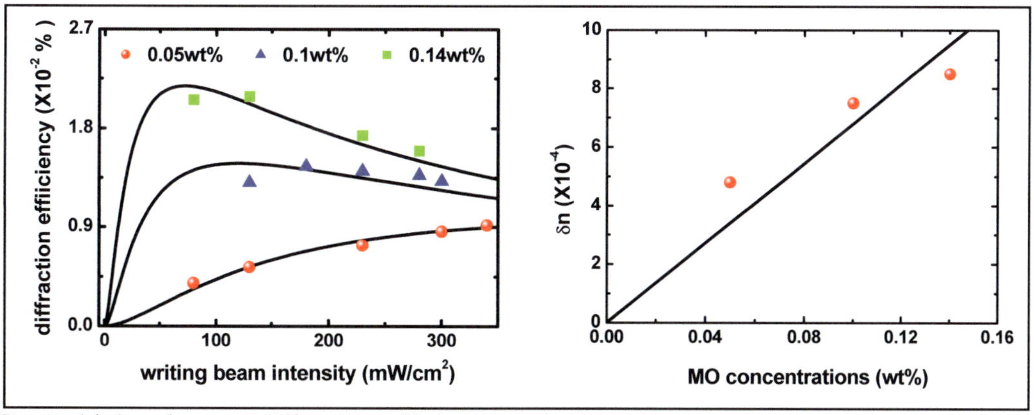

Fig. 7. (a) Steady-state diffraction efficiency against writing beam intensity for various MO concentrations and (b) maximum nonlinear refractive index change versus MO concentration. The solid lines are theoretical curves.

3.3 Determinations of nonlinear characteristics by photoinduced anisotropy

In order to measure the photoinduced birefringence kinetics of MO/PVA film we performed the pump-probe experiment. Figure 8 shows the experimental geometry for pump-probe technique to measure the PIB kinetics. We used a linearly polarized Ar-ion laser beam of 488nm wavelength as a pump beam and a linearly polarized He-Ne laser beam of 633nm wavelength as a probe beam. The wavelength of the probe beam is far away from strong absorption region as shown in Fig. 4 and that the probe beam intensity is taken to be so weak (about 5 mW/cm^2) that it cannot influence the optical properties of the sample.

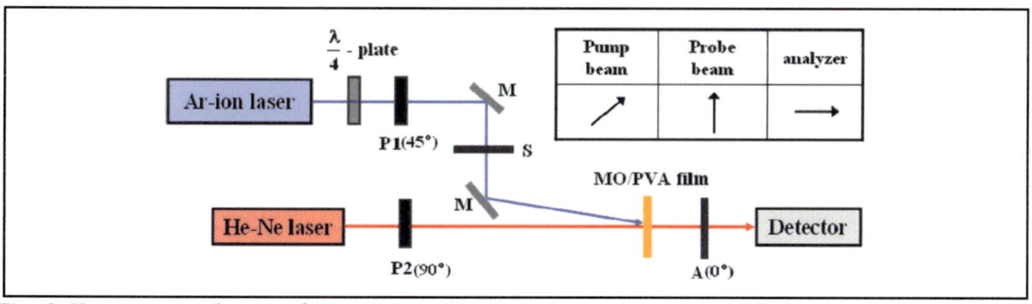

Fig. 8. Experimental setup for pump probe technique to measure PIB kinetics (M: mirror, P1, P2: polarizers, A: analyzer, S: shutter).

The polarization direction of the pump beam is controlled by a quarter wave plate and a polarizer. The intensity of the probe beam transmitted through the analyzer is adjusted to be zero (*i.e.,* to be crossed) without the sample. The film is then placed between the crossed polarizer and analyzer in the path of the probe beam. When the polarizer and analyzer are crossed, the transmittance of the probe beam intensity is given by (Kwak et al., 1992; Yang et al., 2009)

$$T = \frac{\exp\left(-\bar{\alpha}L\right)}{2}\sin^2 2\phi\left[\cosh\left(\frac{\Delta\alpha L}{2}\right) - \cos\left(\frac{2\pi\Delta n L}{\lambda'}\right)\right] \quad (15)$$

where $\bar{\alpha} = \left(\alpha_x + \alpha_y\right)/2$ is the average absorption coefficient, measured with an unpolarized probe light, $\Delta\alpha = \alpha_x - \alpha_y$ represents the photoinduced dichroism (PID), $\Delta n = n_x - n_y$ is the photoinduced birefringence (PIB), ϕ is the relative polarization angle between the linearly polarized probe beam and pump beam, L is the sample thickness and λ' is the wavelength of the probe beam. If one neglects PID of the sample (*i.e.*, $\Delta\alpha = 0$), Eq. (15) provides the PIB transmittance readout by a linearly polarized probe beam:

$$T = \exp\left(-\bar{\alpha}L\right)\sin^2 2\phi \sin^2\left(\frac{\pi\Delta n L}{\lambda'}\right) \quad (16)$$

For maximal readout the PIB the relative polarization angle between the linearly polarized probe and pump beams is chosen as $\phi = \pi/4$. Furthermore, if one may neglect the average absorption coefficient at the wavelength of the probe beam, the PIB kinetics can readily be described by the simple formula of $\Delta n(t) = (\lambda'/\pi L)\sin^{-1}\sqrt{T}$ with the theoretical one, $\Delta n(t) = \mathrm{Re}\left[\delta\chi_{xx} - \delta\chi_{yy}\right]/2n_0(\omega') \equiv \delta n_x(t) - \delta n_y(t)$. Actually, it has experimentally shown that the PID signal was seldom or never detected at 633nm wavelength of the probe beam.

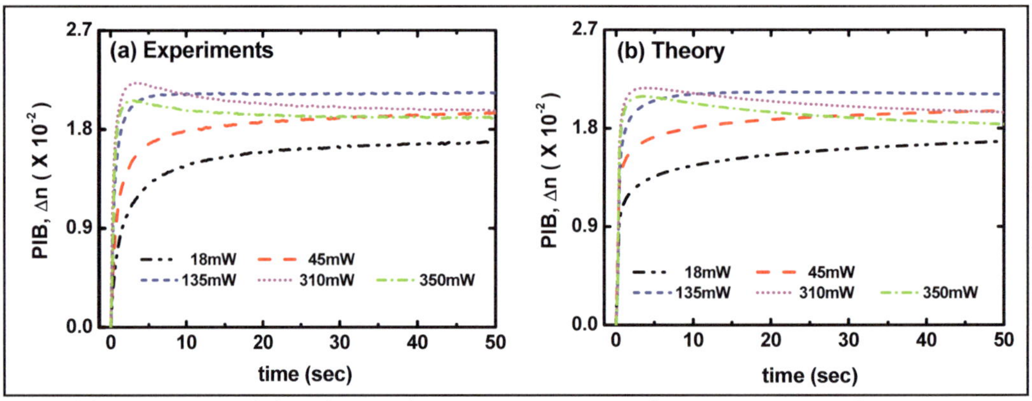

Fig. 9. (a) Experimental data for PIB kinetics against time for various pump beam intensities at MO concentration of 0.08wt% and (b) the corresponding theoretical curves fitted by using Eq. (16) with $\beta = 0.34 \pm 0.04$.

Figure 9 represents the time-dependent PIB data obtained at 0.08wt% MO/PVA film for various pump beam intensities with theoretical predictions of Eq. (16), showing excellent agreements with the experiments. As clearly seen in Fig. 9(a), the PIB kinetics cannot be described by a single exponential kinetics. The stretched exponential PIB kinetics seems to be quite good predictions for the entire time range, indicating the amorphous nature of MO/PVA. From the best curve fitting to the data, we estimated the following physical parameters as: for 0.08wt% of MO concentration, the maximum PIB change,

$\delta n = -(7.4 \pm 0.6) \times 10^{-2}$, the saturation intensity, $I_S = 31 \pm 2 \text{mW/cm}^2$, the characteristic lifetime, $\tau = 75 \pm 5 \sec$ and the stretched exponent, $\beta = 0.34 \pm 0.04$.

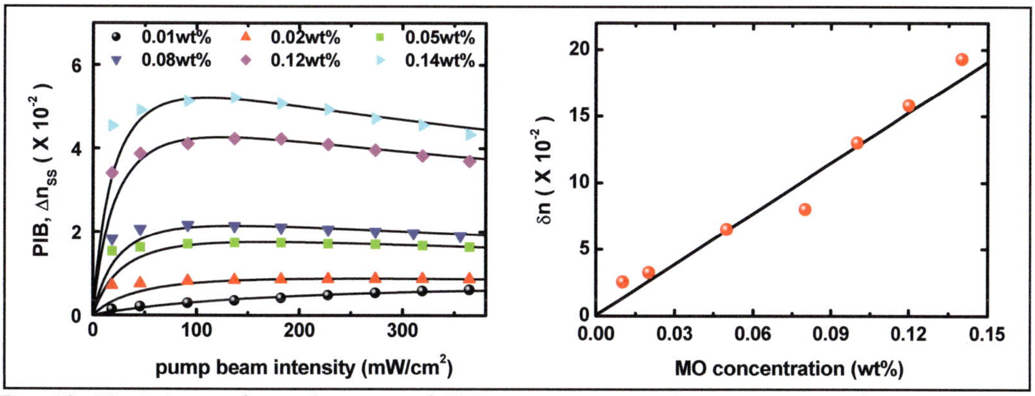

Fig. 10. Variations of steady state of PIB against pump beam intensity for various MO concentrations. Solid lines are the theoretical curves.

Figure 10(a) represents the steady state values of PIB as a function of pump beam intensities for various concentrations of azo dye (MO). As increasing the pump beam intensity the steady state values of PIB for a MO concentration rapidly increase to its maximum value and then gradually decrease. The steady-state value of PIB is uniquely determined by $\Delta n_{ss} = B_0 \delta n = \delta n \left(2 + S - 2\sqrt{1+S}\right) / \left(2S\sqrt{1+S}\right)$ as theoretically predicted. The solid lines are the theoretical predictions. Figure 10(b) shows that the maximum PIB change, δn against the concentration of MO. As described above, $\delta n \propto (\alpha_C - \alpha_T) N_0$, is linearly proportional to the total number density of azo dye N_0.

3.4 On the sign of the optical nonlinearities in azo dye doped polymer

In the previous sections, we have measured only the magnitudes of the optical nonlinearities by means of the holographic gratings (*i.e.*, scalar effects) and the photoinduced birefringence (*i.e.*, vectorial effects) in azo dye doped polymers. One of the simplest ways to determine the sign of the optical nonlinearities is the Z-scan method (Sheik-Bahae et al., 1990). The Z-scan technique is a simple, highly sensitive single beam method that uses the principle of spatial beam distortion to measure both the sign and the magnitude of the optical nonlinearities of materials. The optical material is scanned along the z-axis in the back focal region of an external lens and measures the far-field on-axis (*i.e.*, closed aperture) transmittance and the whole (*i.e.*, open aperture) transmittance as a function of the scan distance z. We have performed the Z-scan experiment by using a He-Ne laser beam, whose photon energy corresponds to the transparent region, as shown in Fig. 4. Figure 11 represent the typical experimental data for Z-scan in azo dye doped polymer films. It is obvious from Fig. 11 that the peak followed by a valley transmittance obtained from the closed aperture Z-scan data indicates the sign of the nonlinear refractivity is negative (*i.e.*, self-defocusing), and that the sign of the nonlinear absorption coefficient is also negative from the open aperture Z-scan (*i.e.*, photobleaching). It is also noted that the closed aperture Z-scan data shows severe asymetric behaviors, revealing the large nonlinear phase shifts (Kwak et al., 1999). Asymmetric behaviors of closed aperture Z-scan data cannot be described by the conventional Z scan theory (Sheik-Bahae et al., 1990).

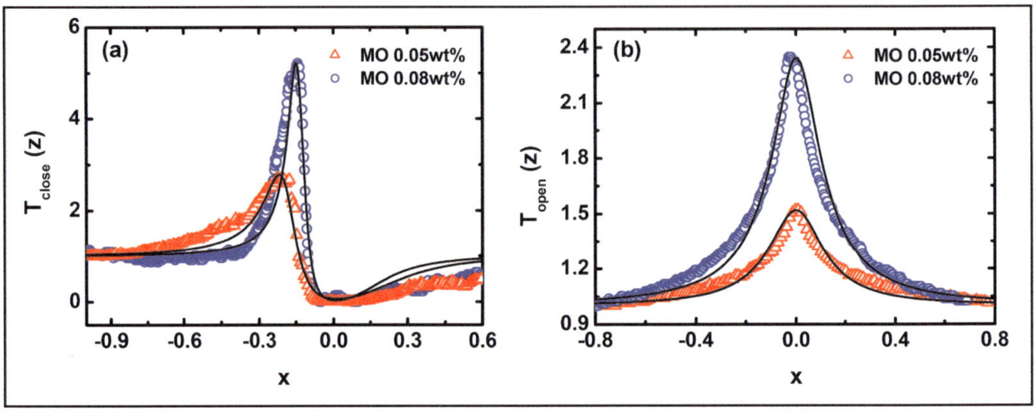

Fig. 11. Typical experimental results of (a) closed aperture Z-scan and (b) the open aperture Z-scan with the theoretical curves.

By employing the complex beam parameter formulation, we have the large phase shift closed aperture Z-scan transmittance, including both of the effects of nonlinear absorption and nonlinear refraction as follows (Kwak et al., 1999):

$$T_{close}(z) = \frac{1}{1 - \dfrac{4(x-\eta)}{(1+x^2)^2(1+q_0)}\Delta\Phi_0 + \dfrac{4(1+\eta^2)}{(1+x^2)^3(1+q_0)^2}\Delta\Phi_0^{\ 2}} \tag{17}$$

where $x = z/z_o$ is the dimensionless distance from a focus of an external lens, z_o is the Rayleigh diffraction length, $\Delta\Phi_o = k\Delta n_o L_{eff}$ is the on-axis nonlinear phase shift at focus, k is the wave number, $\Delta n_o = \gamma I_o$ is the nonlinear refractive index change, γ is the nonlinear refraction coefficient, I_o is the on-axis intensity at focus, L is the sample thickness, $L_{eff} = [1-\exp(-\alpha_o L)]/\alpha_o$ is the effective length of the sample, α_o is the linear absorption coefficient. Here, the coupling factor, $\eta = \beta/2k\gamma$ is the ratio of the imaginary part to the real part of the complex nonlinearity, which is inversely proportional to the figure of merit (FOM), defined as $\mathrm{FOM} = \gamma/\beta\lambda$ (Lenz *et al.*, 2000), where β is the nonlinear absorption coefficient. The nonlinear absorptive and refractive contributions to the closed aperture Z-scan transmittance are coupled in terms of η or FOM. For a CW laser beam, the open aperture Z-scan transmittance is given by (Kwak et al., 1999):

$$T_{open}(z) = \frac{\ln(1+q_0)}{q_0} \tag{18}$$

where $q_o = \beta I_o L_{eff}/(1+x^2)$. Once the nonlinear absorption coefficient β is unambiguously extracted from an open aperture Z-scan, one can use the closed aperture Z-scan transmittance to determine the remaining unknown coefficient γ or $\Delta\Phi_o$ from Eq. (17).

The solid lines in Fig. 11 depict the theoretical curves, showing excellent agreements with experimental data. We have obtained the nonlinear coefficients for several azo dye concentrations:

$\gamma(0.05wt\%) = -2\times10^{-5}\, cm^2/W$, $\beta(0.05wt\%) = -2.75\, cm/W$ and $\Delta\Phi_o(0.05wt\%) = -1.99$,

$\gamma(0.08wt\%) = -4.8\times10^{-5}\, cm^2/W$, $\beta(0.08wt\%) = -4.02\, cm/W$ and $\Delta\Phi_o(0.08wt\%) = -4.78$.

4. Conclusion

We have presented on the determinations of the optical nonlinearities of azo dye doped polymer film by means of the holographic gratings as a scalar effect and the photoinduced birefringence as a vector effect. We have measured the diffraction efficiency of the holographic gratings and the photoinduced birefringence caused by a linear polarized pump beam as a function of time for various laser beam intensities and azo dye concentrations. It is found that the real time behaviors of both of the diffraction efficiencies and the photoinduced birefringence reveal the stretched exponential kinetics. A three state model for photoisomerization is proposed to analyse the stretched exponential kinetic behaviors. Theoretical predictions are in good agreements with the experimental data.

To determine the sign of the optical nonlinearities we have conducted the Z-scan experiments and found that the sign of the nonlinear refractivity of azo dye doped polymer (MO/PVA) film is negative (*i.e.*, self-defocusing) from the closed aperture Z-scan, and that the sign of the nonlinear absorption coefficient is also negative (*i.e.*, photobleaching) from the open aperture Z-scan.

5. References

Benatar, L. E.; Redfield, D. & Bube, R. (1993). Interpretation of the activation energy derived from a stretched-exponential description of defect density kinetics in hydrogenated amorphous silicon. *J. Appl. Phys.*, Vol. 73, Issue 12, 8659-8661, ISSN : 0021-8979

Dureiko, R. D.; Schuele, D. E. & Singer, K. D. (1998). Modeling relaxation processes in poled electro-optic polymer films. *J. Opt. Soc. Am. B*, Vol. 15, Issue 1, 338-350, ISSN : 0740-3224

Egami, C.; Suzuki, Y.; Sugihara, O.; Okamoto, N.; Fujimura, H.; Nakagawa, H. & Fujiwara, H. (1997). Third-order resonant optical nonlinearity from trans–cis photoisomerization of an azo dye in a rigid matrix. *Appl. Phys.* Vol. B 64, Issue 4, 471-478, ISSN : 1432-0649

Fragnito, H. L.; Pereira, S. F. & Kiel, A. (1987). Self-diffraction in population gratings. *J. Opt. Soc. Am. B*, Vol. 4, Issue 8, 1309-1315, ISSN : 0740-3224

Fujiwara, H. & Nakagawa, K. (1985). Phase conjugation in fluorescein film by degenerate four-wave mixing and holographic process. *Opt. Comm.*, Vol. 55, Issue 6, 386-390, ISSN : 0030-4018

Huang, T. & Wagner, K. H. (1993). Holographic diffraction in photoanisotropic organic materials. *J. Opt. Soc. Am. A*. Vol. 10, Issue 2, 306-315, ISSN : 0740-3232

Johanson, R. E.; Kowalyshen, M.; DeForrest, D.; SHimakawa, K. & Kasap, S. O. (2007). The kinetics of photo-induced dichroism in thin films of amorphous arsenic triselenide. *J. Mater. Sci: Mater. Electron.*, Vol. 18, S127-S130, ISSN : 1573-482X

Kogelnik, H. (1969). Coupled wave theory for thick hologram gratings. *Bell System Tech. J.*, Vol. 48, Issue 9, 2909-2947, ISBN : 0005-8580

Kwak, C. H.; Kim, J. T. & Lee, S. S. (1988). Scalar and vector holographic gratings recorded in a photoanisotropic amorphous As_2S_3 thin film. *Opt. Lett.*, Vol. 13, Issue 6, 437-439, ISSN : 0146-9592

Kwak, C. H.; Kim, J. T. & Lee, S. S. (1989). Nonlinear optical image processing in photoanisotropic amorphous As_2S_3 thin film. *Appl. Opt.*, Vol. 28. Issue 4, 737-739, ISSN : 0003-6935

Kwak, C. H.; Park, S. Y; Kim, H. M.; Lee, E. H. & Kim, C. M. (1992). Dammann gratings for multispot array generation by using photoinduced anisotropic materials. *Opt. Comm.*, Vol. 88, 249-257, ISSN : 0030-4018

Kwak, C. H.; Lee, Y. L. & Kim, S. G. (1999). Analysis of asymmetric Z-scan measurement for large optical nonlinearities in an amorphous As_2S_3 thin film. *J. Opt. Soc. Am. B,* Vol. 16, Issue 4, 600-604, ISSN : 0740-3224

Lenz, G.; Zimmermann, J.; Katsufuji, T.; Lines, M. E.; Hwang, H. Y.; Spalter, S.; Slusher, R. E.; Cheong, S. W.; Sangher, J. S. & Aggarwal, I. D. (2000). Large Kerr effect in bulk Se-based chalcogenide glasses. *Opt. Lett.*, Vol. 25, Issue 4, 254-256, ISSN : 0146-9592

Nikolova, L.; Markovsky, P.; Tomova, N.; Dragostinova, V. & Tateva, N. (1988). Optically controlled photo-induced birefringence in photo anisotropy materials. *J. Mod. Opt*, Vol. 35, Issue 11, 1789-1799, ISSN : 0950-0340

Palmer, R. G.; Stein, D. L.; Abrahams, E. & Anderson, P. W. (1984). Models of hierarchically constructed dynamics for glassy relaxation. *Phys. Rev. Lett.*, Vol. 53, Issue 10, 958-961, ISSN : 0031-9007

Pham, V. P.; Galstyan, T.; Granger, A. & Lessard, R. A. (1997). Novel azo dye-doped poly (methyl methacrylate) film as optical data storage media. *Jpn. J. Appl. Phys.*, Vol. 36, Issue 1, 429-438, ISSN : 0021-4922

Sharma, K. K.; Divakara Rao, K. & Ravindra Kumar, G. (1994). Nonlinear optical interactions in dye-doped solids. *Opt. & Quantum Electro.,* Vol. 26, Issue 1, 1-23, ISSN : 1572-817X

Sekkat, Z. & Knoll, W. (1995). Creation of second-order nonlinear optical effects by photoisomerization of polar azo dyes in polymeric films: theoretical study of steady-state and transient properties. *J. Opt. Soc. Am.*, Vol. 12, Issue 10, 1855-1867, ISSN : 0740-3232

Sheik-Bahae, M.; Said, A. A.; Wei, T. H.; Hagan, D. J. & VanStryland, E. W. (1990). Sensitive measurement of optical nonlinearities using a single beam. *IEEE J. Quantum Electron.*, Vol. 26, Issue 4, 760-769, ISSN : 0018-9197

Yaroshchuk, O. V.; Kiselev, A. D.; Azkrevskyy, Yu.; Bidna, T.; Kelly, J.; Chien, L. –C. & Lindau, J. (2003). Photoinduced three-dimension orientational order in side chain liquid crystalline azopolymers. *Phys. Rev. E,* Vol. 68, Issue 11, 118031-1180315, ISSN : 1550-2376

Yang, H. R.; Kim, E. J.; Lee, S. J.; Kim, G. Y. & Kwak, C. H. (2009). Stretched exponential kinetics for photoinduced birefringence in azo dye doped PVA films. *Opt. Comm.*, Vol. 282, 1902-1908, ISSN : 0030-4018